21st Century Canon of Lunar Eclipses

Full Color Edition

Fred Espenak

Edition 1.0
September 2020

21st Century Canon of Lunar Eclipses – Full Color Edition

Astropixels Publishing
P.O. Box 16197
Portal, AZ 85632

www.astropixels.com/pubs

Printed in the United States of America

This book may be ordered at: *www.astropixels.com/pubs/21CCLE.html*

More about lunar eclipses of the 21st Century can be found at:

www.eclipsewise.com/lunar/LEcatalog/21CCLEcat.html

Astropixels Publication Number: AP020

ISBN 978-1-941983-23-2

First Edition (Version 1.0a)

Front Cover: A time sequence shows the partial phases and totality during the total lunar eclipse of 2000 July 16 from Maui, Hawaii (Copyright ©2000 by Fred Espenak).

Back Cover: Portrait of Fred Espenak (Copyright ©2016 by Fred Espenak).

Table of Contents

Photo 1 – Five minutes before the start of totality (2018 Jan 31), the Moon is bathed in an orange-red light. The narrow rim outside the umbra and still in sunlight appears brilliant white . ©2018 F. Espenak

Photo 2 – Various phases of the total lunar eclipse of 2001 Jan 21. ©2001 F. Espenak

Section 1: Lunar Eclipse Fundamentals

1.1 Introduction

The Moon orbits Earth once every 29.5306 days with respect to the Sun. Over the course of its orbit, the Moon's changing position relative to the Sun results in its familiar phases: New Moon > New Crescent > First Quarter > Waxing Gibbous > Full Moon > Waning Gibbous > Last Quarter > Old Crescent > New Moon. The New Moon phase is not visible because the illuminated side of the Moon then points away from Earth. The other phases are easily seen as the Moon cycles through them month after month.

During Full Moon, the Moon appears opposite the Sun in the sky. It rises as the Sun sets and is visible throughout the night. The Full Moon sets in the morning just as the Sun rises. This unique geometry occurs when the Moon is 180° from the Sun as seen from Earth. It is the same direction that Earth casts its shadow into space.

The Moon's orbit is tilted about 5.1° to Earth's orbit around the Sun. As seen from Earth, the points where the two orbits appear to cross are called the nodes. When the Full Moon occurs near one of these nodes, Earth's shadow can fall on some portion of the Moon and a lunar eclipse takes place.

Earth's shadow is composed of two cone-shaped components, one nested inside the other. The outer or penumbral shadow is a zone where the Sun's rays are partially blocked, while the inner or umbral shadow is a region where direct rays from the Sun are completely blocked.

1.2 Classification of Lunar Eclipses

There are three basic types of lunar eclipses:

1. **Penumbral Lunar Eclipse** — The Moon passes through Earth's faint penumbral shadow. Penumbral eclipses are subtle and quite difficult to observe. On rare occasions, the Moon's entire disk can be enveloped within the penumbral shadow. Such events are called total penumbral eclipses.

2. **Partial Lunar Eclipse** — A portion of the Moon passes through Earth's dark umbral shadow. The remaining part of the Moon appears bright even though it lies deep within the penumbra. Partial eclipses are easy to see, even with the unaided eye.

3. **Total Lunar Eclipse** — The entire Moon passes through Earth's umbral shadow. Total eclipses are quite striking for the vibrant range of colors the Moon can take on during the total phase, referred to as totality.

Figure 1–1. Geometry of a Total Lunar Eclipse

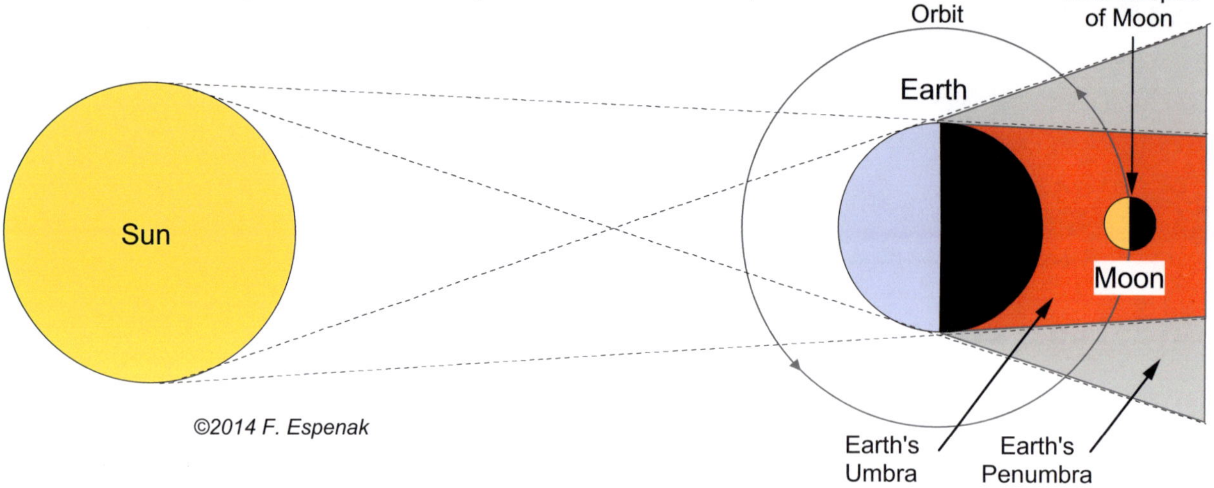

Figure 1–1 illustrates the geometry of a total lunar eclipse. A partial eclipse is visible if only part of the Moon enters Earth's umbral shadow. If the Moon passes through the penumbral shadow but misses the umbral shadow, then a penumbral eclipse occurs.

Figure 1–2. Types of Lunar Eclipses

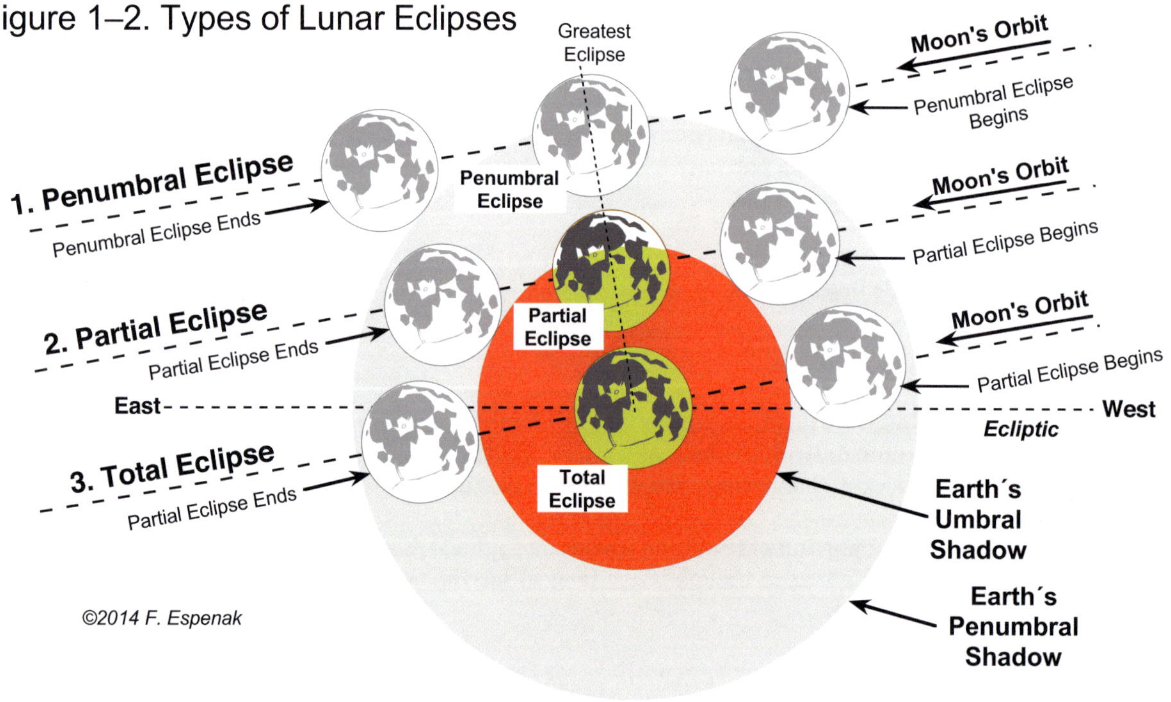

Figure 1–2 illustrates the three types of lunar eclipses as seen from Earth. 1) A penumbral eclipse occurs when the Moon passes through the penumbra but completely misses the umbra. 2) A partial eclipse occurs if some portion of the Moon enters the umbra. 3) A total lunar eclipse takes place when the entire disk of the Moon enters the umbral shadow.

Photo 3 – The visual appearance of penumbral, partial, and total lunar eclipses is illustrated. ©2020 F. Espenak

1.3 Visual Appearance of Penumbral Lunar Eclipses

The visual appearance of penumbral, partial and total lunar eclipses differs significantly. While penumbral eclipses are pale and difficult to see, partial eclipses are easy naked-eye events while total eclipses are colorful and dramatic.

Earth's penumbral shadow forms a diverging cone that expands into space in the direction away from the Sun. From within this zone, Earth blocks part but not the Sun's entire disk. Thus, some portion of the Sun's direct rays continues to reach the Moon during a penumbral eclipse.

The early and late stages of a penumbral eclipse are completely invisible to the eye. It is only when about 2/3 of the Moon's disk has entered the penumbral shadow that a skilled observer can detect a faint shading across the Moon.

Even when 90% of the Moon is immersed in the penumbra, about 10% of the Sun's rays still reach the deepest edge of the Moon. Under such conditions, the Moon remains relatively bright with only a subtle shadow gradient across its disk.

1.4 Visual Appearance of Partial Lunar Eclipses

Compared to penumbral eclipses, partial eclipses are easy to see with the naked eye. The lunar limb that extends into the umbral shadow appears very dark or black. This is due to a contrast effect since the remaining portion of the Moon in the penumbra may be 500x brighter. Because the umbral shadow's diameter is about 2.7x times the Moon's diameter, it appears as though a semi-circular bite has been taken out of the Moon.

Aristotle (384–322 BCE) first proved that Earth was round using the curved umbral shadow seen at partial eclipses. In comparing observations of several eclipses, he noted that Earth's shadow was round no matter where the eclipse took place, whether the Moon was high in the sky or low near the horizon. Aristotle correctly reasoned that only a sphere casts a round shadow from every angle.

1.5 Visual Appearance of Total Lunar Eclipses

A total lunar eclipse is the most dramatic and visually compelling type of lunar eclipse. The Moon's appearance can vary enormously throughout the period of totality and from one eclipse to the next. The geometry of the Moon's path through the umbra plays a significant role in determining the appearance of totality. The effect that Earth's atmosphere has on a total eclipse is not as apparent. Although the physical mass of Earth blocks all direct sunlight from the umbra, the planet's atmosphere filters, attenuates and bends some of the Sun's rays into the shadow.

The molecules in Earth's atmosphere scatter short wavelength light (i.e., yellow, green, blue) more than long wavelength light (i.e., orange, red). The same process responsible for making sunsets red also gives total lunar eclipses their characteristic ruddy color. The exact appearance can vary considerably in both hue and brightness.

Because the lowest layers of the atmosphere are the thickest, they absorb more sunlight and refract it through larger angles. About 75% of the atmosphere's mass is concentrated in the bottom 10 kilometers (troposphere) as well as most of the water vapor, which can form massive clouds that block even more light. Just above the troposphere lies the stratosphere (10 to 50 kilometers), a rarified zone above most of the planet's weather systems. The stratosphere is subject to important photochemical reactions due to the high level of solar ultraviolet radiation that penetrates the region. The troposphere and stratosphere act together as a ring-shaped lens that refracts heavily reddened sunlight into Earth's umbral shadow. Since the higher stratospheric layers contain less gas, they refract sunlight through progressively smaller angles into the outer parts of the umbra. In contrast, denser tropospheric layers refract sunlight through larger angles to reach the inner parts of the umbra.

Photo 4 – The beginning, middle and end of totality during the total lunar eclipse of 2004 October 28. ©2004 F. Espenak

Because of this lensing effect, the amount of light refracted into the umbra tends to increase radially from center to edge. Inhomogeneities from varying amounts of cloud and dust at differing latitudes can cause significant variations in brightness throughout the umbra.

Besides water (cloud, mist, precipitation), Earth's atmosphere also contains aerosols or tiny particles of organic debris, meteoric dust, volcanic ash and photochemical droplets. This material attenuates sunlight before it is refracted into the umbra. For instance, major volcanic eruptions in 1963 (Agung) and 1982 (El Chichon) each dumped huge quantities of gas and ash into the stratosphere and were followed by several years of very dark eclipses (Keen, 1983).

The 1991 eruption of Pinatubo in the Philippines had a similar effect. While most of the solid ash fell to Earth several days after circulating through the troposphere, a sizable volume of sulphur dioxide (SO_2) reached the stratosphere where it combined with water vapor to produce sulphuric acid (H_2SO_4). This high-altitude volcanic haze layer severely attenuates sunlight that must travel several hundred kilometers horizontally through the layer before being refracted into the umbral shadow. Thus, total eclipses following large volcanic eruptions are particularly dark. The total lunar eclipse of 1992 Dec 09 (1½ years after Pinatubo) was so dark that it was difficult to see the Moon's dull gray disk with the naked eye.

All total eclipses begin with penumbral and partial phases. After the total phase, the eclipse ends with partial followed by penumbral phases. While *solar* eclipses require special filters for safe viewing, no such precautions are needed to watch *lunar* eclipses. The best views of a lunar eclipse are with binoculars and the naked eye.

1.6 Danjon Scale of Lunar Eclipse Brightness

The French astronomer A. Danjon proposed a five-point scale to evaluate the visual appearance and brightness of the Moon during a total lunar eclipse. The L values for various luminosities are defined in Table 1–1.

Table 1–1. Danjon Brightness Scale for Total Lunar Eclipses

Danjon Value	Visual Description
L=0	Very dark eclipse (Moon is almost invisible, especially at mid-totality)
L=1	Dark eclipse, grey or brownish in coloration (details are distinguishable only with difficulty)
L=2	Deep red or rust-colored eclipse (very dark central shadow, while outer umbra is relatively bright)
L=3	Brick-red eclipse (umbral shadow usually has a bright or yellow rim)
L=4	Very bright copper-red or orange eclipse (umbral shadow has a bluish, very bright rim)

The Danjon scale is a useful tool to characterize the range of colors and brightness the Moon takes on during a total lunar eclipse. Assigning an L value is best done with the naked eye or binoculars near the time of mid-totality. The Moon's appearance should also be evaluated just after the start and before the end of totality. The Moon is then near the edge of the shadow, providing an opportunity to assign an L value to the outer umbra. In making such determinations, the instrumentation and the time should be recorded.

Photo 5 – Various phases of the total lunar eclipse of 2014 Apr 15 are captured in this multiple exposure sequence. ©2014 F. Espenak

Section 2: Lunar Eclipse Predictions

2.1 Lunar Eclipse Contacts

During the course of a lunar eclipse, the instants when the Moon's disk becomes tangent to Earth's shadows are known as eclipse contacts. They mark the primary stages or phases of a lunar eclipse (see Figure 2–1).

Penumbral lunar eclipses have two major contacts although neither of these events is observable.

P1 — Instant of first exterior tangency of the Moon with the Penumbra
(Penumbral Eclipse Begins)
P4 — Instant of last exterior tangency of the Moon with the Penumbra
(Penumbral Eclipse Ends)

Partial lunar eclipses have four contacts. As the Moon's limb enters and exits the umbral shadow, contacts U1 and U4 mark the instants when the partial eclipse phase begins and ends, respectively.

P1 — Instant of first exterior tangency of the Moon with the Penumbra
(Penumbral Eclipse Begins)
U1 — Instant of first exterior tangency of the Moon with the Umbra
(Partial Umbral Eclipse Begins)
U4 — Instant of last exterior tangency of the Moon with the Umbra
(Partial Umbral Eclipse Ends)
P4 — Instant of last exterior tangency of the Moon with the Penumbra
(Penumbral Eclipse Ends)

Total lunar eclipses have six contacts. Contacts U2 and U3 mark the instants when the Moon's entire disk is first and last internally tangent to the umbra. These are the times when the total phase of the eclipse begins and ends, respectively.

P1 — Instant of first exterior tangency of the Moon with the Penumbra
(Penumbral Eclipse Begins)
U1 — Instant of first exterior tangency of the Moon with the Umbra
(Partial Umbral Eclipse Begins)
U2 — Instant of first interior tangency of the Moon with the Umbra
(Total Umbral Eclipse Begins)
U3 — Instant of last interior tangency of the Moon with the Umbra
(Total Umbral Eclipse Ends)
U4 — Instant of last exterior tangency of the Moon with the Umbra
(Partial Umbral Eclipse Ends)
P4 — Instant of last exterior tangency of the Moon with the Penumbra
(Penumbral Eclipse Ends)

The instant of greatest eclipse occurs when the Moon passes closest to the shadow axis. This corresponds to the maximum phase of the eclipse when the Moon is at its deepest position within either the penumbral or umbral shadow.

Figure 2–1. Lunar Eclipse Contacts

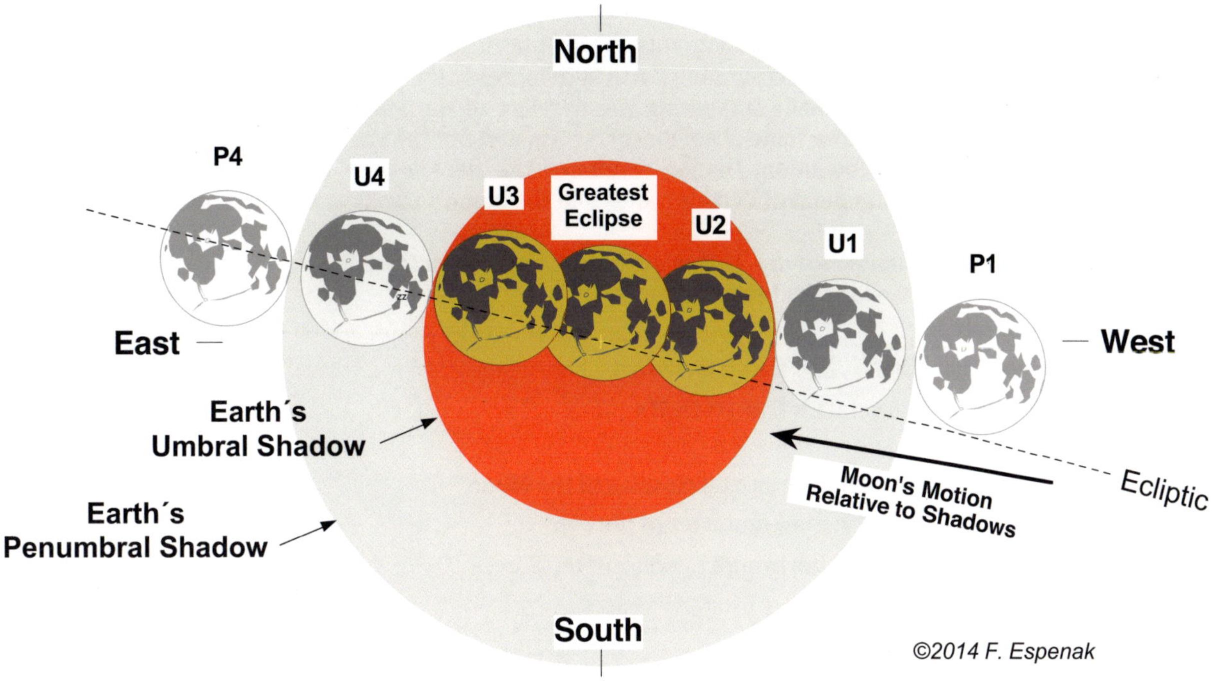

Figure 2–1 illustrates the six contacts for a total lunar eclipse. These correspond to the instants when the Moon's disk is externally tangent to the penumbra (P1 and P4), or either externally or internally tangent to the umbra (U1, U2, U3, and U4). Partial eclipses do not have contacts U2 and U3, while penumbral eclipses only have contacts P1 and P4.

2.2 Enlargement of Earth's Shadows

In 1707, Philippe de La Hire made a curious observation about Earth's umbra. The predicted radius of the shadow needed to be enlarged by about 1/41 in order to fit timings made during a lunar eclipse. Additional observations over the next two centuries revealed that the shadow enlargement was somewhat variable from one eclipse to the next. According to William Chauvenet (1891):

"This fractional increase of the breath of the shadow was given by Lambert as 1/40, and by Mayer as 1/60. Beer and Maedler found 1/50 from a number of observations of eclipses of lunar spots in the very favorable eclipse of December 26, 1833."

Chauvenet adopted a value of 1/50, which has become the standard enlargement factor for lunar eclipse predictions published by many national institutes worldwide. The enlargement enters into the definitions of the penumbral and umbral shadow radii as follows.

penumbral radius: $$Rp = 1.02 \times (0.998340 \times \pi m + Ss + \pi s) \quad (2\text{–}1)$$

umbral radius: $$Ru = 1.02 \times (0.998340 \times \pi m - Ss + \pi s) \quad (2\text{–}2)$$

Where:
πm = Equatorial horizontal parallax of the Moon
Ss = Geocentric semi-diameter of the Sun
πs = Equatorial horizontal parallax of the Sun

The factor 1.02 is the enlargement of the shadows by 1/50. Earth's true figure approximates that of an oblate ellipsoid with a flattening of ~1/300. Furthermore, the degree of axial tilt of the planet towards or away from the Sun throughout the year means the shape of the penumbral and umbral shadows varies by a small amount. A mean radius of Earth at latitude 45° is used to approximate the departure from perfectly circular shadows. The

Astronomical Almanac uses a factor of 0.998340 to scale the Moon's equatorial horizontal parallax to account for this (0.998340 ≈ 1.0 – 0.5 × 1/300).

Some authorities dispute Chauvenet's shadow enlargement convention. Danjon (1951) notes the only reasonable way of accounting for a layer of opaque air surrounding Earth is to increase the planet's radius by the altitude of the layer. This is accomplished by proportionally increasing the parallax of the Moon. The radii of the umbral and penumbral shadows are then subject to the same absolute correction and not the same relative correction employed in the traditional Chauvenet 1/50 convention. Danjon estimates the thickness of the occulting layer to be 75 kilometers and this results in an enlargement of Earth's radius and the Moon's parallax of about 1/85.

In 1951, the French almanac *Connaissance des Temps* adopted Danjon's method for the enlargement of Earth's shadows in their eclipse predictions as shown below.

Penumbral radius: $$Rp = 1.01 \times \pi m + Ss + \pi s \quad (2\text{–}3)$$

Umbral radius: $$Ru = 1.01 \times \pi m - Ss + \pi s \quad (2\text{–}4)$$

Where:

πm = Equatorial horizontal parallax of the Moon

Ss = Geocentric semi-diameter of the Sun

πs = Equatorial horizontal parallax of the Sun

And $1.01 \approx 1 + 1/85 - 1/594$

The factor 1.01 combines the 1/85 shadow enlargement term with a 1/594 term to correct for Earth's oblateness at a latitude of 45°.

Danjon's method correctly models the geometric relationship between an enlargement of Earth's radius and the corresponding increase in the size of its shadows. Meeus and Mucke (1979) and Espenak & Meeus (2009a) both use Danjon's method. However, the resulting umbral and penumbral eclipse magnitudes are smaller by approximately 0.006 and 0.026, respectively, as compared to predictions using the traditional Chauvenet convention of 1/50.

For instance, the umbral magnitude of the partial lunar eclipse of 2008 Aug 16 was 0.813 according to the *Astronomical Almanac for 2008* (2008) using Chauvenet's method, but only 0.8076 according to Espenak & Meeus (2009a) using Danjon's method.

Chauvenet's method is still used by the *Astronomical Almanac* (published jointly by the USNO and HMNAO) to calculate lunar eclipse circumstances, while Danjon's method is used by Meeus and Mucke (1979), Espenak and Meeus (2009a) and *Connaissance des Temps* (published by the Bureau des Longitudes).

2.3 Earth's Elliptical Shadows

Both the Chaunenet and Danjon methods of accounting for the enlargement of Earth's two shadows assume circular shadows scaled at 45° latitude. However, Earth is flattened at the poles and bulges at the Equator, so an oblate spheroid more closely represents its shape. The projection of each of the planet's shadows is an ellipse rather than a circle. Furthermore, Earth's axial tilt towards or away from the Sun throughout the year means the elliptical shape of the penumbral and umbral shadows varies as well.

Herald and Sinnott performed an analysis of 22,539 observations made at 94 lunar eclipses from 1842 to 2011 (Herald and Sinnott, 2014). This is the largest collection of crater and contact timings ever compiled. The authors define the height of a 'notional eclipse-forming layer' in Earth's atmosphere (abbreviated as NEL) corresponding to the occulting layer height used by Danjon. Given the size and consistency of their dataset, they refine the NEL height to 87 kilometers (compared to Danjon's value of 75 kilometers).

Herald and Sinnott find that size and shape of the umbra are consistent with an oblate spheroid at the time of each eclipse, enlarged by the empirically determined NEL that uniformly surrounds Earth. They conclude that future lunar eclipse predictions should be based on a Danjon-like approach with full allowance for an oblate Earth, with the umbral radius r_u being computed using equation 2–5.

$$r_u = R_\oplus \pi_m - S_s + \pi_s \qquad (2\text{–}5)$$

Where: $R_\oplus$ = Radius of Earth, where $R_\oplus = 1 + h - 0.003353 \sin^2 \psi \cos^2 (\delta_s + f \sin \psi)$

h = 0.0136 is the height of the NEL in Earth radii (h = 87 / 6378.137)

ψ = Angular position angle (measured from the east–west direction, positive to the north) of the relevant contact point about the edge of the umbra

$f = S_s - R_\oplus \pi_s$

δ_s = Declination of the Sun

π_m = Equatorial horizontal parallax of the Moon

S_s = Geocentric semi-diameter of the Sun

π_s = Equatorial horizontal parallax of the Sun

The calculation of r_u requires a single iteration between $R_\oplus$ and f to generate mutually consistent values for a given ψ. Similar adjustments can be made for the penumbral radius r_p although the resulting effects are not observable.

The Herald and Sinnott method of calculating Earth's shadow enlargement is the most rigorous and accurate procedure to date. It is superior to the methods of Chaunenet and Danjon because it uses a better determined value of the NEL and an elliptical cross section for Earth's shadow. The *21st Century Canon of Lunar Eclipses* uses the Herald and Sinnott method in the lunar eclipse predictions presented here.

2.4 Solar and Lunar Coordinates

The coordinates of the Sun and the Moon used in the eclipse predictions presented here have been calculated with the JPL DE430 (Jet Propulsion Laboratory Developmental Ephemeris 430). The DE430 is based upon the International Celestial Reference Frame (ICRF), the adopted reference frame of the IAU. The DE430 includes both nutation or libration and has an absolute accuracy of several kilometers for planetary positions. In most cases this corresponds to a small fraction of an arc-second.

The Moon's center of figure does not coincide with its center of mass. To compensate, an empirical correction is sometimes added to the Moon's center of mass position. Unfortunately, the large variation in lunar libration from one eclipse to the next minimizes the effectiveness of this empirical correction. Because of this, no correction has been made to the Moon's center of mass position in the *21st Century Canon.*

2.5 Measurement of Time

The most natural form of time measurement is the solar day (usually measured from solar noon to solar noon). The length of the solar day varies during the year because of the eccentricity of Earth's orbit around the Sun. Mean solar time resolves this problem by using an average to define the mean solar day.

In 1884, Greenwich Mean Time (GMT) — the mean solar time on the Greenwich Meridian (0° longitude) — was adopted as the standard reference time for clocks around the world. A fundamental basis of GMT is the assumption that Earth's rotation on its axis is constant. It wasn't until the mid-twentieth century that astronomers realized the rotation period is gradually increasing. Earth is slowing down because of tidal friction with the Moon.

For purposes of orbital calculations, time using Earth's rotation was abandoned for a more uniform time scale based on Earth's orbit about the Sun. In 1952, Ephemeris Time was introduced to address the problem. The ephemeris second was defined as a fraction of the tropical year[1] for 1900 Jan 01 as calculated from Newcomb's *Tables of the Sun* (1895). Ephemeris Time was used for Solar System ephemeris calculations until 1979.

Terrestrial Dynamical Time (TD) is the modern replacement for Ephemeris Time and is used in theories of planetary motion in the Solar System. TD is based on International Atomic Time (TAI), which is a high-precision standard using several hundred atomic clocks worldwide. To ensure continuity with Ephemeris Time, TD was defined to match ET for the date 1977 Jan 01. In 1991, the IAU refined the definition of TD to make it more precise. It was also renamed Terrestrial Time (TT) although the author prefers to use the older name Terrestrial Dynamical Time.

Civilian time used throughout the world is still based on mean solar time, although indirectly. While Greenwich Mean Time was determined though observations of the Sun, its modern day replacement, Universal Time (actually UT1) is based on Earth's rotation using observations of distant quasars. UT1 is a nonuniform time because Earth is gradually slowing down at an irregular rate. At present (2020), the accumulated error in the rotation of Earth in the course of one year is ~0.3 seconds.

Coordinated Universal Time (UTC) is derived from International Atomic Time (TAI). The length of the UTC second is defined in terms of an atomic transition of cesium and is accurate to approximately one nanosecond (billionth of a second) per day. UTC was defined to closely parallel UT1. However, the two time systems are intrinsically incompatible since UTC is uniform while UT1 is based on Earth's rotation, which is gradually slowing. In order to keep the two times within 0.9 seconds of each other, a leap second is added to UT1 as needed (currently once every few years).

Today, UTC is the time standard used to define time zones around the world. It is the time reference for GPS satellites and aviation, and is used to synchronize the clocks of computers across the Internet.

2.6 ΔT (Delta T)

The orbital positions of the Sun and the Moon, required by eclipse predictions, are calculated using Terrestrial Dynamical Time (TD) because it is a uniform time scale. However, world time zones and daily life are based on Universal Time[2] (UT1). In order to convert predictions from TD to UT1, the difference between these two time scales must be known. The parameter ΔT (Delta T) is the arithmetic difference, in seconds, between the two as:

$$\Delta T = TD - UT1 \qquad (2\text{–}1)$$

Past values of ΔT can be deduced from historical records. In spite of their relatively low precision, these data represent the only evidence for the value of ΔT prior to 1600. In the centuries following the introduction of the telescope (circa 1609), thousands of high quality observations have been made of lunar occultations of stars, affording valuable data with increased accuracy in the determination of ΔT.

In modern times, the determination of ΔT is made using atomic clocks and radio observations of quasars. From 1955 to 2010, the average 1-year change in ΔT ranges from 0.18 seconds to 1.06 seconds. Future changes in ΔT are unknown since theoretical models of the physical causes are imprecise. Extrapolations from the table weighted by the long period trend from tidal braking of the Moon offer estimates of +71 seconds in 2024, +85 seconds in 2050, and +128 seconds in 2100.

Polynomial expressions for ΔT based on this data can be found at: *eclipsewise.com/help/deltatpoly2014.html*

[1] The tropical year is the length of time that the Sun takes to return to the same position in the cycle of seasons, as seen from Earth (e.g., the time from vernal equinox to vernal equinox).
[2] World time zones are actually based on Coordinated Universal Time (UTC). It is an atomic time synchronized and adjusted to stay within a second of astronomically determined Universal Time (UT1) through the addition of an occasional "leap second" to compensate for the gradual slowing of Earth's rotation.

2.7 Date Format

There are a number of ways to write the calendar date through variations in the order of day, month, and year. The International Organization for Standardization's (ISO) 8601 advises a numeric date representation, which organizes the elements from the largest to the smallest. The exact format is YYYY–MM–DD where YYYY is the calendar year, MM is the month of the year between 01 (January) and 12 (December), and DD is the day of the month between 01 and 31. For example, the 27th day of April in the year 1943 would then be expressed as 1943–04–27. The ISO convention is adopted here, but the month number has been replaced with the three-letter English abbreviation of the month name for additional clarity. From the previous example, the date then is expressed as 1943 Apr 27.

Photo 6 – Thirteen phases of the total lunar eclipse of 2000 January 21 appear in this composite image. ©2000 F. Espenak

Section 3: Lunar Eclipse Statistics

3.1 Statistical Distribution of Lunar Eclipse Types

There are three types of lunar eclipses:

1. **Penumbral** — The Moon passes partially or completely into Earth's penumbral shadow
2. **Partial** — The Moon passes through Earth's penumbral shadow and partially into the umbral shadow
3. **Total** — The Moon passes through Earth's penumbral shadow and completely into the umbral shadow

During the 100-year period from 2001 to 2100, Earth experiences 228 eclipses of the Moon. The statistical distribution of the three eclipse types over this interval is shown in Table 3–1.

Table 3–1: Distribution of Basic Lunar Eclipse Types During 21st Century

Eclipse Type	Abbreviation	Number	Percent
All Eclipses	—	228	100.0%
Penumbral	N	86	37.7%
Partial	P	57	25.0%
Total	T	85	37.3%

During most penumbral eclipses, only part of the Moon passes through Earth's penumbral shadow. Examples of such *partial* penumbral eclipses include: 2023 May 05, 2024 Mar 25, 2027 Feb 20 and 2027 Aug 17. It is also possible to have a penumbral eclipse in which the Moon passes completely within Earth's penumbral shadow while not entering the inner umbral shadow. Such *total* penumbral eclipses are quite rare compared the typical *partial* penumbral eclipses. During the 21st century, there are 81 *partial* penumbral eclipses, but only 5 *total* penumbral eclipses (5.8%): 2006 Mar 14, 2053 Aug 29, 2070 Apr 25, 2082 Aug 08, and 2099 Sep 29.

While there are no special classifications for partial eclipses, total lunar eclipses through Earth's umbral shadow can be categorized into two types:

1. **Central** — The Moon passes through the central axis of Earth's umbral shadow
2. **Non-Central** — The Moon misses the central axis of Earth's umbral shadow

Table 3–2: Statistics of Total Eclipses During 21st Century

Eclipse Type	Number	Percent
All Total Eclipses	85	100.0%
Central Total	24	28.2%
Non-Central Total	61	71.8%

Examples of central total eclipses include: 2022 May 16, 2022 Nov 08, 2029 Jun 26 and 2036 Aug 07. Several examples of non-central total eclipses are: 2025 Mar 14, 2025 Sep 07, 2026 Mar 03 and 2028 Dec 31. The track of central total eclipses through Earth's umbral shadow is longer than for non-central total eclipses, so the total phase durations of centrals are longer than non-centrals.

3.2 Lunar Eclipse Frequency and the Calendar Year

There are two to five lunar eclipses in every calendar year. Table 3–3 shows the distribution in the number of eclipses per year for the 100 years covered in the *21st Century Canon*.

Table 3–3: Number of Lunar Eclipses per Year

Number of Eclipses Per Year	Number of Years	Percent
2	78	78%
3	16	16%
4	6	6%
5	0	0%

When two eclipses occur in one calendar year, they can be in any combination of penumbral, partial, or total (N, P, or T, respectively).

The 16 years in which three lunar eclipses occur include: 2001, 2002, 2009, 2013, 2027, 2028, 2031, 2048, 2049, 2060, 2066, 2067, 2074, 2075, 2078, 2092, and 2094.

The 6 years in which four lunar eclipses occur include: 2009, 2020, 2038, 2056, 2085, and 2096.

During the 1000-year period 1501 to 2500, there are 7 years containing 5 lunar eclipses: 1676, 1694, 1749, 1879, 2132, 2262, and 2400.

3.3 Extremes in Duration and Eclipse Magnitude

The longest and shortest eclipses of the century appear in Table 3–4.

Table 3–4: Extremes in Duration

Extrema Type	Date (Dynamical Time)	Eclipse Duration	Eclipse Magnitude
Longest Total	2018 Jul 27	01h43m47s	1.6100
Shortest Total	2015 Apr 04	00h07m16s	1.0019
Longest Partial	2021 Nov 19	03h29m09s	0.9760
Shortest Partial	2082 Feb 13	00h27m20s	0.0153
Longest Penumbral	2099 Apr 05	04h49m05s	1.0360*
Shortest Penumbral	2027 Jul 18	00h17m39s	0.0032*

*penumbral eclipse magnitude

The largest and smallest eclipses (in umbral eclipse magnitude) appear in Table 3–5.

Table 3–5: Extremes in Eclipse Magnitude

Extrema Type	Date (Dynamical Time)	Eclipse Duration	Eclipse Magnitude
Largest Total	2029 Jun 26	01h42m40s	1.8452
Smallest Total	2015 Apr 04	00h07m16s	1.0019
Largest Partial	2086 Nov 20	03h08m52s	0.9877
Smallest Partial	2082 Feb 13	00h27m20s	0.0153
Largest Penumbral	2070 Apr 25	04h47m43s	1.0527*
Smallest Penumbral	2027 Jul 18	00h17m39s	0.0032*

*penumbral eclipse magnitude

3.4 Lunar Eclipse Tetrads

When four consecutive lunar eclipses are all total, the group is termed a tetrad. They occur because of the eccentricity of Earth's orbit in conjunction with the timing of eclipse seasons (Section 3.18). During the 21st Century the first eclipse of every tetrad occurs between January and June.

The umbral magnitudes of the total eclipses making up a tetrad are all relatively small. For the 300-year period 1901 to 2200, the largest umbral magnitude of a tetrad eclipse is 1.4251 on 1949 Apr 13. Table 3-6 lists the date of the first total eclipse in each of the 27 tetrads from 1501 to 2500.

Table 3-6. Date of First Total Eclipse in Lunar Eclipse Tetrads: 1501 to 2500

1504 Mar 01	1927 Jun 15	2043 Mar 25	2137 Mar 07
1515 Jan 30	1949 Apr 13	2050 May 06	2155 Mar 19
1522 Mar 12	1967 Apr 24	2061 Apr 04	2448 Jun 17
1533 Feb 09	1985 May 04	2072 Mar 04	2466 Jun 28
1562 Jan 20	2003 May 16	2090 Mar 15	2477 May 28
1580 Jan 31	2014 Apr 15	2101 Feb 14	2495 Jun 08
1909 Jun 04	2032 Apr 25	2119 Feb 25	

3.5 Eclipse Seasons

Because of its ~5.1° inclination, the Moon's orbit crosses the ecliptic at two points or nodes. If Full Moon takes place within approximately 17° of a node,[3] then a lunar eclipse will be visible from some region of Earth.

The Sun makes one complete circuit of the ecliptic in 365.24 days, so its average angular velocity is 0.99° per day. At this rate, it takes 34.5 days for the Sun and, at the opposite node, Earth's umbral and penumbral shadows to cross the 34° wide eclipse zone centered on each node. Because the Moon's orbit with respect to the Sun has a mean duration of 29.53 days, there will always be one and possibly two lunar eclipses during each 34.5-day interval when the Sun (and Earth's shadows) pass through the nodal eclipse zones. These time periods are called eclipse seasons.

The mid-point of each eclipse season is separated by 173.3 days because this is the mean time for the Sun to travel from one node to the next. The period is a little less than half a calendar year because the lunar nodes regress westward by 19.3° per year.

3.6 Quincena

The mean time interval between New Moon and Full Moon is 14.77 days. This is less than half the duration of an eclipse season. As a consequence, the same Sun–node alignment geometry responsible for producing a lunar eclipse always results in a complementary solar eclipse within a fortnight. The solar eclipse may either precede or succeed the lunar eclipse. In either case, the pair of eclipses is referred to here as a Quincena[4]. The QSE (Quincena Solar Eclipse parameter) identifies the type of the solar eclipse and whether it precedes or succeeds a particular lunar eclipse.

---

[3] The exact angular distance from the node depends of the distances of the Sun and Moon from Earth, which determine their angular diameters as well as the diameters of Earth's shadows.

[4] Quincena is a Spanish word meaning *a period of fifteen days*. This also happens to be the time, rounded to the nearest day, between New Moon and Full Moon, or Full Moon and New Moon. So *quincena* is a convenient and appropriate term for describing a pair of eclipses (one solar and one lunar) separated by this period.

Solar eclipses can be classified into four different types:

1. partial solar eclipse (p) — The Moon's penumbral shadow traverses Earth; the umbral/antumbral shadow misses Earth
2. annular solar eclipse (a) — The Moon's antumbral shadow traverses Earth; the Moon is too far to completely cover the Sun
3. total solar eclipse (t) — The Moon's umbral shadow traverses Earth; the Moon is close enough to completely cover the Sun
4. hybrid solar eclipse (h) — The Moon's umbral and antumbral shadows traverse different parts of Earth; the eclipse appears either total or annular along different sections of its path; hybrid eclipses are also known as annular-total eclipses

The QSE is a two-character string consisting of one or more of the above solar eclipse types. The first character in the QSE identifies the type of solar eclipse preceding a lunar eclipse. The second character identifies the type of solar eclipse succeeding a lunar eclipse. In most instances, one of the two characters is "–" indicating no solar eclipse occurs. For example, a QSE of "–p" means that no solar eclipse precedes the lunar eclipse, but the lunar eclipse is followed by a partial solar eclipse 15 days later.

On rare occasions, a double Quincena occurs in which a lunar eclipse is both preceded and succeeded by a solar eclipse. In every case in the *Canon*, the lunar eclipse is always total.

Photo 7 – A time lapse sequence captures the phases of a partial lunar eclipse on 2012 Jun 04 from Portal, AZ. ©2012 F. Espenak

Section 4: Explanation of Lunar Eclipse Catalog in Appendix A

4.1 Introduction

Earth experiences 228 eclipses of the Moon during the 21st Century. The catalog in *Appendix A* summarizes the principal characteristics of each eclipse and complements the figure in *Appendices B and C.*

Each line in the tables corresponds to a single lunar eclipse and provides parameters to characterize the eclipse. The calendar date and Dynamical Time of greatest eclipse (when the Moon passes closest to the axis of Earth's shadows) are given along with the adopted value of Delta T (ΔT). The lunation number and the Saros series are listed along with the eclipse type (N=Penumbral, P=Partial, or T=Total). Gamma is the distance from the Moon's center to the axis of Earth's shadow cones at greatest eclipse, while the penumbral and umbral eclipse magnitudes are defined as the fraction of the Moon's diameter immersed in each shadow at that instant. The duration of the penumbral, partial, and total eclipse phases are given in minutes. Finally, the geographic latitude and longitude are given for the location where the Moon lies in the zenith at greatest eclipse. A more detailed description of each field in the catalog appears in the following sections.

4.2 Cat Num (Catalog Number)

The catalog number is the sequential number assigned to each eclipse from 1 to 228.

4.3 Calendar Date

Gregorian calendar date at greatest eclipse is given in the ISO order of YEAR-MONTH-DAY.

4.4 TD of Greatest Eclipse (Terrestrial Dynamical Time of Greatest Eclipse)

The instant of greatest eclipse occurs when the distance between the axis of the Moon's shadow cone and the center of Earth reaches a minimum. This instant differs slightly from the instant of greatest magnitude due to ellipticity of Earth's shadows.

Greatest eclipse is given in Terrestrial Dynamical Time or TD (Sect. 2.5), which is a time system based on International Atomic Time. As such, TD is the modern equivalent to its predecessor Ephemeris Time and is used in theories of planetary motion in the Solar System. To determine the geographic visibility of an eclipse, TD is converted to Universal Time (UT1) using the parameter Delta T.

4.5 ΔT (Delta T)

ΔT (Delta T) is the arithmetic difference, in seconds, between Terrestrial Dynamical Time (TD) and Universal Time (UT1). For more information on ΔT, see Section 2.6.

4.6 Luna Num (Lunation Number)

The lunation number is the number of synodic months or lunations since New Moon on 2000 Jan 06. It can be converted to the Brown Lunation Number[5] by adding 953.

4.7 Saros Num (Saros Series Number)

Each lunar eclipse belongs to a Saros series using a numbering system first introduced by van den Bergh (1955). The eclipses with an even Saros number take place at the ascending node of the Moon's orbit; those with an odd Saros number take place at the descending node. This relationship is reversed for *solar* eclipses.

---

[5] The *Brown Lunation Number* defines lunation 1 as beginning at the first New Moon of 1923, the year when Ernest W. Brown's lunar theory was introduced in the major national astronomical almanacs.

The Saros is a period of 223 synodic months (~ 18 years, 11 days, and 8 hours). Eclipses separated by this period belong to the same Saros series and share similar geometry and characteristics.

4.8 Ecl Type (Lunar Eclipse Type)

The first character in this 2-character parameter gives the lunar eclipse type. The three basic types of lunar eclipses are:

1. Penumbral Lunar Eclipse (N) — The Moon passes through Earth's penumbral shadow
2. Partial Lunar Eclipse (P) — The Moon passes partly into Earth's umbral shadow
3. Total Lunar Eclipse (T) — The Moon passes completely into Earth's umbral shadow

The second character of the lunar eclipse type is a qualifier defined as follows.

1. m = Middle eclipse of Saros series
2. + = Central total eclipse (Moon's center passes north of shadow axis)
3. – = Central total eclipse (Moon's center passes south of shadow axis)
4. * = Total penumbral eclipse
5. b = Saros series begins (first penumbral eclipse in a Saros series)
6. e = Saros series ends (last penumbral eclipse in a Saros series)

Qualifiers 1 through 3 are used exclusively with total lunar eclipses while qualifiers 4 through 6 are exclusively used with penumbral eclipses.

4.9 QSE (Quincena Solar Eclipse Parameter)

A solar eclipse always occurs within 15 days of a lunar eclipse. The Quincena Solar Eclipse parameter (QSE) identifies the type of the solar eclipse (partial, annular, total, or hybrid) and whether it precedes or succeeds a particular lunar eclipse.

The QSE is a two-character string consisting of one or more of the above solar eclipse types (p, a, t, or h). The first character in the QSE identifies a solar eclipse preceding a lunar eclipse, while the second character identifies a solar eclipse succeeding a lunar eclipse. In most instances, one of the two characters is "–" indicating no solar eclipse occurs. On rare occasions, a double Quincena occurs in which a lunar eclipse is both preceded and succeeded by solar eclipses. See Section 3.6 for more information on the Quincena.

4.10 Gamma

Gamma is the minimum distance from the center of the Moon to the axis of Earth's umbral shadow cone in units of Earth's equatorial radius. This distance is positive or negative, depending on whether the Moon passes north or south, respectively, of the shadow cone axis.

4.11 Um Mag (Umbral Eclipse Magnitude)

The umbral magnitude is defined as the fraction of the Moon's diameter immersed in Earth's umbral shadow. During a partial lunar eclipse, some portion of the Moon's disk enters the umbral shadow. The umbral magnitude for partial eclipses in the *Canon* ranges from 0.0153 to 0.9760 (Sect. 3.3). Of course, the Moon also passes through the penumbra during a partial eclipse, so the penumbral magnitude is usually greater than 1.

In the case of a total lunar eclipse, the Moon's entire disk passes through Earth's umbral shadow, so the umbral magnitude is always equal to or greater than 1.0. During totality, the Moon can take on a range of colors from bright orange, to deep red, dark brown, or even very dark gray (Sect. 1.5). The only light reaching the Moon at this time is heavily filtered and attenuated by Earth's atmosphere. The umbral magnitude for total eclipses in the *Canon* ranges from 1.0019 to 1.8452 (Sect. 3.3).

4.12 Pen Mag (Penumbral Eclipse Magnitude)

The penumbral eclipse magnitude is defined as the fraction of the Moon's diameter immersed in Earth's penumbral shadow.

The penumbral eclipse magnitude of penumbral eclipses in the *Canon* ranges from 0.0032 to 1.0360 (Sect. 3.3). For most penumbral eclipses the penumbral magnitude is less than 1, meaning only a fraction of the Moon's disk enters the penumbra. When the penumbral magnitude is greater than or equal to 1, the Moon's entire disk is immersed in the penumbra and the event is termed a total penumbral eclipse. Penumbral eclipses are subtle events (Sect. 1.3). They cannot be detected visually (with or without optical aid) unless the eclipse magnitude is greater than ~0.6. The umbral eclipse magnitude of a penumbral eclipse is always negative. It is a measure of the distance of the Moon's limb to the edge of the umbral shadow in units of the Moon's diameter.

4.13 Phase Durations (Pen, Par & Total)

The duration of a penumbral eclipse "*Pen*" is the time between first and last external tangencies of the Moon with the penumbral shadow (= P4 – P1). Similarly, the duration of a partial eclipse "*Par*" is the time between first and last external tangencies of the Moon with the umbral shadow (= U4 – U1). Finally, the duration of a total eclipse "*Total*" is the time between first and last internal tangencies of the Moon with the umbral shadow (= U3 – U2). The position of the Moon at each contact is illustrated in Figure 2–1. The duration given for each eclipse phase is given to the nearest 0.1 of a minute.

4.14 Greatest in Zenith Lat Long (Latitude & Longitude)

The geographic latitude and longitude where the Moon appears in the zenith at greatest eclipse.

4.15 EclipseWise.com and the 21st Century Canon

Much of the content of the *21st Century Canon of Lunar Eclipses* is based on the eclipse predictions website *www.EclipseWise.com*. A plain text file containing the entire lunar eclipse catalog in *Appendix A* can be downloaded at: *www.EclipseWise.com/pubs/21CCLE.html*.

A web-based version of the catalog is available at: *www.eclipsewise.com/lunar/LEcatalog/21CCLEcat.html*

Section 5: Explanation of Lunar Eclipse Figures in Appendices B and C

5.1 Explanation of Small Eclipse Figures in Appendix B

Earth experiences 228 eclipses of the Moon during the 21st Century. An individual diagram and visibility map for each eclipse appears in *Appendix B*.

The figure for each eclipse consists of two diagrams. The top one depicts the Moon's path through Earth's penumbral and umbral shadows with Celestial North directed up. The second diagram is a map showing the global visibility of each eclipse phase. All salient features in these diagrams are identified in figure 5–1, which serves as a key.

Figure 5–1: Key to Lunar Eclipse Figures

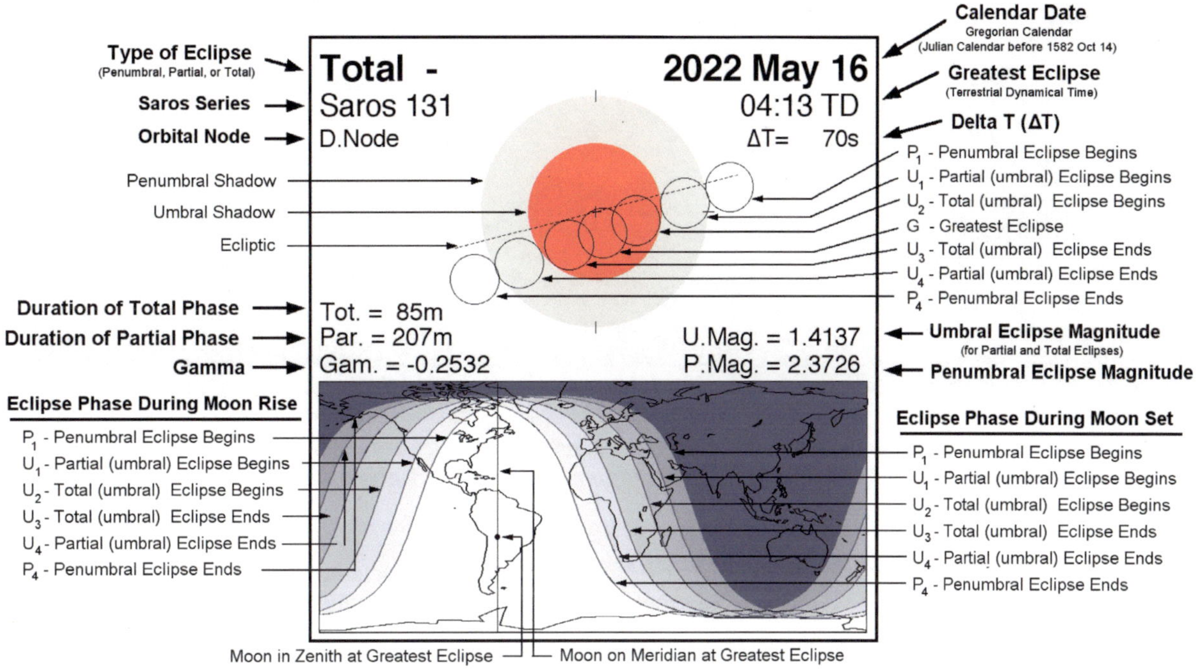

In the diagram, the Moon's orbital motion with respect to the shadows is from west to east (right to left). Each phase of the eclipse is defined by the instant when the Moon's limb is externally or internally tangent to the penumbra or umbra. The six primary contacts of the Moon with the penumbral and umbral shadows are defined as follows.

P1 — Instant of first exterior tangency of the Moon with the Penumbra
(Penumbral Eclipse Begins)
U1 — Instant of first exterior tangency of the Moon with the Umbra
(Partial Umbral Eclipse Begins)
U2 — Instant of first interior tangency of the Moon with the Umbra
(Total Umbral Eclipse Begins)
U3 — Instant of last interior tangency of the Moon with the Umbra
(Total Umbral Eclipse Ends)
U4 — Instant of last exterior tangency of the Moon with the Umbra
(Partial Umbral Eclipse Ends)
P4 — Instant of last exterior tangency of the Moon with the Penumbra
(Penumbral Eclipse Ends)

Penumbral lunar eclipses have two primary contacts: P1 and P4, but neither is observable because the edge of the penumbra is indistinct and extremely faint.

Photo 7 – The beginning, middle and end of the total lunar eclipse of 2015 September 28. ©2015 F. Espenak

In addition to the penumbral contacts, partial lunar eclipses have two more contacts as the Moon's limb enters and exits the umbral shadow: U1 and U4, respectively. These two contacts mark the instants when the partial phase of the eclipse begins and ends.

Total lunar eclipses undergo all six contacts. The two additional umbral contacts are the instants when the Moon's entire disk is first and last internally tangent to the umbra: U2 and U3, respectively. They mark the times when the total phase of the eclipse begins and ends.

The Moon passes closest to the shadow axis at the instant of greatest eclipse. This corresponds to the maximum phase of the eclipse and the Moon's position at this instant is also depicted in the path diagrams.

The bottom diagram in each figure is an equidistant cylindrical projection map of Earth showing the geographic region of visibility at each phase of the eclipse. This is accomplished using a series of curves showing where Moonrise and Moonset occur at each eclipse contact. The map is shaded to indicate eclipse visibility. The entire eclipse is visible from the zone with no shading. Conversely, none of the eclipse can be seen from the zone with the darkest shading.

At greatest eclipse, the Moon is deepest in Earth's shadow. A vertical line running through the middle of the clear zone (complete eclipse visibility) indicates the meridian or line of longitude that the Moon is crossing. An observer positioned on this line would then see the Moon at its highest point in the sky due south or due north, depending on the observer's latitude and the Moon's declination. The geographic location where the Moon appears in the zenith at greatest eclipse is shown by a black dot on the meridian. All salient features of the eclipse figures are identified in Figure 5–1, which serves as the key to the figures in *Appendix B*.

Data relevant to the eclipse appear in the corners of each figure. To the top left are the eclipse type (penumbral, partial or total), the Saros series of the eclipse, and the node of the Moon's orbit where the eclipse occurs. To the top right are the Gregorian calendar date (Julian calendar dates are used prior to 1582 Oct 15), the time of greatest eclipse (Terrestrial Dynamical Time), and the value of ΔT (Delta T). The lower left corner lists the duration of the major phases of the eclipse in minutes. Depending on the eclipse type, the duration of the penumbral, partial or total phases are given. Beneath the eclipse durations is the quantity gamma, the minimum distance of the Moon's center from the axis of Earth's shadow cones at the instant of greatest eclipse. The umbral and penumbral eclipse magnitudes are given to the lower right. The list below summarizes the parameters appearing in each eclipse figure.

Lunar Eclipse Type – One of three basic types of lunar eclipses: Penumbral, Partial, or Total.

Saros Series – The Saros series to which the eclipse belongs. Eclipses with an odd number take place at the Moon's ascending node, while those with an even number are at the descending node.

Node – The orbital node near which the eclipse takes place. The ascending node (A. Node) is the point where the Moon travels from south to north through Earth's orbital plane. Similarly, the descending node (D. Node) is the point where the Moon travels from north to south.

Calendar Date – Gregorian calendar date at greatest eclipse is given in the ISO order of YEAR-MONTH-DAY.

Greatest Eclipse – The instant (Terrestrial Dynamical Time) when the distance between the Moon and the axis of Earth's shadow cone reaches a minimum.

ΔT (Delta T) – Difference, (seconds) between Terrestrial Dynamical Time (TD) and Universal Time (UT1).

Gamma – Minimum distance from the Moon's center to Earth's shadow axis (units of Earth's equatorial radius).

Duration – The duration of a penumbral eclipse (**Pen.**), a partial eclipse (**Par.**), and a total eclipse (**Tot.**).

Umbral Eclipse Magnitude (U. Mag) – The fraction of the Moon's diameter immersed in the umbra at greatest eclipse.

Penumbral Eclipse Magnitude (P. Mag) – The fraction of the Moon's diameter immersed in the penumbra at greatest eclipse.

Photo 8 – The setting Moon in total eclipse on 2018 January 31. ©2018 F. Espenak

5.2 Explanation of Full Page Lunar Eclipse Figures in Appendix C

The heart of the *21st Century Canon of Lunar Eclipses – Color Edition* is the set of 112 full-page figures appearing in Appendix C — one map for every lunar eclipse from 2021 through 2070[6]. This feature is similar to the *Fifty Year Canon of Lunar Eclipses: 1986 – 2035* (Espenak 1989) and makes *21st Century Canon* its modern successor.

Each lunar eclipse has two diagrams associated with it along with data about the eclipse. The top figure shows the path of the Moon through Earth's penumbral and umbral shadows. Above this figure is listed the instant of greatest eclipse, expressed in Terrestrial Dynamical Time and Universal Time (UT1). The *Penumbral Magnitude* and *Umbral Magnitude* are defined as the fraction of the Moon's diameter immersed in each shadow at greatest eclipse. *Gamma* is the minimum distance of the Moon's center from Earth's shadow axis in Earth radii, and *Axis* is the same parameter expressed in degrees. The *Saros Series* of the eclipse is listed next. *Saros Member* consists of a pair of numbers. The first is the eclipse's relative position in the Saros series; the second is the total number of eclipses in the series.

In the upper left and right corners are the geocentric coordinates of the Sun and the Moon, respectively, at the instant of greatest eclipse. They are:

R.A. - Right Ascension
Dec. - Declination
S.D. - Apparent Semi-Diameter
H.P. - Horizontal Parallax

To the lower left are the durations of the penumbral, umbral (partial), and total eclipses. Below them are the solar/lunar ephemerides used in the predictions, the applied shadow *Rule*, and the extrapolated value of ΔT. To the lower right are the contact times of the Moon with Earth's penumbral and umbral shadows, as follows:

P1 - Instant of first exterior tangency of Moon with Penumbra. (Penumbral Eclipse Begins)
U1 - Instant of first exterior tangency of Moon with Umbra. (Partial Umbral Eclipse Begins)
U2 - Instant of first interior tangency of Moon with Umbra. (Total Umbral Eclipse Begins)
U3 - Instant of last interior tangency of Moon with Umbra. (Total Umbral Eclipse Ends)
U4 - Instant of last exterior tangency of Moon with Umbra (Partial Umbral Eclipse Ends)
P4 - Instant of last exterior tangency of Moon with Penumbra. (Penumbral Eclipse Ends)

The bottom figure is a cylindrical equidistant projection map of Earth that shows the regions of visibility for each stage of the eclipse. In particular, the moonrise/moonset terminator is plotted for each contact and is labeled accordingly. An asterisk indicates the point where the Moon is at the zenith at greatest eclipse. The region that is unshaded will observe the entire eclipse, while the darkest shaded area will witness no eclipse. The remaining lightly shaded areas will experience moonrise or moonset while the eclipse is in progress. The shaded zones east of the asterisk (*) will witness moonset before the eclipse ends, and the shaded zones west will witness moonrise after the eclipse has begun.

[6] Note that the *Deluxe Edition* contains 228 full-page figures – one figure for every lunar eclipse from 2001 through 2100.

The salient features of these figures are identified in Figure 5–2, which serves as a key.

Figure 5–2: Key to Full Page Lunar Eclipse Figures (Appendix C)

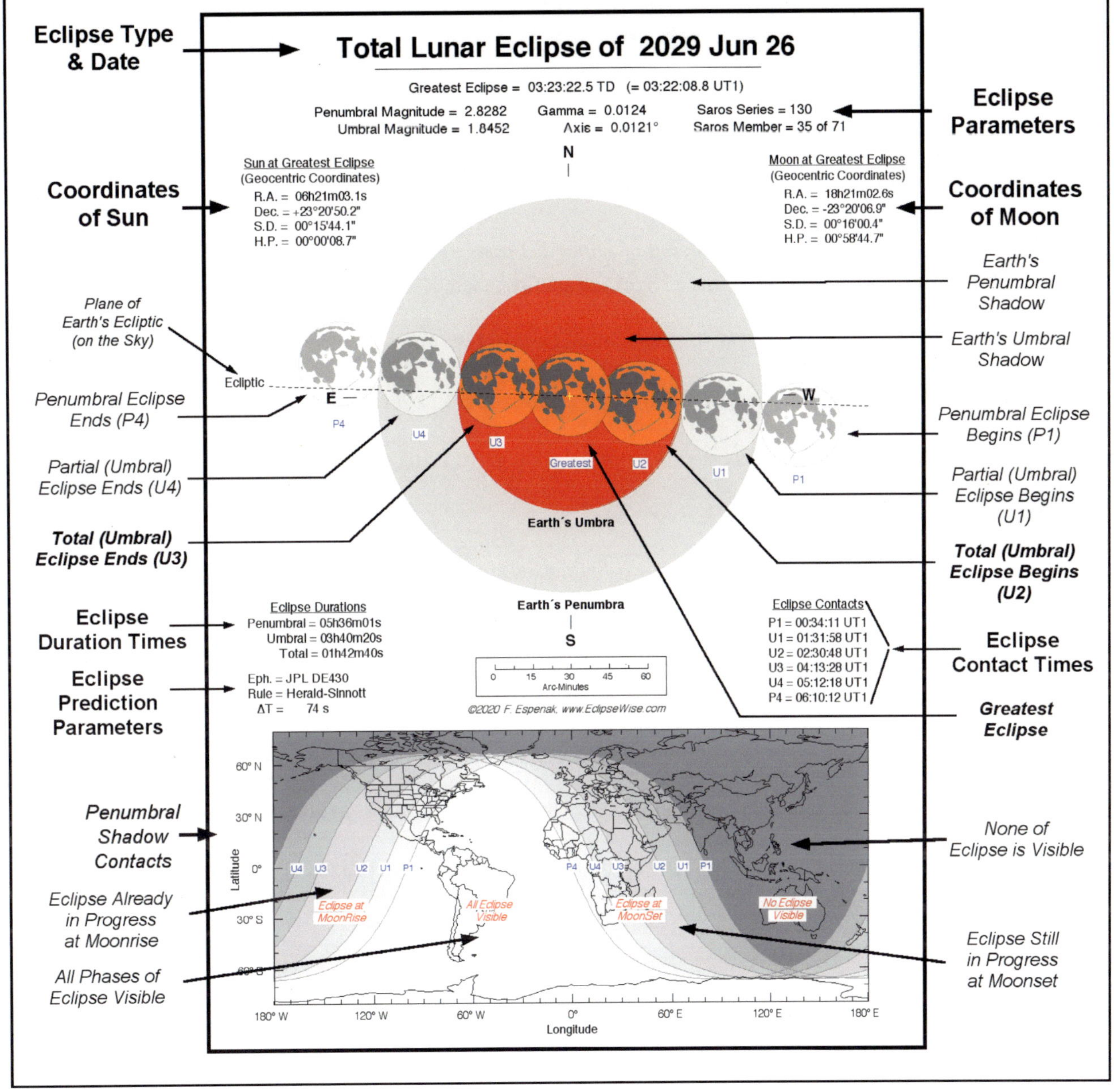

References

Astronomical Almanac for 2011, Washington: US Government Printing Office; London: HM Stationery Office (2010).

Chauvenet, W.A., *Manual of Spherical and Practical Astronomy*, Vol.1, edition of 1891, (Dover reprint, New York, 1960).

Danjon, A., "Les éclipses de Lune par la pénombre en 1951," *L'Astronomie*, 65, 51–53 (1951).

Espenak, F., *Fifty Year Canon of Lunar Eclipses: 1986–2035*, Sky Publishing Corp., Cambridge, Massachusetts (1989).

Espenak, F., *Thousand Year Canon of Lunar Eclipses: 1501 to 2500*, Astropixels Publishing, Portal, Arizona (2014).

Espenak, F., and Meeus, J., *Five Millennium Canon of Lunar Eclipses: –1999 to +3000 (2000 BCE to 3000 CE)*, NASA Tech. Pub. 2006–214172, NASA Goddard Space Flight Center, Greenbelt, Maryland (2009a).

Espenak, F., and Meeus, J., *Five Millennium Catalog of Lunar Eclipses: –1999 to +3000 (2000 BCE to 3000 CE)*, NASA Tech. Pub. 2006–214173, NASA Goddard Space Flight Center, Greenbelt, Maryland (2009b).

Explanatory Supplement to the Ephemeris, H.M. Almanac Office, London (1974).

Herald, D., and Sinnott, R. W., "Analysis of Lunar Crater Timings, 1842–2011," *J. Br. Astron. Assoc.*, **124**, 5 (2014)

Keen, R. A., "Volcanic Aerosols and Lunar Eclipses", *Science*, vol. 222, p. 1011-1013, Dec. 2, 1983.

Lahire, P., *Tabulae Astronomicae* (Paris 1707).

Link, F., *Eclipse Phenomena in Astronomy*, Springer-Verlag, New York (1969).

Liu, Bao-Lin and Fiala, A. D., *Canon of Lunar Eclipses 1500 B.C.-A.D. 3000*, Willmann-Bell, p215 (1992).

Meeus, J. and Mucke, H., *Canon of Lunar Eclipse: -2002 to +2526*, Astronomisches Buro, Wein (1979).

Newcomb, S., "Tables of the Motion of the Earth on its Axis Around the Sun," *Astron. Papers Amer. Eph.,* Vol. 6, Part I (1895).

van den Bergh, *Periodicity and Variation of Solar (and Lunar) Eclipses*, Tjeenk Willink, and Haarlem, Netherlands (1955).

von Oppolzer, T.R., *Canon der Finsternisse*, Wien, (1887); Gingerich, O., (Translator) *Canon of Eclipses*, Dover Publications, New York (1962).

Appendix A

Lunar Eclipse Catalog: 2001 to 2100

Key to Lunar Eclipse Catalog

Cat Num — sequential Catalog Number assigned to each eclipse from 1 to 228

Calendar Date — Gregorian date of Greatest Eclipse

TD of Greatest Eclipse — Terrestrial Dynamical Time of Greatest Eclipse

ΔT — arithmetic difference between Terrestrial Dynamical Time Universal Time (seconds)

Luna Num — number of synodic months, or lunations, since New Moon on 2000 Jan 06

Saros Num — Saros Series Number of eclipse

Ecl Type — Lunar Eclipse Type

N = Penumbral Lunar Eclipse
P = Partial Lunar Eclipse
T = Total Lunar Eclipse
m = Middle eclipse of Saros series
+ = Central total eclipse (Moon's center passes north of shadow axis)
– = Central total eclipse (Moon's center passes south of shadow axis)
* = Total penumbral eclipse
b = Saros series begins (first penumbral eclipse in a Saros series)
e = Saros series ends (last penumbral eclipse in a Saros series)

QSE — Quincena Solar Eclipse Parameter

p = Partial Solar Eclipse
a = Annular Solar Eclipse
t = Total Solar Eclipse
h = Hybrid Solar Eclipse

Gamma — minimum distance from center of Moon to axis of Earth's umbral shadow cone

Pen Mag — Penumbral Magnitude; fraction of the Moon's diameter immersed in penumbra

Um Mag — Umbral Magnitude; fraction of the Moon's diameter immersed in umbra

Phase Durations
Pen — elapsed time from contact P1 to P4 (minutes)
Par — elapsed time from contact U1 to U4 (minutes)
Total — elapsed time from contact U2 to U3 (minutes)

Greatest in Zenith
Lat & Long — latitude and longitude where Moon appears in zenith at greatest eclipse

Cat Num	Calendar Date	TD of Greatest Eclipse	ΔT s	Luna Num	Saros Num	Ecl Type	QSE	Gamma	Pen Mag	Um Mag	Phase Durations Pen m	Phase Durations Par m	Phase Durations Total m	Greatest in Zenith Lat	Greatest in Zenith Long
001	2001 Jan 09	20:21:40	64	12	134	T	p-	0.3720	2.1631	1.1902	311.9	197.1	61.7	22N	57E
002	2001 Jul 05	14:56:23	64	18	139	P	t-	-0.7287	1.5489	0.4961	326.0	160.0	-	23S	137E
003	2001 Dec 30	10:30:22	64	24	144	N	a-	1.0732	0.8948	-0.1141	244.3	-	-	24N	157W
004	2002 May 26	12:04:27	64	29	111	N	-a	1.1759	0.6910	-0.2871	217.3	-	-	20S	179E
005	2002 Jun 24	21:28:13	64	30	149	N	a-	-1.4440	0.2110	-0.7910	129.8	-	-	25S	39E
006	2002 Nov 20	01:47:41	64	35	116	N	-t	-1.1127	0.8618	-0.2246	265.1	-	-	19N	30W
007	2003 May 16	03:41:13	64	41	121	T	-a	0.4123	2.0765	1.1294	307.3	194.6	52.1	19S	56W
008	2003 Nov 09	01:19:38	64	47	126	T	-t	-0.4319	2.1157	1.0197	364.0	212.2	23.2	16N	24W
009	2004 May 04	20:31:17	65	53	131	T	p-	-0.3132	2.2645	1.3054	316.5	203.9	76.1	17S	51E
010	2004 Oct 28	03:05:11	65	59	136	T	p-	0.2846	2.3656	1.3100	354.6	219.4	81.1	13N	50W
011	2005 Apr 24	09:55:54	65	65	141	N	h-	-1.0885	0.8669	-0.1417	246.4	-	-	14S	150W
012	2005 Oct 17	12:04:27	65	71	146	P	a-	0.9796	1.0605	0.0645	260.5	57.0	-	10N	175E
013	2006 Mar 14	23:48:35	65	76	113	Nx	-t	1.0211	1.0320	-0.0583	288.2	-	-	3N	6E
014	2006 Sep 07	18:52:25	65	82	118	P	-a	-0.9262	1.1349	0.1857	255.0	91.8	-	7S	77E
015	2007 Mar 03	23:21:59	65	88	123	T	-p	0.3175	2.3208	1.2347	366.2	221.8	74.0	7N	13E
016	2007 Aug 28	10:38:27	65	94	128	T-	-p	-0.2146	2.4545	1.4777	328.0	212.8	90.6	10S	159W
017	2008 Feb 21	03:27:09	66	100	133	T	a-	-0.3992	2.1470	1.1081	339.7	206.1	50.5	10N	48W
018	2008 Aug 16	21:11:12	66	106	138	P	t-	0.5646	1.8385	0.8095	331.3	188.8	-	13S	43E
019	2009 Feb 09	14:39:21	66	112	143	N	a-	-1.0640	0.9013	-0.0863	239.5	-	-	14N	144E
020	2009 Jul 07	09:39:44	66	117	110	N	-t	-1.4916	0.1578	-0.9116	122.4	-	-	24S	143W
021	2009 Aug 06	00:40:16	66	118	148	N	t-	1.3572	0.4038	-0.6642	190.6	-	-	16S	9W
022	2009 Dec 31	19:23:46	66	123	115	P	-a	0.9766	1.0572	0.0779	251.8	60.8	-	24N	70E
023	2010 Jun 26	11:39:34	66	129	120	P	-t	-0.7091	1.5789	0.5383	323.0	163.6	-	24S	174W
024	2010 Dec 21	08:18:04	66	135	125	T	-p	0.3214	2.2821	1.2576	336.0	209.5	73.0	24N	125W
025	2011 Jun 15	20:13:43	67	141	130	T+	pp	0.0897	2.6883	1.7014	337.0	220.1	101.0	23S	57E
026	2011 Dec 10	14:32:57	67	147	135	T	p-	-0.3882	2.1875	1.1076	357.3	213.1	51.8	23N	140E
027	2012 Jun 04	11:04:20	67	153	140	P	a-	0.8248	1.3198	0.3718	270.8	127.2	-	22S	166W
028	2012 Nov 28	14:34:07	67	159	145	N	t-	-1.0869	0.9168	-0.1859	276.8	-	-	20N	139E
029	2013 Apr 25	20:08:37	67	164	112	P	-a	-1.0121	0.9878	0.0160	248.4	28.1	-	14S	57E
030	2013 May 25	04:11:07	67	165	150	Nb	a-	1.5351	0.0170	-0.9322	35.1	-	-	19S	63W
031	2013 Oct 18	23:51:25	67	170	117	N	-h	1.1508	0.7660	-0.2706	239.8	-	-	11N	2W
032	2014 Apr 15	07:46:48	67	176	122	T	-a	-0.3017	2.3193	1.2918	344.8	215.5	78.5	10S	116W
033	2014 Oct 08	10:55:44	67	182	127	T	-p	0.3827	2.1467	1.1670	318.9	200.2	59.4	6N	167W
034	2015 Apr 04	12:01:24	68	188	132	T	t-	0.4460	2.0802	1.0019	358.4	209.8	7.3	5S	179W
035	2015 Sep 28	02:48:17	68	194	137	T	p-	-0.3296	2.2307	1.2774	311.5	200.6	72.5	2N	44W
036	2016 Mar 23	11:48:22	68	200	142	N	t-	1.1592	0.7758	-0.3107	256.1	-	-	0S	175W
037	2016 Sep 16	18:55:27	68	206	147	N	a-	-1.0549	0.9091	-0.0624	239.9	-	-	3S	75E
038	2017 Feb 11	00:45:03	68	211	114	N	-a	-1.0255	0.9896	-0.0342	259.9	-	-	13N	8W
039	2017 Aug 07	18:21:38	69	217	119	P	-t	0.8669	1.2898	0.2477	301.7	115.9	-	15S	86E
040	2018 Jan 31	13:31:00	69	223	124	T	-p	-0.3014	2.2954	1.3167	318.0	203.5	76.7	17N	161E
041	2018 Jul 27	20:22:54	69	229	129	T+	pp	0.1168	2.6805	1.6100	374.8	235.4	103.8	19S	56E
042	2019 Jan 21	05:13:27	69	235	134	T	p-	0.3684	2.1697	1.1966	312.3	197.5	62.6	20N	75W
043	2019 Jul 16	21:31:55	69	241	139	P	t-	-0.6430	1.7050	0.6544	334.6	178.7	-	22S	39E
044	2020 Jan 10	19:11:11	69	247	144	N	a-	1.0727	0.8969	-0.1146	245.3	-	-	23N	74E
045	2020 Jun 05	19:26:14	70	252	111	N	-a	1.2406	0.5699	-0.4036	198.9	-	-	21S	69E
046	2020 Jul 05	04:31:12	70	253	149	N	a-	-1.3639	0.3560	-0.6422	165.7	-	-	24S	66W
047	2020 Nov 30	09:44:02	70	258	116	N	-t	-1.1309	0.8302	-0.2602	261.8	-	-	21N	148W
048	2021 May 26	11:19:53	70	264	121	T	-a	0.4774	1.9558	1.0112	302.8	188.1	15.9	21S	170W
049	2021 Nov 19	09:04:06	70	270	126	P	-t	-0.4552	2.0738	0.9760	362.4	209.2	-	19N	139W
050	2022 May 16	04:12:42	70	276	131	T-	p-	-0.2532	2.3743	1.4154	319.5	207.9	85.5	19S	64W
051	2022 Nov 08	11:00:22	71	282	136	T+	p-	0.2570	2.4161	1.3607	354.7	220.6	85.7	17N	169W
052	2023 May 05	17:24:04	71	288	141	N	h-	-1.0350	0.9655	-0.0438	258.3	-	-	17S	98E
053	2023 Oct 28	20:15:17	71	294	146	P	a-	0.9472	1.1200	0.1239	265.3	78.1	-	14N	52E
054	2024 Mar 25	07:14:00	71	299	113	N	-t	1.0610	0.9577	-0.1304	279.9	-	-	1S	106W
055	2024 Sep 18	02:45:26	71	305	118	P	-a	-0.9792	1.0392	0.0869	246.9	63.7	-	3S	42W
056	2025 Mar 14	06:59:56	72	311	123	T	-p	0.3485	2.2615	1.1804	363.4	218.9	66.1	3N	102W
057	2025 Sep 07	18:12:58	72	317	128	T	-p	-0.2752	2.3459	1.3638	327.4	210.0	82.7	6S	87E
058	2026 Mar 03	11:34:52	72	323	133	T	a-	-0.3765	2.1858	1.1526	339.4	207.8	59.0	6N	171W
059	2026 Aug 28	04:14:04	72	329	138	P	t-	0.4964	1.9664	0.9319	338.5	198.8	-	9S	63W
060	2027 Feb 20	23:14:05	73	335	143	N	a-	-1.0480	0.9286	-0.0549	241.7	-	-	10N	15E

Cat Num	Calendar Date	TD of Greatest Eclipse	ΔT s	Luna Num	Saros Num	Ecl Type	QSE	Gamma	Pen Mag	Um Mag	Phase Durations Pen m	Phase Durations Par m	Phase Durations Total m	Greatest in Zenith Lat	Greatest in Zenith Long
061	2027 Jul 18	16:04:10	73	340	110	Ne	-t	-1.5759	0.0032	-1.0662	17.7	-	-	22S	121E
062	2027 Aug 17	07:14:58	73	341	148	N	t-	1.2797	0.5476	-0.5234	219.4	-	-	12S	108W
063	2028 Jan 12	04:14:13	73	346	115	P	-a	0.9818	1.0485	0.0679	251.4	56.9	-	23N	61W
064	2028 Jul 06	18:20:57	73	352	120	P	-t	-0.7904	1.4282	0.3908	311.5	142.2	-	23S	86E
065	2028 Dec 31	16:53:15	73	358	125	T	-p	0.3258	2.2758	1.2478	337.1	209.6	72.0	23N	108E
066	2029 Jun 26	03:23:22	74	364	130	T+	pp	0.0124	2.8282	1.8452	336.0	220.3	102.7	23S	50W
067	2029 Dec 20	22:43:12	74	370	135	T	p-	-0.3811	2.2023	1.1190	358.9	214.1	54.4	23N	19E
068	2030 Jun 15	18:34:33	74	376	140	P	a-	0.7535	1.4495	0.5040	279.0	145.0	-	23S	82E
069	2030 Dec 09	22:28:51	74	382	145	N	t-	-1.0732	0.9430	-0.1613	280.1	-	-	22N	21E
070	2031 May 07	03:52:01	75	387	112	N	-a	-1.0695	0.8827	-0.0892	238.1	-	-	18S	59W
071	2031 Jun 05	11:45:17	75	388	150	N	a-	1.4732	0.1306	-0.8185	96.3	-	-	21S	176W
072	2031 Oct 30	07:46:44	75	393	117	N	-h	1.1774	0.7173	-0.3193	232.5	-	-	15N	121W
073	2032 Apr 25	15:14:51	75	399	122	T	-a	-0.3558	2.2204	1.1925	343.3	212.0	66.2	14S	131E
074	2032 Oct 18	19:03:40	75	405	127	T	-p	0.4169	2.0841	1.1039	316.2	196.6	47.7	10N	71E
075	2033 Apr 14	19:13:51	76	411	132	T	t-	0.3954	2.1722	1.0955	362.1	215.8	49.8	9S	72E
076	2033 Oct 08	10:56:23	76	417	137	T	p-	-0.2889	2.3068	1.3508	313.4	203.1	79.4	6N	167W
077	2034 Apr 03	19:07:00	76	423	142	N	t-	1.1144	0.8557	-0.2263	266.2	-	-	5S	75E
078	2034 Sep 28	02:47:37	76	429	147	P	a-	-1.0110	0.9922	0.0155	249.4	27.8	-	1N	44W
079	2035 Feb 22	09:06:11	76	434	114	N	-a	-1.0367	0.9663	-0.0523	256.4	-	-	9N	133W
080	2035 Aug 19	01:12:15	77	440	119	P	-t	0.9434	1.1519	0.1049	290.6	77.2	-	12S	17W
081	2036 Feb 11	22:13:06	77	446	124	T	-p	-0.3110	2.2762	1.3006	316.9	202.7	75.1	14N	31E
082	2036 Aug 07	02:52:32	77	452	129	T+	pp	0.2004	2.5279	1.4556	373.1	232.2	96.1	16S	41W
083	2037 Jan 31	14:01:38	77	458	134	T	p-	0.3619	2.1815	1.2086	312.9	198.2	64.3	18N	153E
084	2037 Jul 27	04:09:53	78	464	139	P	t-	-0.5582	1.8596	0.8108	341.7	193.2	-	20S	60W
085	2038 Jan 21	03:49:52	78	470	144	N	a-	1.0711	0.9009	-0.1127	246.5	-	-	21N	54W
086	2038 Jun 17	02:45:02	78	475	111	N	-a	1.3083	0.4438	-0.5259	177.0	-	-	22S	41W
087	2038 Jul 16	11:35:56	78	476	149	N	a-	-1.2838	0.5012	-0.4938	193.2	-	-	23S	172W
088	2038 Dec 11	17:45:00	79	481	116	N	-t	-1.1449	0.8062	-0.2876	259.3	-	-	22N	93E
089	2039 Jun 06	18:54:25	79	487	121	P	-a	0.5460	1.8288	0.8863	297.5	180.0	-	22S	77E
090	2039 Nov 30	16:56:28	79	493	126	P	-t	-0.4721	2.0435	0.9443	361.0	206.8	-	21N	103E
091	2040 May 26	11:46:22	79	499	131	T-	p-	-0.1872	2.4955	1.5365	322.2	211.5	92.9	21S	177W
092	2040 Nov 18	19:04:40	80	505	136	T+	p-	0.2361	2.4543	1.3991	354.4	221.2	88.6	20N	70E
093	2041 May 16	00:43:02	80	511	141	P	t-	-0.9747	1.0765	0.0663	270.5	59.4	-	20S	12W
094	2041 Nov 08	04:35:04	80	517	146	P	a-	0.9212	1.1675	0.1714	268.7	91.1	-	18N	73W
095	2042 Apr 05	14:30:12	80	522	113	N	-t	1.1080	0.8700	-0.2156	269.2	-	-	5S	144E
096	2042 Sep 29	10:45:47	81	528	118	N	-a	-1.0262	0.9548	-0.0010	239.2	-	-	2N	163W
097	2043 Mar 25	14:32:04	81	534	123	T	-t	0.3849	2.1920	1.1161	360.0	215.3	54.1	2S	144E
098	2043 Sep 19	01:51:50	81	540	128	T	-a	-0.3316	2.2452	1.2575	326.4	206.7	72.3	2S	29W
099	2044 Mar 13	19:38:33	82	546	133	T	a-	-0.3496	2.2322	1.2050	339.1	209.7	67.0	2N	68E
100	2044 Sep 07	11:20:44	82	552	138	T	t-	0.4318	2.0879	1.0476	344.8	206.9	34.8	5S	171W
101	2045 Mar 03	07:43:25	82	558	143	N	a-	-1.0274	0.9643	-0.0148	244.6	-	-	6N	113W
102	2045 Aug 27	13:54:49	82	564	148	N	t-	1.2061	0.6845	-0.3899	242.4	-	-	9S	151E
103	2046 Jan 22	13:02:37	83	569	115	P	-a	0.9886	1.0365	0.0550	250.7	51.4	-	21N	168E
104	2046 Jul 18	01:06:05	83	575	120	P	-t	-0.8692	1.2824	0.2478	299.0	115.3	-	22S	14W
105	2047 Jan 12	01:26:14	83	581	125	T	-p	0.3317	2.2665	1.2358	338.1	209.6	70.7	22N	19W
106	2047 Jul 07	10:35:45	84	587	130	T-	pp	-0.0636	2.7326	1.7529	334.3	219.3	101.6	23S	157W
107	2048 Jan 01	06:53:55	84	593	135	T	p-	-0.3746	2.2158	1.1297	360.4	215.1	56.7	23N	102W
108	2048 Jun 26	02:02:28	84	599	140	P	a-	0.6797	1.5841	0.6404	286.5	159.8	-	23S	30W
109	2048 Dec 20	06:27:48	84	605	145	N	t-	-1.0624	0.9632	-0.1420	282.4	-	-	22N	97W
110	2049 May 17	11:26:38	85	610	112	N	-a	-1.1337	0.7650	-0.2073	225.0	-	-	21S	172W
111	2049 Jun 15	19:14:12	85	611	150	N	a-	1.4069	0.2526	-0.6970	132.7	-	-	22S	72E
112	2049 Nov 09	15:52:11	85	616	117	N	-h	1.1965	0.6821	-0.3541	226.8	-	-	18N	118E
113	2050 May 06	22:32:02	85	622	122	T	-h	-0.4181	2.1064	1.0779	340.9	206.8	43.8	17S	21E
114	2050 Oct 30	03:21:47	86	628	127	T	-p	0.4435	2.0356	1.0549	313.9	193.6	35.1	14N	54W
115	2051 Apr 26	02:16:28	86	634	132	T	p-	0.3371	2.2785	1.2034	365.7	221.7	70.3	13S	34W
116	2051 Oct 19	19:11:50	86	640	137	T-	p-	-0.2542	2.3719	1.4130	315.0	205.0	84.2	10N	69E
117	2052 Apr 14	02:18:06	87	646	142	N	t-	1.0629	0.9478	-0.1294	276.8	-	-	9S	34W
118	2052 Oct 08	10:45:58	87	652	147	P	a-	-0.9727	1.0653	0.0832	257.3	63.9	-	5N	164W
119	2053 Mar 04	17:22:09	87	657	114	N	-a	-1.0531	0.9334	-0.0796	251.8	-	-	5N	102E
120	2053 Aug 29	08:05:50	87	663	119	Nx	-t	1.0165	1.0203	-0.0319	278.5	-	-	8S	121W

Cat Num	Calendar Date	TD of Greatest Eclipse	ΔT s	Luna Num	Saros Num	Ecl Type	QSE	Gamma	Pen Mag	Um Mag	Phase Durations Pen m	Phase Durations Par m	Phase Durations Total m	Greatest in Zenith Lat	Greatest in Zenith Long
121	2054 Feb 22	06:51:27	88	669	124	T	-p	-0.3242	2.2502	1.2780	315.5	201.6	72.7	10N	99W
122	2054 Aug 18	09:26:30	88	675	129	T	pp	0.2806	2.3817	1.3074	370.4	227.4	83.7	13S	140W
123	2055 Feb 11	22:46:17	88	681	134	T	p-	0.3526	2.1982	1.2258	313.7	199.1	66.6	14N	22E
124	2055 Aug 07	10:53:18	89	687	139	P	t-	-0.4769	2.0081	0.9606	347.2	204.2	-	17S	161W
125	2056 Feb 01	12:26:06	89	693	144	N	a-	1.0682	0.9069	-0.1084	248.0	-	-	18N	177E
126	2056 Jun 27	10:03:09	89	698	111	N	-a	1.3770	0.3158	-0.6504	150.6	-	-	22S	150W
127	2056 Jul 26	18:43:25	89	699	149	N	a-	-1.2048	0.6448	-0.3477	215.2	-	-	20S	81E
128	2056 Dec 22	01:48:56	90	704	116	N	-t	-1.1560	0.7872	-0.3093	257.3	-	-	22N	27W
129	2057 Jun 17	02:26:20	90	710	121	P	-a	0.6168	1.6982	0.7570	291.4	170.0	-	23S	36W
130	2057 Dec 11	00:53:38	90	716	126	P	-t	-0.4853	2.0194	0.9197	359.7	204.8	-	23N	15W
131	2058 Jun 06	19:15:48	91	722	131	T-	pp	-0.1181	2.6226	1.6628	324.5	214.1	98.0	23S	71E
132	2058 Nov 30	03:16:18	91	728	136	T+	p-	0.2208	2.4819	1.4277	353.9	221.5	90.4	22N	52W
133	2059 May 27	07:55:34	91	734	141	P	t-	-0.9098	1.1963	0.1846	282.5	97.9	-	22S	119W
134	2059 Nov 19	13:01:36	92	740	146	P	a-	0.9004	1.2055	0.2097	271.2	99.9	-	20N	161E
135	2060 Apr 15	21:37:05	92	745	113	N	-t	1.1622	0.7694	-0.3136	255.7	-	-	9S	37E
136	2060 Oct 09	18:53:33	92	751	118	N	-a	-1.0671	0.8816	-0.0779	232.0	-	-	6N	74E
137	2060 Nov 08	04:04:13	93	752	156	Nb	a-	1.5332	0.0286	-0.9356	45.3	-	-	18N	65W
138	2061 Apr 04	21:54:05	93	757	123	T	-t	0.4300	2.1064	1.0360	355.8	210.3	30.9	6S	33E
139	2061 Sep 29	09:38:13	93	763	128	T	-a	-0.3810	2.1576	1.1640	325.5	203.1	59.6	2N	146W
140	2062 Mar 25	03:33:50	94	769	133	T	p-	-0.3150	2.2925	1.2715	339.1	212.0	75.3	2S	52W
141	2062 Sep 18	18:34:02	94	775	138	T	p-	0.3736	2.1979	1.1515	350.0	213.1	60.2	1S	80E
142	2063 Mar 14	16:05:49	94	781	143	P	a-	-1.0008	1.0108	0.0363	248.5	41.9	-	1N	121E
143	2063 Sep 07	20:41:11	95	787	148	N	t-	1.1375	0.8121	-0.2657	261.2	-	-	5S	49E
144	2064 Feb 02	21:48:57	95	792	115	P	-a	0.9969	1.0215	0.0395	249.8	43.7	-	18N	37E
145	2064 Jul 28	07:52:48	95	798	120	P	-t	-0.9473	1.1378	0.1055	285.0	76.5	-	20S	116W
146	2065 Jan 22	09:58:58	96	804	125	T	-p	0.3371	2.2579	1.2248	339.0	209.7	69.4	20N	146W
147	2065 Jul 17	17:48:40	96	810	130	T-	pp	-0.1402	2.5907	1.6138	331.9	217.0	97.7	21S	95E
148	2066 Jan 11	15:04:47	96	816	135	T	p-	-0.3687	2.2276	1.1395	361.6	216.0	58.7	21N	136E
149	2066 Jul 07	09:30:29	97	822	140	P	a-	0.6056	1.7196	0.7770	293.1	172.0	-	22S	141W
150	2066 Dec 31	14:30:09	97	828	145	N	t-	-1.0540	0.9789	-0.1264	284.2	-	-	22N	143E
151	2067 May 28	18:56:07	97	833	112	N	-a	-1.2013	0.6416	-0.3316	209.2	-	-	23S	76E
152	2067 Jun 27	02:41:06	97	834	150	N	a-	1.3394	0.3770	-0.5736	160.5	-	-	22S	39W
153	2067 Nov 21	00:04:42	98	839	117	N	-h	1.2107	0.6557	-0.3798	222.2	-	-	21N	4W
154	2068 May 17	05:42:17	98	845	122	P	-t	-0.4852	1.9839	0.9545	337.5	199.8	-	20S	86W
155	2068 Nov 09	11:47:00	99	851	127	T	-p	0.4645	1.9974	1.0161	312.0	190.9	19.2	18N	180E
156	2069 May 06	09:09:57	99	857	132	T+	pp	0.2717	2.3977	1.3242	369.1	227.0	85.0	17S	138W
157	2069 Oct 30	03:35:06	99	863	137	T-	p-	-0.2263	2.4247	1.4628	316.3	206.4	87.4	14N	57W
158	2070 Apr 25	09:21:25	100	869	142	Nx	t-	1.0044	1.0527	-0.0197	287.7	-	-	12S	140W
159	2070 Oct 19	18:51:12	100	875	147	P	a-	-0.9406	1.1270	0.1395	263.9	82.3	-	9N	74E
160	2071 Mar 16	01:31:09	100	880	114	N	-a	-1.0757	0.8890	-0.1183	245.8	-	-	1N	21W
161	2071 Sep 09	15:05:40	101	886	119	N	-t	1.0835	0.9000	-0.1575	265.9	-	-	4S	133E
162	2072 Mar 04	15:23:07	101	892	124	T	-p	-0.3431	2.2137	1.2452	313.9	200.1	69.0	6N	132E
163	2072 Aug 28	16:05:42	102	898	129	T	-t	0.3563	2.2439	1.1673	366.9	221.1	64.8	9S	119E
164	2073 Feb 22	07:24:53	102	904	134	T	p-	0.3389	2.2230	1.2514	314.6	200.4	69.8	10N	107W
165	2073 Aug 17	17:42:41	102	910	139	T	t-	-0.3998	2.1490	1.1024	351.3	212.3	50.8	13S	96E
166	2074 Feb 11	20:55:58	103	916	144	N	a-	1.0612	0.9203	-0.0960	250.2	-	-	15N	50E
167	2074 Jul 08	17:21:37	103	921	111	N	-a	1.4457	0.1884	-0.7751	117.3	-	-	21S	101E
168	2074 Aug 07	01:56:04	103	922	149	N	a-	-1.1291	0.7826	-0.2079	232.9	-	-	17S	27W
169	2075 Jan 02	09:55:03	104	927	116	N	-t	-1.1643	0.7729	-0.3256	255.8	-	-	22N	147W
170	2075 Jun 28	09:55:36	104	933	121	P	-a	0.6897	1.5639	0.6235	284.2	157.6	-	23S	148W
171	2075 Dec 22	08:55:55	104	939	126	P	-t	-0.4945	2.0024	0.9028	358.6	203.4	-	23N	134W
172	2076 Jun 17	02:39:47	105	945	131	T-	pp	-0.0452	2.7570	1.7959	326.2	215.9	100.9	23S	39W
173	2076 Dec 10	11:34:51	105	951	136	T+	p-	0.2102	2.5006	1.4476	353.1	221.4	91.6	23N	175W
174	2077 Jun 06	14:59:52	106	957	141	P	t-	-0.8388	1.3274	0.3139	294.4	125.7	-	24S	135E
175	2077 Nov 29	21:35:53	106	963	146	P	a-	0.8855	1.2326	0.2372	272.8	105.6	-	23N	33E
176	2078 Apr 27	04:35:45	106	968	113	N	-t	1.2223	0.6577	-0.4227	239.0	-	-	13S	69W
177	2078 Oct 21	03:08:04	107	974	118	N	-a	-1.1022	0.8191	-0.1442	225.5	-	-	10N	50W
178	2078 Nov 19	12:40:02	107	975	156	N	a-	1.5148	0.0633	-0.9028	67.2	-	-	21N	166E
179	2079 Apr 16	05:10:45	107	980	123	P	-t	0.4800	2.0119	0.9471	350.9	204.1	-	10S	77W
180	2079 Oct 10	17:30:30	108	986	128	T	-a	-0.4246	2.0806	1.0811	324.5	199.3	43.1	7N	95E

Cat Num	Calendar Date	TD of Greatest Eclipse	ΔT s	Luna Num	Saros Num	Ecl Type	QSE	Gamma	Pen Mag	Um Mag	Phase Durations Pen m	Phase Durations Par m	Phase Durations Total m	Greatest in Zenith Lat	Greatest in Zenith Long
181	2080 Apr 04	11:23:38	108	992	133	T	p-	-0.2751	2.3626	1.3479	339.0	214.2	82.8	6S	170W
182	2080 Sep 29	01:52:42	109	998	138	T	p-	0.3203	2.2986	1.2462	354.5	218.0	74.4	3N	30W
183	2081 Mar 25	00:22:01	109	1004	143	P	a-	-0.9688	1.0673	0.0973	253.1	68.0	-	3S	4W
184	2081 Sep 18	03:35:25	109	1010	148	N	t-	1.0747	0.9291	-0.1524	276.4	-	-	1S	55W
185	2082 Feb 13	06:29:20	110	1015	115	P	-a	1.0101	0.9974	0.0153	247.9	27.3	-	14N	93W
186	2082 Aug 08	14:46:42	110	1021	120	Nx	-t	-1.0204	1.0030	-0.0275	270.5	-	-	17S	141E
187	2083 Feb 02	18:26:46	111	1027	125	T	-p	0.3464	2.2418	1.2070	339.7	209.5	67.2	17N	87E
188	2083 Jul 29	01:05:34	111	1033	130	T-	pp	-0.2143	2.4537	1.4791	328.8	213.6	91.1	19S	14W
189	2084 Jan 22	23:13:00	111	1039	135	T	p-	-0.3610	2.2425	1.1531	362.9	217.0	61.3	19N	15E
190	2084 Jul 17	16:58:51	112	1045	140	P	a-	0.5313	1.8557	0.9136	298.9	182.1	-	20S	107E
191	2085 Jan 10	22:32:29	112	1051	145	N	t-	-1.0453	0.9944	-0.1102	285.7	-	-	21N	24E
192	2085 Jun 08	02:17:36	113	1056	112	N	-a	-1.2746	0.5079	-0.4668	189.2	-	-	24S	34W
193	2085 Jul 07	10:04:39	113	1057	150	N	a-	1.2695	0.5064	-0.4461	184.2	-	-	21S	150W
194	2085 Dec 01	08:25:35	113	1062	117	N	-a	1.2190	0.6400	-0.3944	219.2	-	-	23N	129W
195	2086 May 28	12:43:47	114	1068	122	P	-t	-0.5585	1.8499	0.8193	332.9	190.2	-	22S	169E
196	2086 Nov 20	20:19:42	114	1074	127	P	-p	0.4800	1.9692	0.9877	310.4	188.9	-	20N	52E
197	2087 May 17	15:55:20	115	1080	132	T+	pp	0.1999	2.5289	1.4568	371.9	231.5	95.9	19S	121E
198	2087 Nov 10	12:05:33	115	1086	137	T-	p-	-0.2043	2.4667	1.5018	317.3	207.3	89.6	17N	175E
199	2088 May 05	16:16:50	115	1092	142	P	t-	0.9388	1.1708	0.1032	298.7	77.8	-	16S	116E
200	2088 Oct 30	03:03:20	116	1098	147	P	a-	-0.9147	1.1773	0.1843	269.3	94.2	-	13N	49W
201	2089 Mar 26	09:34:13	116	1103	114	N	-a	-1.1039	0.8343	-0.1670	238.4	-	-	4S	142W
202	2089 Sep 19	22:11:16	117	1109	119	N	-t	1.1448	0.7904	-0.2726	252.9	-	-	0N	26E
203	2090 Mar 15	23:48:31	117	1115	124	T	-p	-0.3675	2.1670	1.2023	312.0	198.1	63.6	1N	5E
204	2090 Sep 08	22:52:29	118	1121	129	T	-t	0.4257	2.1178	1.0387	362.9	213.9	32.5	5S	17E
205	2091 Mar 05	15:58:22	118	1127	134	T	p-	0.3212	2.2548	1.2843	315.8	202.0	73.5	6N	124E
206	2091 Aug 29	00:38:25	119	1133	139	T	t-	-0.3270	2.2821	1.2362	354.3	218.3	73.5	10S	9W
207	2092 Feb 23	05:21:00	119	1139	144	N	a-	1.0509	0.9395	-0.0777	253.1	-	-	11N	76W
208	2092 Jul 19	00:41:57	119	1144	111	Ne	-a	1.5132	0.0634	-0.8979	68.6	-	-	19S	9W
209	2092 Aug 17	09:14:00	120	1145	149	N	a-	-1.0569	0.9143	-0.0746	247.4	-	-	14S	137W
210	2093 Jan 12	18:00:02	120	1150	116	N	-t	-1.1734	0.7567	-0.3430	253.9	-	-	20N	93E
211	2093 Jul 08	17:24:18	120	1156	121	P	-a	0.7632	1.4289	0.4886	276.1	142.5	-	22S	101E
212	2094 Jan 01	17:00:06	121	1162	126	P	-t	-0.5025	1.9872	0.8886	357.4	202.0	-	22N	106E
213	2094 Jun 28	10:01:57	121	1168	131	T+	pp	0.0288	2.7879	1.8249	327.3	216.5	101.4	23S	149W
214	2094 Dec 21	19:56:32	122	1174	136	T+	p-	0.2016	2.5153	1.4642	352.1	221.3	92.4	24N	61E
215	2095 Jun 17	22:00:11	122	1180	141	P	t-	-0.7653	1.4632	0.4474	305.5	147.6	-	24S	31E
216	2095 Dec 11	06:15:02	123	1186	146	P	a-	0.8743	1.2526	0.2581	273.7	109.6	-	24N	95W
217	2096 May 07	11:24:44	123	1191	113	N	-t	1.2897	0.5327	-0.5451	217.7	-	-	16S	171W
218	2096 Jun 06	02:43:40	123	1192	151	Nb	t-	-1.5724	0.0064	-1.0567	24.7	-	-	24S	41W
219	2096 Oct 31	11:30:24	124	1197	118	N	-a	-1.1308	0.7685	-0.1987	220.0	-	-	13N	176W
220	2096 Nov 29	21:22:20	124	1198	156	N	a-	1.5018	0.0879	-0.8799	79.0	-	-	23N	37E
221	2097 Apr 26	12:18:17	124	1203	123	P	-t	0.5377	1.9032	0.8439	344.8	195.9	-	13S	176E
222	2097 Oct 21	01:30:55	125	1209	128	T	-a	-0.4608	2.0171	1.0116	323.8	195.9	16.7	11N	26W
223	2098 Apr 15	19:04:48	125	1215	133	T-	p-	-0.2272	2.4474	1.4389	339.0	216.5	89.7	10S	74E
224	2098 Oct 10	09:19:58	126	1221	138	T	pp	0.2749	2.3850	1.3266	358.2	221.7	83.4	7N	143W
225	2099 Apr 05	08:30:56	126	1227	143	P	a-	-0.9304	1.1353	0.1700	258.4	88.8	-	7S	127W
226	2099 Sep 29	10:36:37	127	1233	148	Nx	t-	1.0175	1.0360	-0.0491	289.1	-	-	3N	162W
227	2100 Feb 24	15:05:12	127	1238	115	N	-a	1.0267	0.9669	-0.0151	245.3	-	-	10N	138E
228	2100 Aug 19	21:44:59	128	1244	120	N	-t	-1.0906	0.8735	-0.1556	254.9	-	-	13S	36E

Appendix B

Small Lunar Eclipse Figures: 2001 to 2100

Key to Small Lunar Eclipse Figures

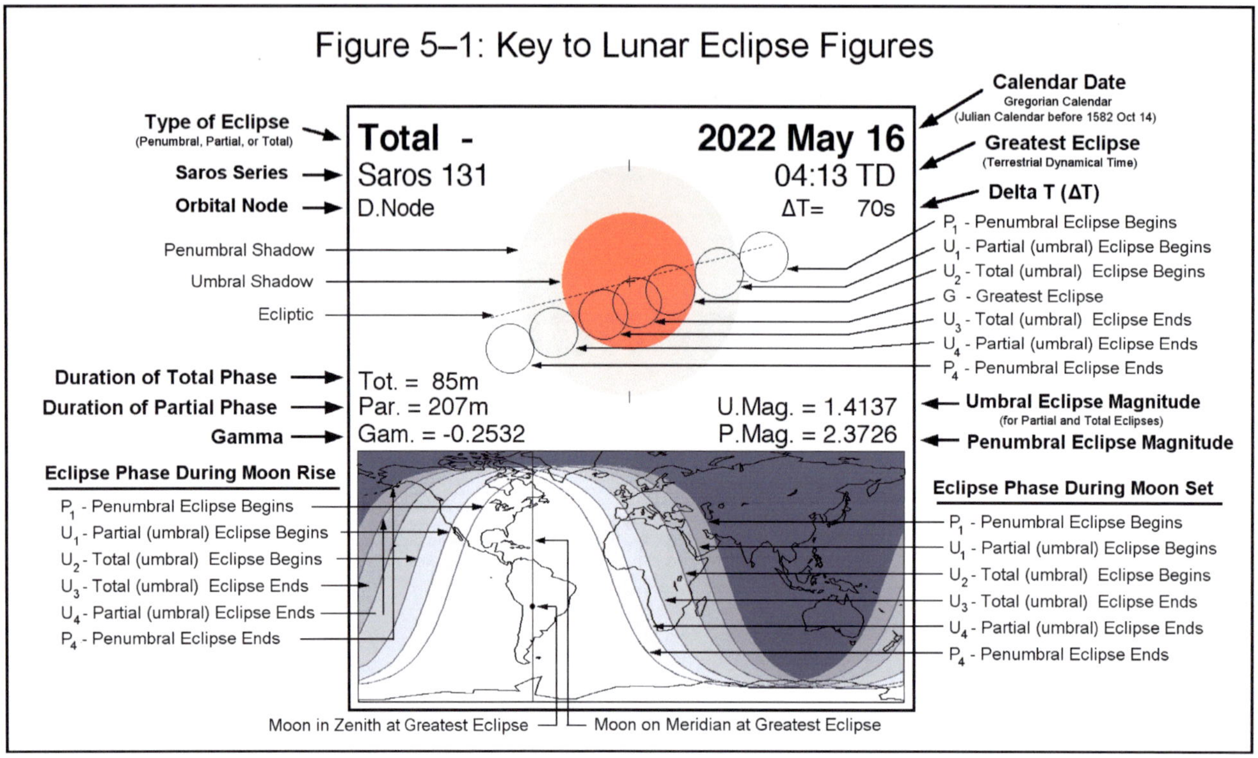

Lunar Eclipse Type – One of three basic types of lunar eclipses: Penumbral, Partial, or Total.

Saros Series – The Saros series that the eclipse belongs to. Eclipses with an even number take place at the Moon's ascending node, while those with an odd number are at the descending node.

Node – The orbital node near which the eclipse takes place. The ascending node (A. Node) is the point where the Moon travels from south to north through Earth's orbital plane. Similarly, the descending node (D. Node) is the point where the Moon travels from north to south.

Calendar Date – Gregorian calendar date at greatest eclipse is given in the ISO order of YEAR-MONTH-DAY.

Greatest Eclipse – The instant (Terrestrial Dynamical Time) when the distance between the Moon and the axis of Earth's shadow cone reaches a minimum.

ΔT (Delta T) – the arithmetic difference, in seconds, between Terrestrial Dynamical Time (TD) and Universal Time (UT1).

Gamma – The minimum distance from the center of the Moon to Earth's shadow axis, in units of Earth's equatorial radius.

Duration – The duration of a penumbral eclipse (**Pen.**), a partial eclipse (**Par.**) , and a total eclipse (**Tot.**).

Umbral Eclipse Magnitude (U. Mag) – The fraction of the Moon's diameter immersed in the umbra at greatest eclipse.

Penumbral Eclipse Magnitude (P. Mag) – The fraction of the Moon's diameter immersed in the penumbra at greatest eclipse.

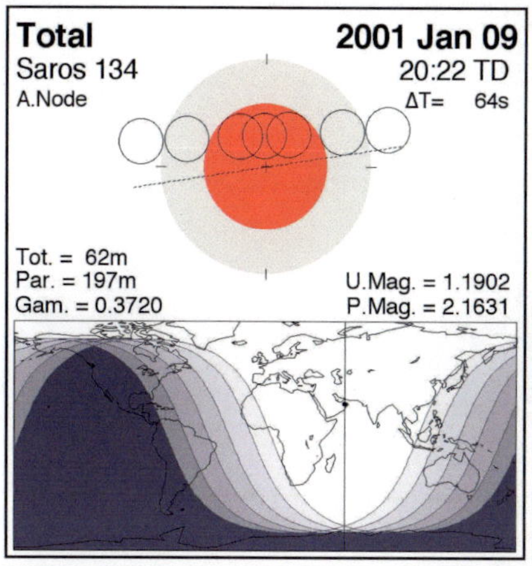
Total
2001 Jan 09
Saros 134
20:22 TD
A.Node
ΔT= 64s
Tot. = 62m
Par. = 197m
Gam. = 0.3720
U.Mag. = 1.1902
P.Mag. = 2.1631

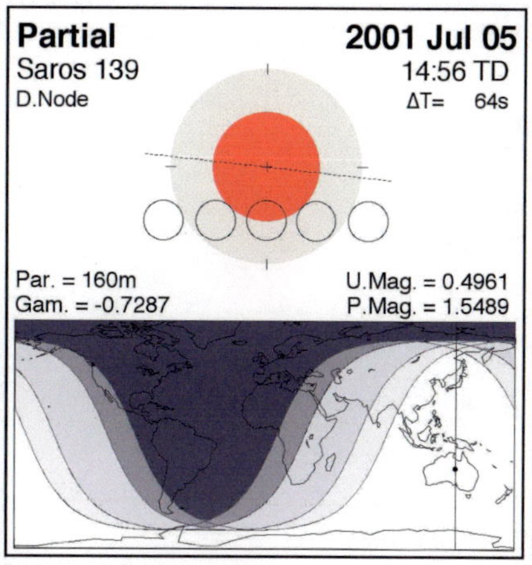
Partial
2001 Jul 05
Saros 139
14:56 TD
D.Node
ΔT= 64s
Par. = 160m
Gam. = -0.7287
U.Mag. = 0.4961
P.Mag. = 1.5489

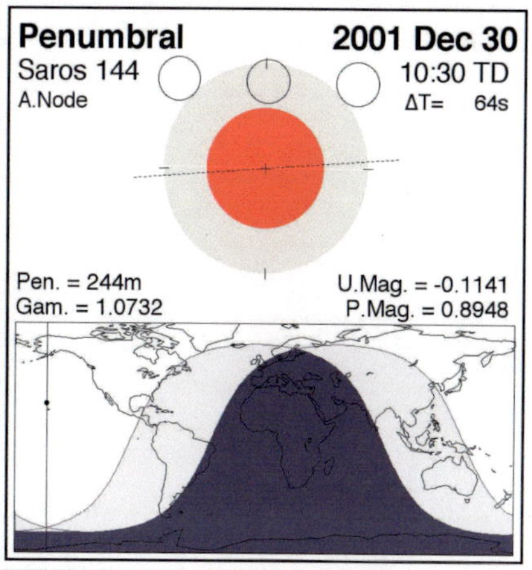
Penumbral
2001 Dec 30
Saros 144
10:30 TD
A.Node
ΔT= 64s
Pen. = 244m
Gam. = 1.0732
U.Mag. = -0.1141
P.Mag. = 0.8948

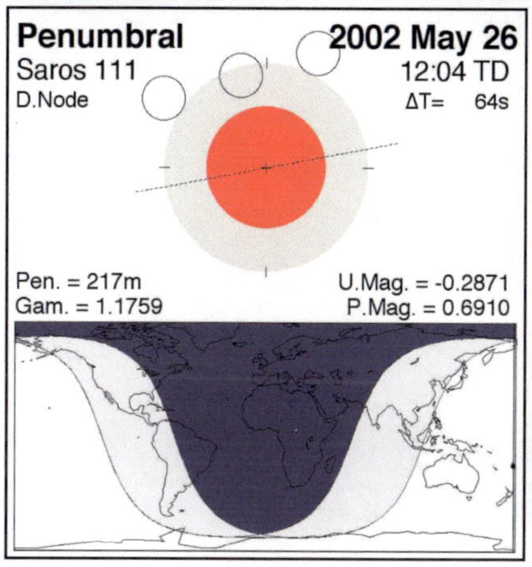
Penumbral
2002 May 26
Saros 111
12:04 TD
D.Node
ΔT= 64s
Pen. = 217m
Gam. = 1.1759
U.Mag. = -0.2871
P.Mag. = 0.6910

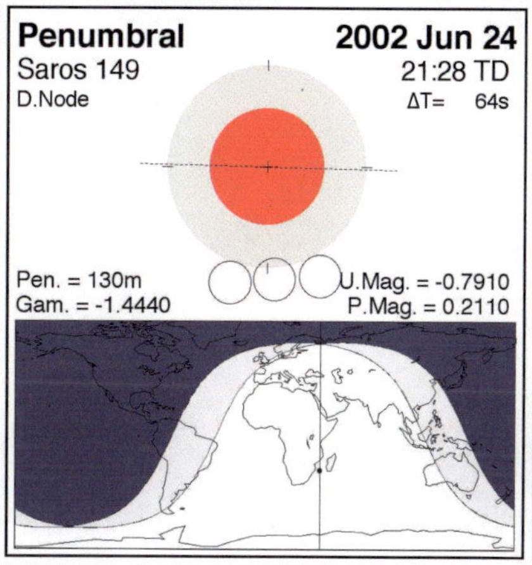
Penumbral
2002 Jun 24
Saros 149
21:28 TD
D.Node
ΔT= 64s
Pen. = 130m
Gam. = -1.4440
U.Mag. = -0.7910
P.Mag. = 0.2110

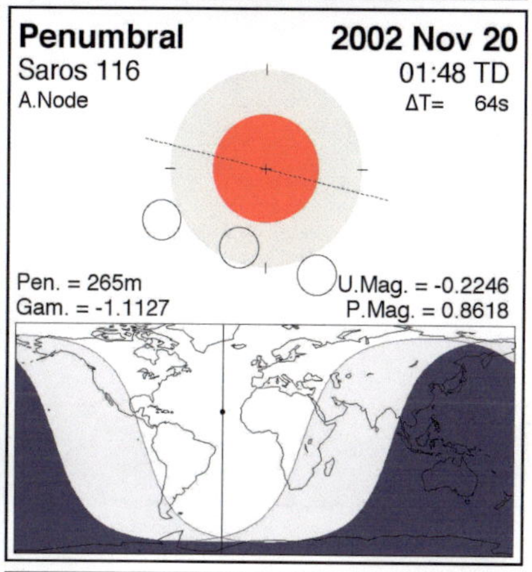
Penumbral
2002 Nov 20
Saros 116
01:48 TD
A.Node
ΔT= 64s
Pen. = 265m
Gam. = -1.1127
U.Mag. = -0.2246
P.Mag. = 0.8618

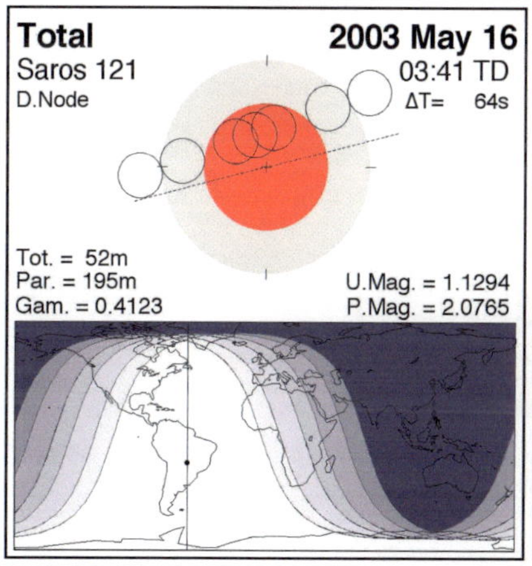
Total
2003 May 16
Saros 121
03:41 TD
D.Node
ΔT= 64s
Tot. = 52m
Par. = 195m
Gam. = 0.4123
U.Mag. = 1.1294
P.Mag. = 2.0765

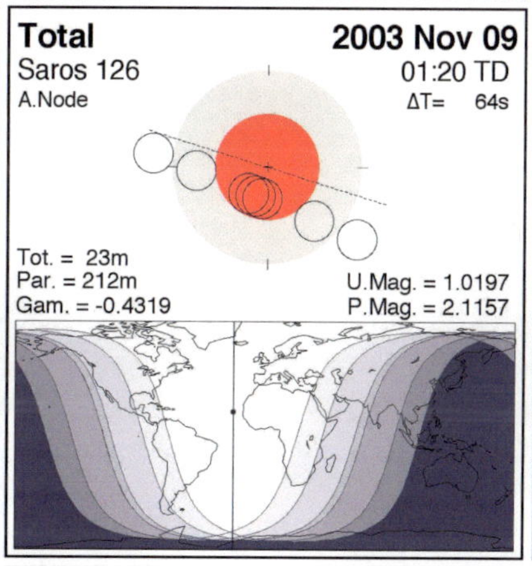
Total
2003 Nov 09
Saros 126
01:20 TD
A.Node
ΔT= 64s
Tot. = 23m
Par. = 212m
Gam. = -0.4319
U.Mag. = 1.0197
P.Mag. = 2.1157

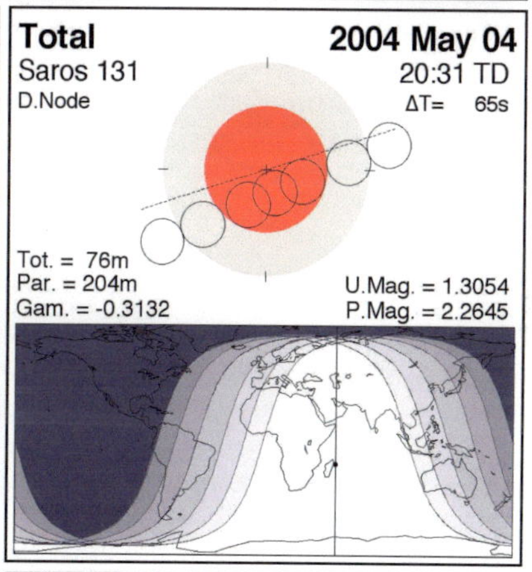
Total
2004 May 04
Saros 131
20:31 TD
D.Node
ΔT= 65s
Tot. = 76m
Par. = 204m
Gam. = -0.3132
U.Mag. = 1.3054
P.Mag. = 2.2645

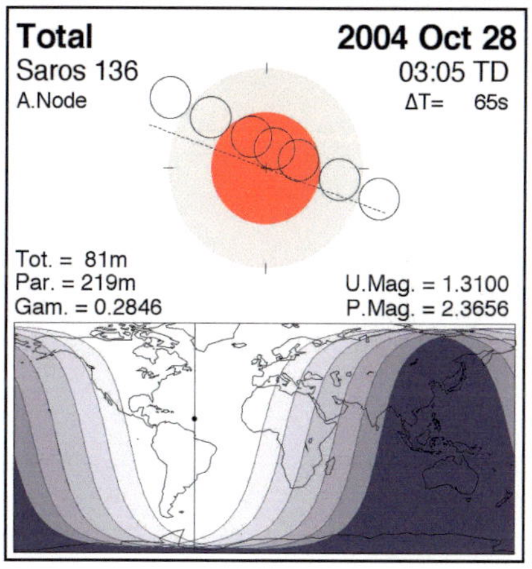
Total
2004 Oct 28
Saros 136
03:05 TD
A.Node
ΔT= 65s
Tot. = 81m
Par. = 219m
Gam. = 0.2846
U.Mag. = 1.3100
P.Mag. = 2.3656

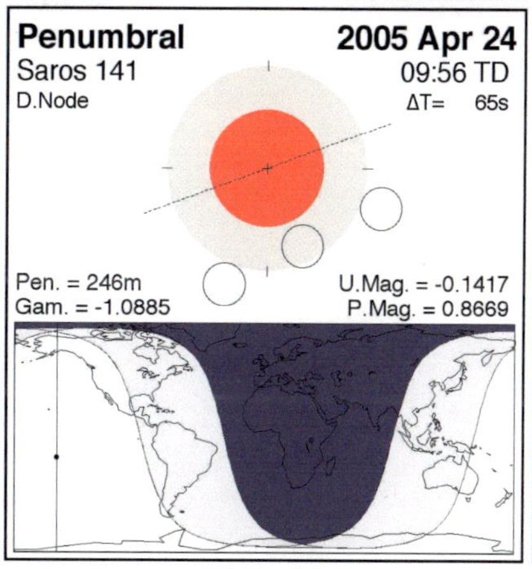
Penumbral
2005 Apr 24
Saros 141
09:56 TD
D.Node
ΔT= 65s
Pen. = 246m
Gam. = -1.0885
U.Mag. = -0.1417
P.Mag. = 0.8669

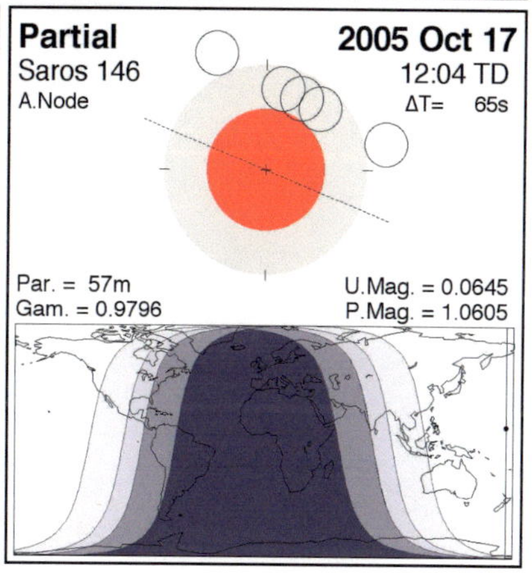
Partial
2005 Oct 17
Saros 146
12:04 TD
A.Node
ΔT= 65s
Par. = 57m
Gam. = 0.9796
U.Mag. = 0.0645
P.Mag. = 1.0605

Plate 001

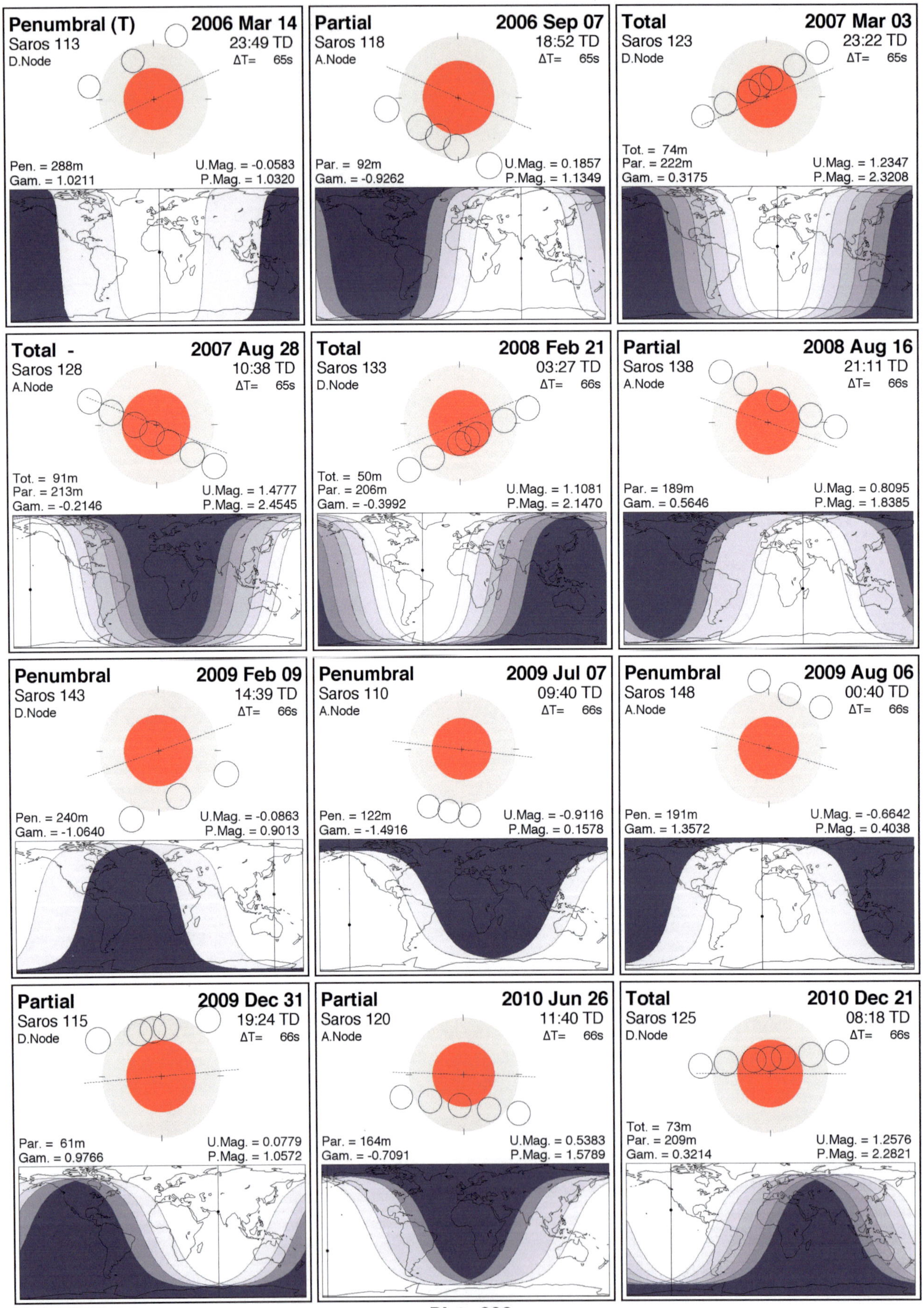
Penumbral (T)
2006 Mar 14
Saros 113
23:49 TD
D.Node
ΔT= 65s
Pen. = 288m
Gam. = 1.0211
U.Mag. = -0.0583
P.Mag. = 1.0320
Partial
2006 Sep 07
Saros 118
18:52 TD
A.Node
ΔT= 65s
Par. = 92m
Gam. = -0.9262
U.Mag. = 0.1857
P.Mag. = 1.1349
Total
2007 Mar 03
Saros 123
23:22 TD
D.Node
ΔT= 65s
Tot. = 74m
Par. = 222m
Gam. = 0.3175
U.Mag. = 1.2347
P.Mag. = 2.3208
Total -
2007 Aug 28
Saros 128
10:38 TD
A.Node
ΔT= 65s
Tot. = 91m
Par. = 213m
Gam. = -0.2146
U.Mag. = 1.4777
P.Mag. = 2.4545
Total
2008 Feb 21
Saros 133
03:27 TD
D.Node
ΔT= 66s
Tot. = 50m
Par. = 206m
Gam. = -0.3992
U.Mag. = 1.1081
P.Mag. = 2.1470
Partial
2008 Aug 16
Saros 138
21:11 TD
A.Node
ΔT= 66s
Par. = 189m
Gam. = 0.5646
U.Mag. = 0.8095
P.Mag. = 1.8385
Penumbral
2009 Feb 09
Saros 143
14:39 TD
D.Node
ΔT= 66s
Pen. = 240m
Gam. = -1.0640
U.Mag. = -0.0863
P.Mag. = 0.9013
Penumbral
2009 Jul 07
Saros 110
09:40 TD
A.Node
ΔT= 66s
Pen. = 122m
Gam. = -1.4916
U.Mag. = -0.9116
P.Mag. = 0.1578
Penumbral
2009 Aug 06
Saros 148
00:40 TD
A.Node
ΔT= 66s
Pen. = 191m
Gam. = 1.3572
U.Mag. = -0.6642
P.Mag. = 0.4038
Partial
2009 Dec 31
Saros 115
19:24 TD
D.Node
ΔT= 66s
Par. = 61m
Gam. = 0.9766
U.Mag. = 0.0779
P.Mag. = 1.0572
Partial
2010 Jun 26
Saros 120
11:40 TD
A.Node
ΔT= 66s
Par. = 164m
Gam. = -0.7091
U.Mag. = 0.5383
P.Mag. = 1.5789
Total
2010 Dec 21
Saros 125
08:18 TD
D.Node
ΔT= 66s
Tot. = 73m
Par. = 209m
Gam. = 0.3214
U.Mag. = 1.2576
P.Mag. = 2.2821

Plate 002

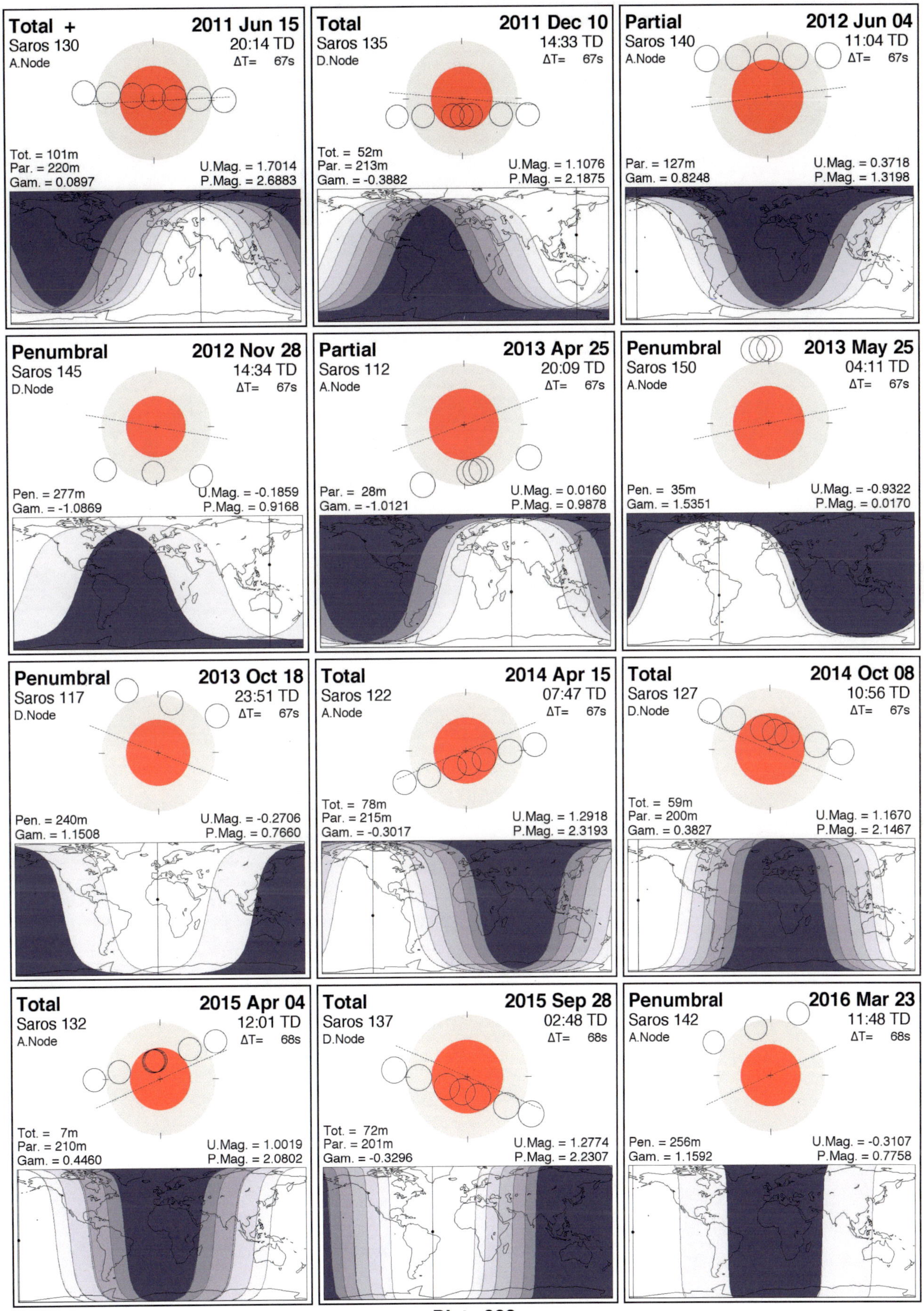
Total +
2011 Jun 15
Saros 130
20:14 TD
A.Node
ΔT= 67s
Tot. = 101m
Par. = 220m
Gam. = 0.0897
U.Mag. = 1.7014
P.Mag. = 2.6883
Total
2011 Dec 10
Saros 135
14:33 TD
D.Node
ΔT= 67s
Tot. = 52m
Par. = 213m
Gam. = -0.3882
U.Mag. = 1.1076
P.Mag. = 2.1875
Partial
2012 Jun 04
Saros 140
11:04 TD
A.Node
ΔT= 67s
Par. = 127m
Gam. = 0.8248
U.Mag. = 0.3718
P.Mag. = 1.3198
Penumbral
2012 Nov 28
Saros 145
14:34 TD
D.Node
ΔT= 67s
Pen. = 277m
Gam. = -1.0869
U.Mag. = -0.1859
P.Mag. = 0.9168
Partial
2013 Apr 25
Saros 112
20:09 TD
A.Node
ΔT= 67s
Par. = 28m
Gam. = -1.0121
U.Mag. = 0.0160
P.Mag. = 0.9878
Penumbral
2013 May 25
Saros 150
04:11 TD
A.Node
ΔT= 67s
Pen. = 35m
Gam. = 1.5351
U.Mag. = -0.9322
P.Mag. = 0.0170
Penumbral
2013 Oct 18
Saros 117
23:51 TD
D.Node
ΔT= 67s
Pen. = 240m
Gam. = 1.1508
U.Mag. = -0.2706
P.Mag. = 0.7660
Total
2014 Apr 15
Saros 122
07:47 TD
A.Node
ΔT= 67s
Tot. = 78m
Par. = 215m
Gam. = -0.3017
U.Mag. = 1.2918
P.Mag. = 2.3193
Total
2014 Oct 08
Saros 127
10:56 TD
D.Node
ΔT= 67s
Tot. = 59m
Par. = 200m
Gam. = 0.3827
U.Mag. = 1.1670
P.Mag. = 2.1467
Total
2015 Apr 04
Saros 132
12:01 TD
A.Node
ΔT= 68s
Tot. = 7m
Par. = 210m
Gam. = 0.4460
U.Mag. = 1.0019
P.Mag. = 2.0802
Total
2015 Sep 28
Saros 137
02:48 TD
D.Node
ΔT= 68s
Tot. = 72m
Par. = 201m
Gam. = -0.3296
U.Mag. = 1.2774
P.Mag. = 2.2307
Penumbral
2016 Mar 23
Saros 142
11:48 TD
A.Node
ΔT= 68s
Pen. = 256m
Gam. = 1.1592
U.Mag. = -0.3107
P.Mag. = 0.7758

Plate 003

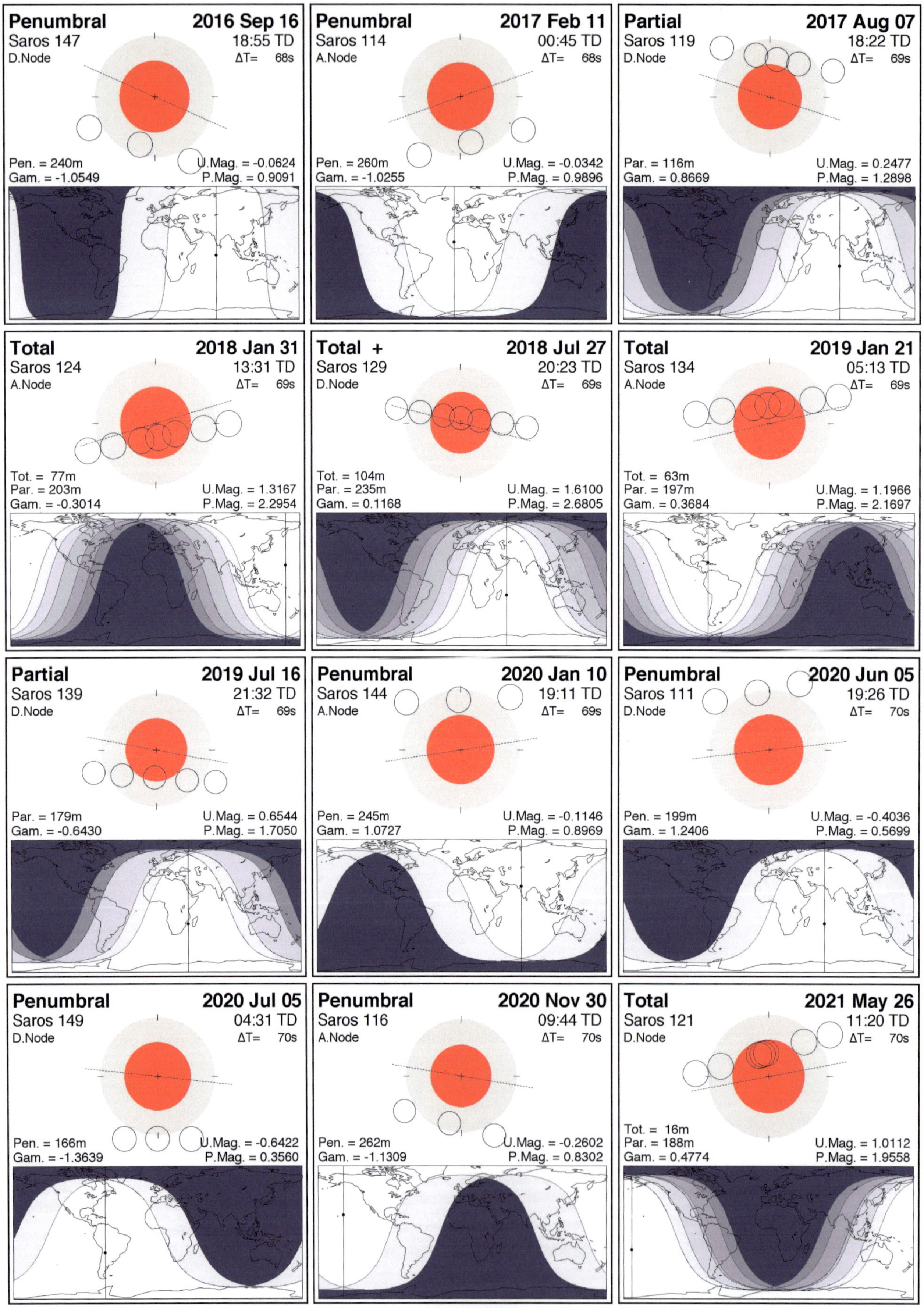
Penumbral
2016 Sep 16
Saros 147
18:55 TD
D.Node
ΔT= 68s
Pen. = 240m
U.Mag. = -0.0624
Gam. = -1.0549
P.Mag. = 0.9091
Penumbral
2017 Feb 11
Saros 114
00:45 TD
A.Node
ΔT= 68s
Pen. = 260m
U.Mag. = -0.0342
Gam. = -1.0255
P.Mag. = 0.9896
Partial
2017 Aug 07
Saros 119
18:22 TD
D.Node
ΔT= 69s
Par. = 116m
U.Mag. = 0.2477
Gam. = 0.8669
P.Mag. = 1.2898
Total
2018 Jan 31
Saros 124
13:31 TD
A.Node
ΔT= 69s
Tot. = 77m
Par. = 203m
U.Mag. = 1.3167
Gam. = -0.3014
P.Mag. = 2.2954
Total +
2018 Jul 27
Saros 129
20:23 TD
D.Node
ΔT= 69s
Tot. = 104m
Par. = 235m
U.Mag. = 1.6100
Gam. = 0.1168
P.Mag. = 2.6805
Total
2019 Jan 21
Saros 134
05:13 TD
A.Node
ΔT= 69s
Tot. = 63m
Par. = 197m
U.Mag. = 1.1966
Gam. = 0.3684
P.Mag. = 2.1697
Partial
2019 Jul 16
Saros 139
21:32 TD
D.Node
ΔT= 69s
Par. = 179m
U.Mag. = 0.6544
Gam. = -0.6430
P.Mag. = 1.7050
Penumbral
2020 Jan 10
Saros 144
19:11 TD
A.Node
ΔT= 69s
Pen. = 245m
U.Mag. = -0.1146
Gam. = 1.0727
P.Mag. = 0.8969
Penumbral
2020 Jun 05
Saros 111
19:26 TD
D.Node
ΔT= 70s
Pen. = 199m
U.Mag. = -0.4036
Gam. = 1.2406
P.Mag. = 0.5699
Penumbral
2020 Jul 05
Saros 149
04:31 TD
D.Node
ΔT= 70s
Pen. = 166m
U.Mag. = -0.6422
Gam. = -1.3639
P.Mag. = 0.3560
Penumbral
2020 Nov 30
Saros 116
09:44 TD
A.Node
ΔT= 70s
Pen. = 262m
U.Mag. = -0.2602
Gam. = -1.1309
P.Mag. = 0.8302
Total
2021 May 26
Saros 121
11:20 TD
D.Node
ΔT= 70s
Tot. = 16m
Par. = 188m
U.Mag. = 1.0112
Gam. = 0.4774
P.Mag. = 1.9558

Plate 004

Partial 2021 Nov 19
Saros 126 09:04 TD
A.Node ΔT= 70s

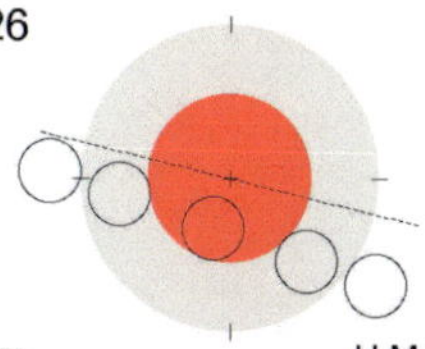

Par. = 209m U.Mag. = 0.9760
Gam. = -0.4552 P.Mag. = 2.0738

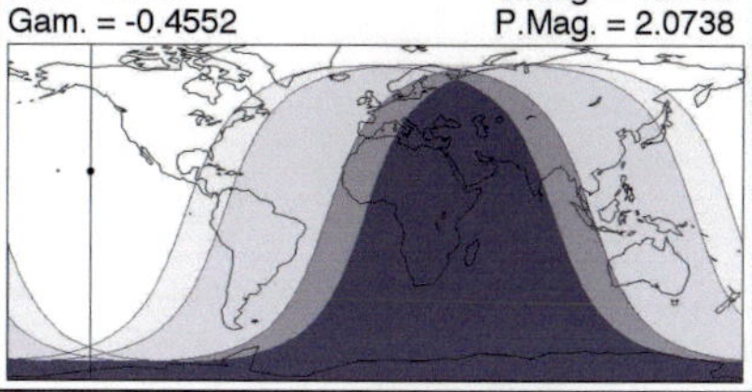

Total - 2022 May 16
Saros 131 04:13 TD
D.Node ΔT= 70s

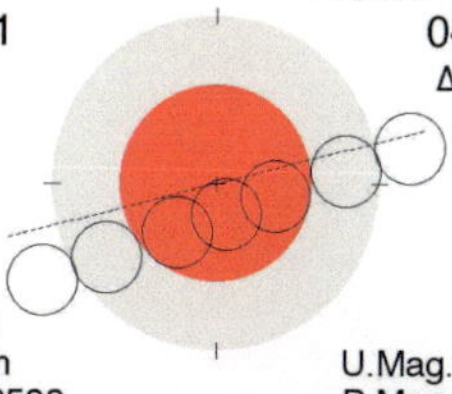

Tot. = 86m
Par. = 208m U.Mag. = 1.4154
Gam. = -0.2532 P.Mag. = 2.3743

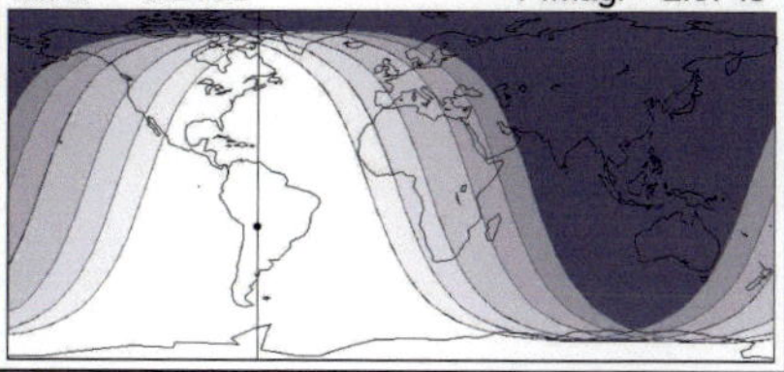

Total + 2022 Nov 08
Saros 136 11:00 TD
A.Node ΔT= 71s

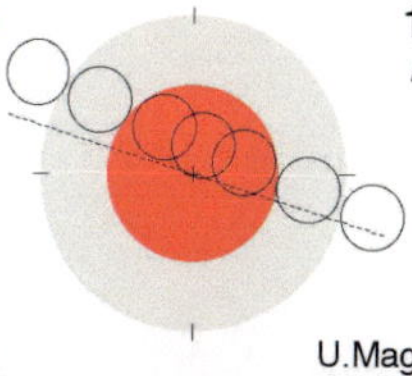

Tot. = 86m
Par. = 221m U.Mag. = 1.3607
Gam. = 0.2570 P.Mag. = 2.4161

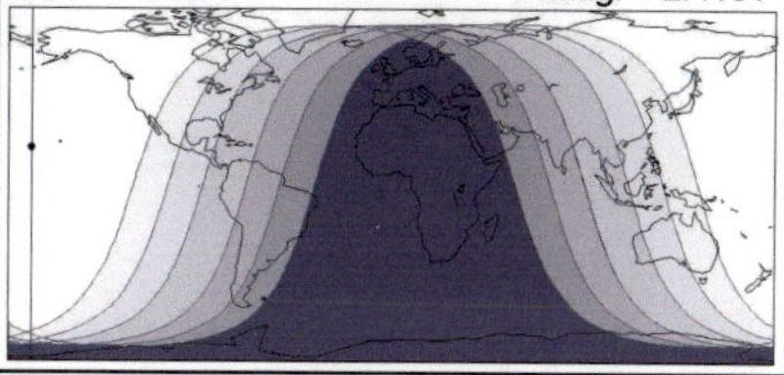

Penumbral 2023 May 05
Saros 141 17:24 TD
D.Node ΔT= 71s

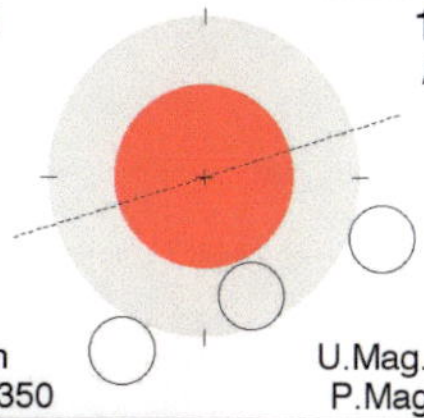

Pen. = 258m U.Mag. = -0.0438
Gam. = -1.0350 P.Mag. = 0.9655

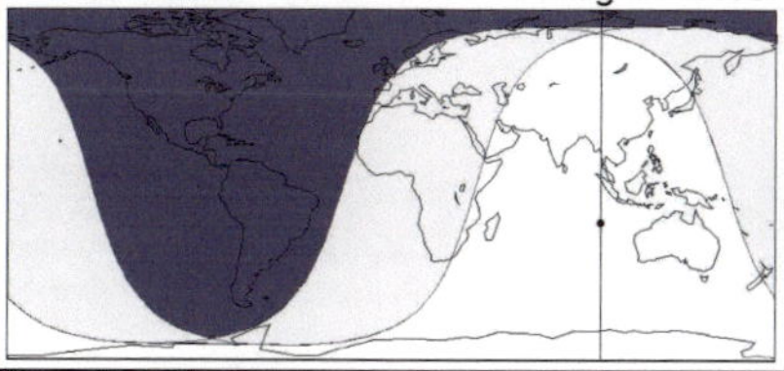

Partial 2023 Oct 28
Saros 146 20:15 TD
A.Node ΔT= 71s

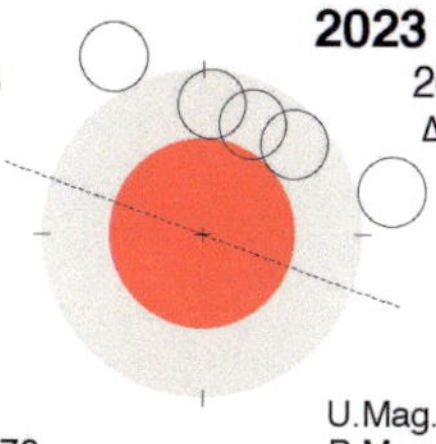

Par. = 78m U.Mag. = 0.1239
Gam. = 0.9472 P.Mag. = 1.1200

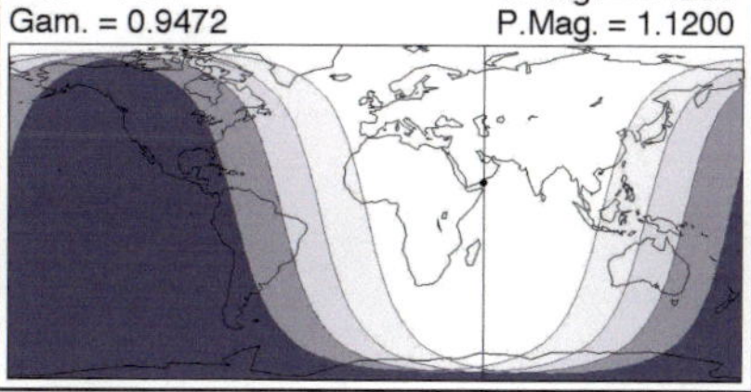

Penumbral 2024 Mar 25
Saros 113 07:14 TD
D.Node ΔT= 71s

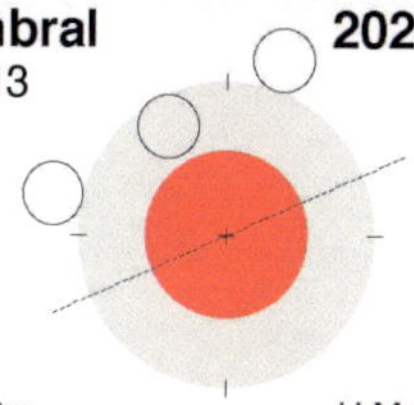

Pen. = 280m U.Mag. = -0.1304
Gam. = 1.0610 P.Mag. = 0.9577

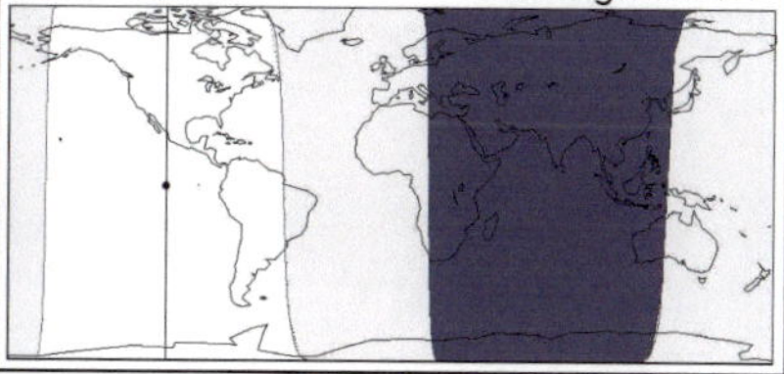

Partial 2024 Sep 18
Saros 118 02:45 TD
A.Node ΔT= 71s

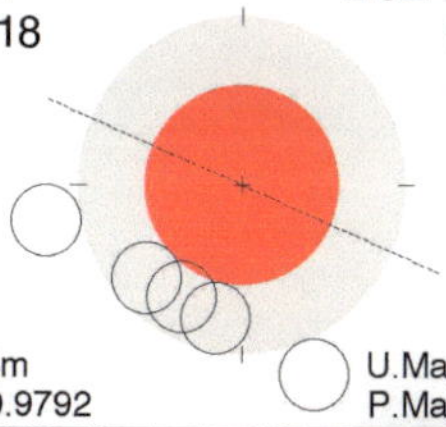

Par. = 64m U.Mag. = 0.0869
Gam. = -0.9792 P.Mag. = 1.0392

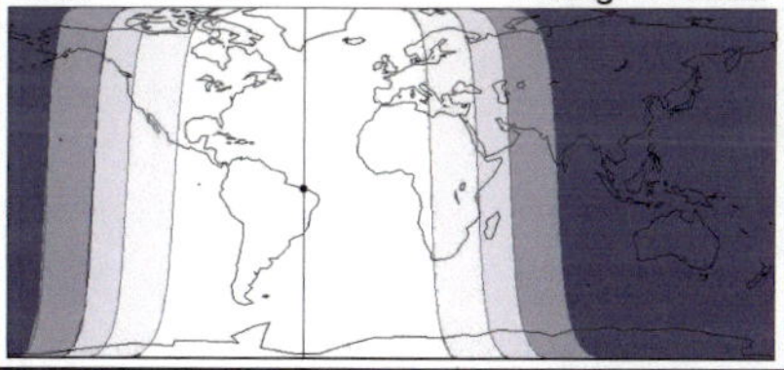

Total 2025 Mar 14
Saros 123 07:00 TD
D.Node ΔT= 72s

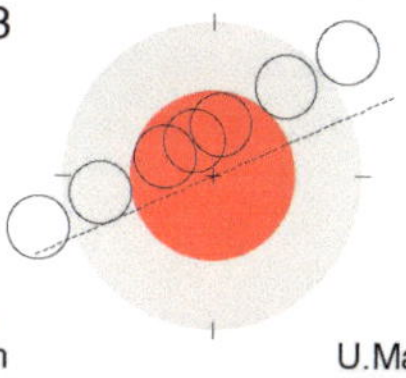

Tot. = 66m
Par. = 219m U.Mag. = 1.1804
Gam. = 0.3485 P.Mag. = 2.2615

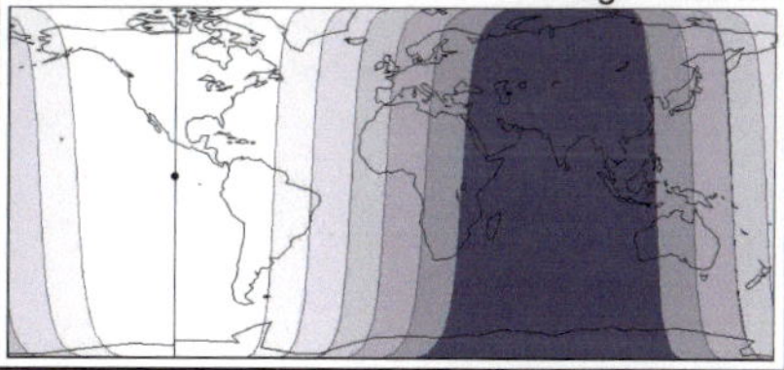

Total 2025 Sep 07
Saros 128 18:13 TD
A.Node ΔT= 72s

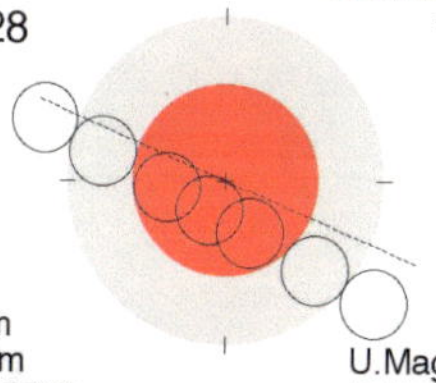

Tot. = 83m
Par. = 210m U.Mag. = 1.3638
Gam. = -0.2752 P.Mag. = 2.3459

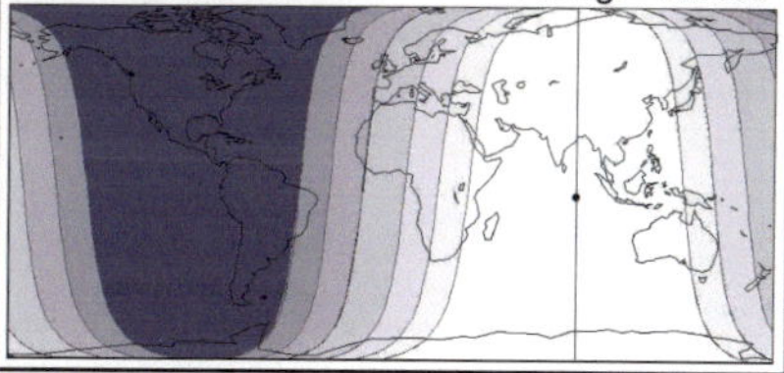

Total 2026 Mar 03
Saros 133 11:35 TD
D.Node ΔT= 72s

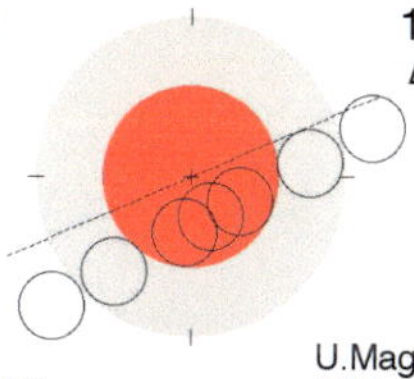

Tot. = 59m
Par. = 208m U.Mag. = 1.1526
Gam. = -0.3765 P.Mag. = 2.1858

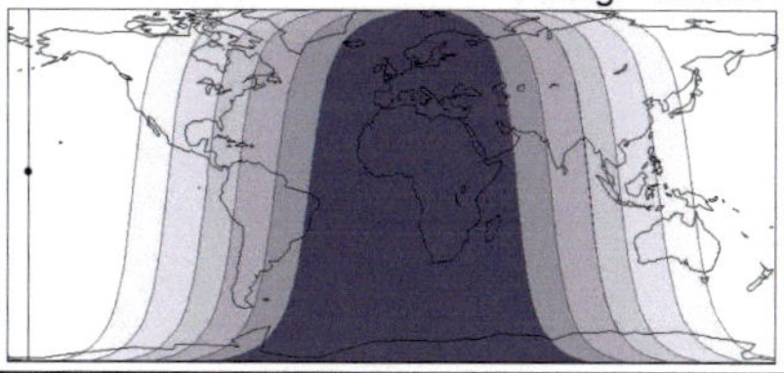

Partial 2026 Aug 28
Saros 138 04:14 TD
A.Node ΔT= 72s

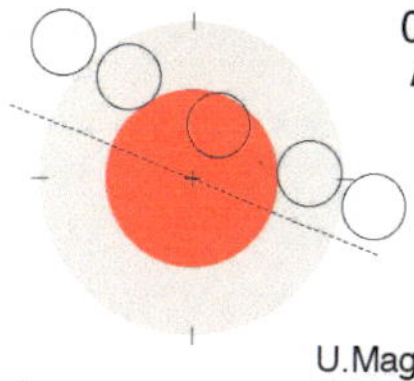

Par. = 199m U.Mag. = 0.9319
Gam. = 0.4964 P.Mag. = 1.9664

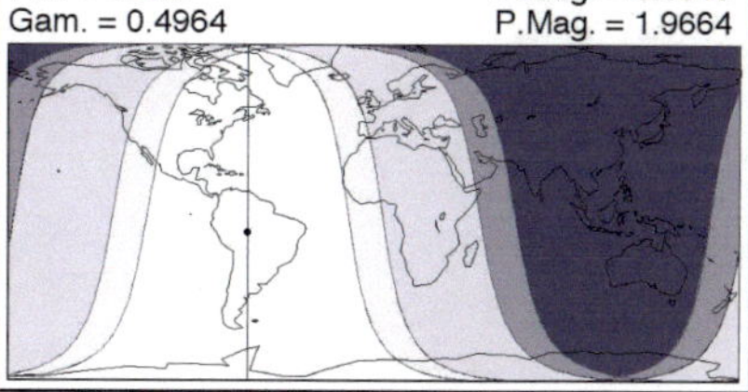

Penumbral 2027 Feb 20
Saros 143 23:14 TD
D.Node ΔT= 73s

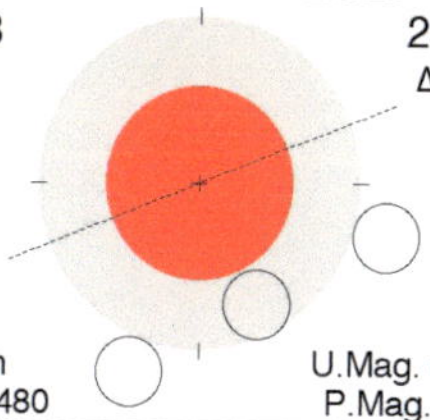

Pen. = 242m U.Mag. = -0.0549
Gam. = -1.0480 P.Mag. = 0.9286

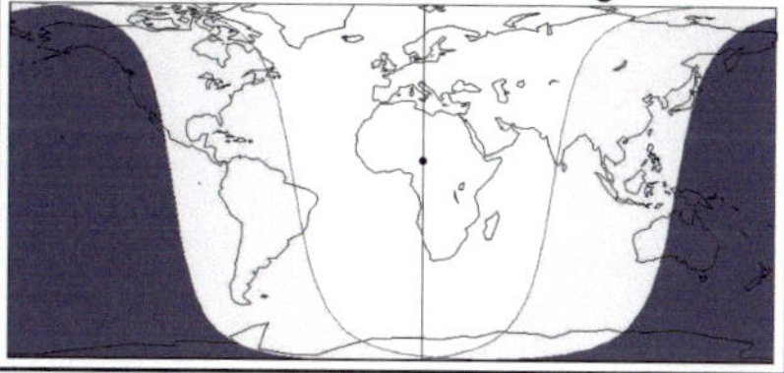

Plate 005

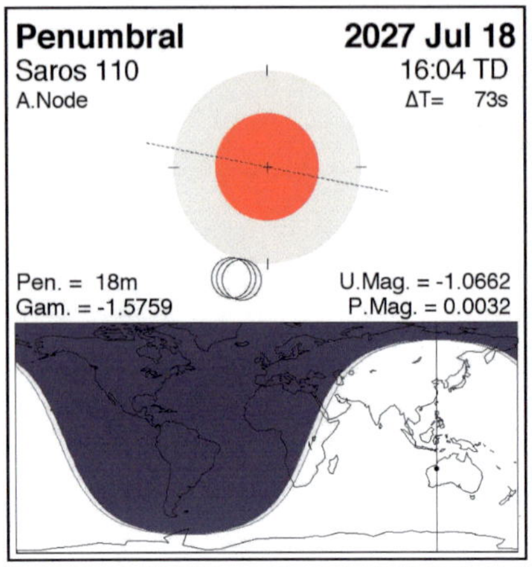
Penumbral
2027 Jul 18
Saros 110
16:04 TD
A.Node
ΔT= 73s
Pen. = 18m
Gam. = -1.5759
U.Mag. = -1.0662
P.Mag. = 0.0032

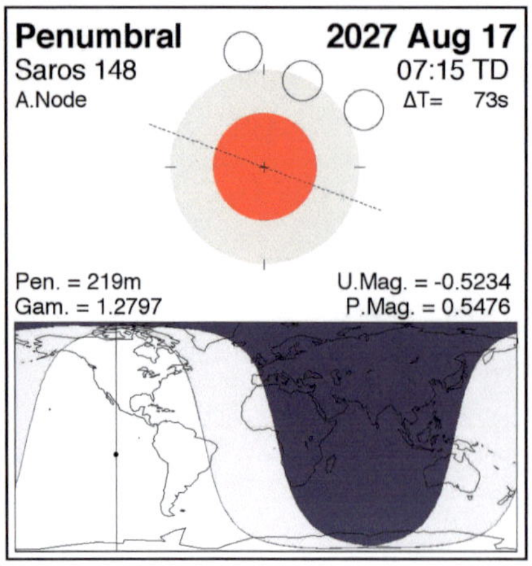
Penumbral
2027 Aug 17
Saros 148
07:15 TD
A.Node
ΔT= 73s
Pen. = 219m
Gam. = 1.2797
U.Mag. = -0.5234
P.Mag. = 0.5476

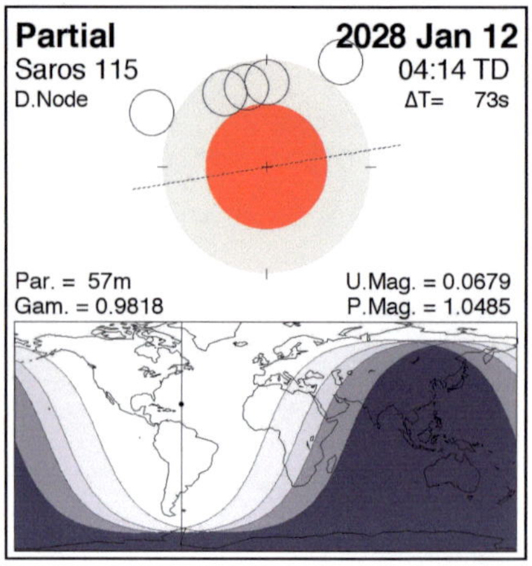
Partial
2028 Jan 12
Saros 115
04:14 TD
D.Node
ΔT= 73s
Par. = 57m
Gam. = 0.9818
U.Mag. = 0.0679
P.Mag. = 1.0485

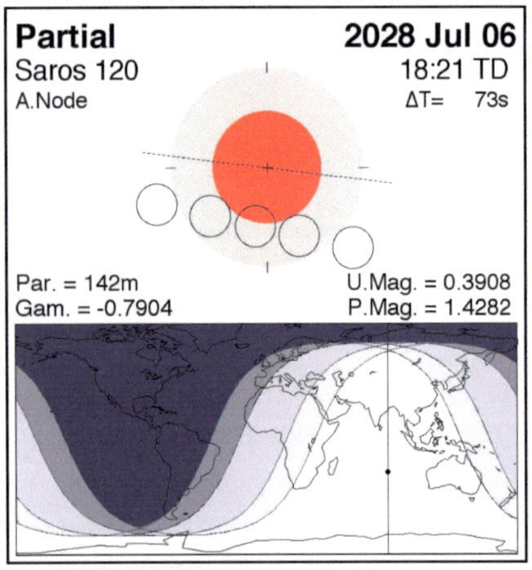
Partial
2028 Jul 06
Saros 120
18:21 TD
A.Node
ΔT= 73s
Par. = 142m
Gam. = -0.7904
U.Mag. = 0.3908
P.Mag. = 1.4282

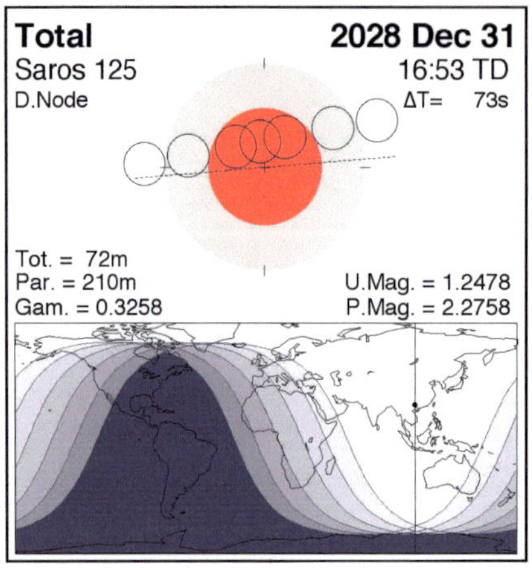
Total
2028 Dec 31
Saros 125
16:53 TD
D.Node
ΔT= 73s
Tot. = 72m
Par. = 210m
Gam. = 0.3258
U.Mag. = 1.2478
P.Mag. = 2.2758

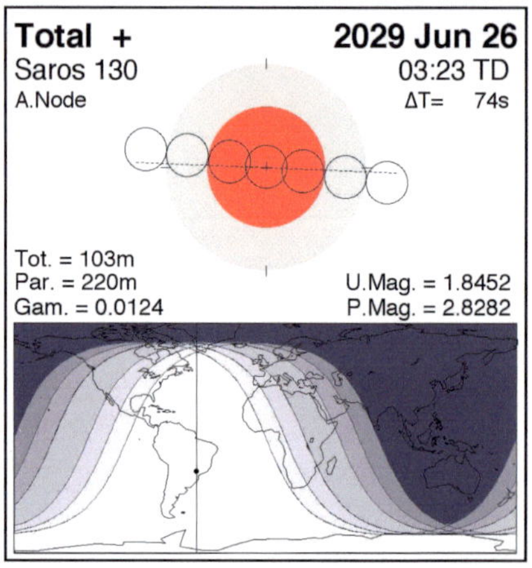
Total +
2029 Jun 26
Saros 130
03:23 TD
A.Node
ΔT= 74s
Tot. = 103m
Par. = 220m
Gam. = 0.0124
U.Mag. = 1.8452
P.Mag. = 2.8282

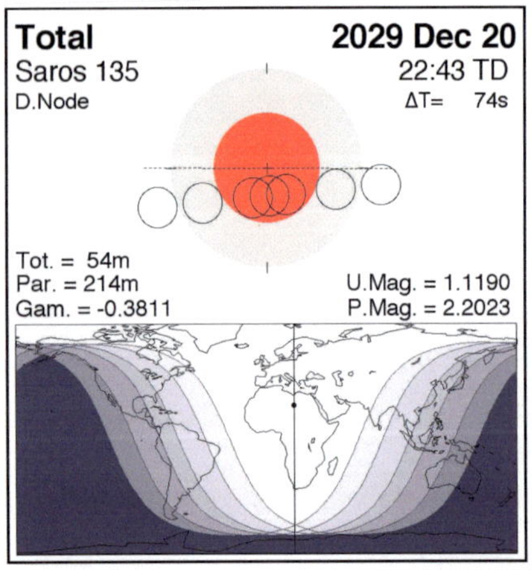
Total
2029 Dec 20
Saros 135
22:43 TD
D.Node
ΔT= 74s
Tot. = 54m
Par. = 214m
Gam. = -0.3811
U.Mag. = 1.1190
P.Mag. = 2.2023

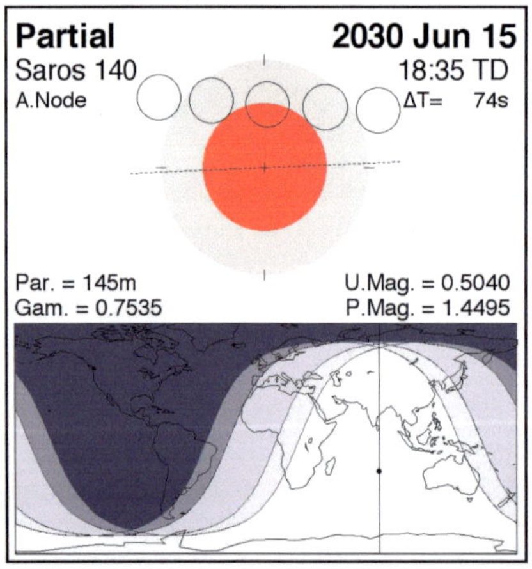
Partial
2030 Jun 15
Saros 140
18:35 TD
A.Node
ΔT= 74s
Par. = 145m
Gam. = 0.7535
U.Mag. = 0.5040
P.Mag. = 1.4495

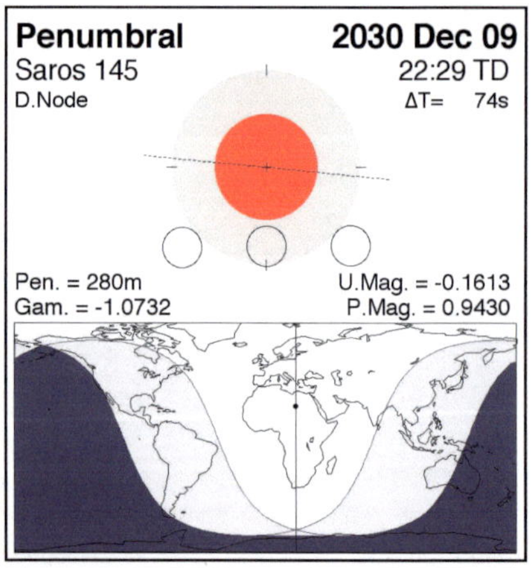
Penumbral
2030 Dec 09
Saros 145
22:29 TD
D.Node
ΔT= 74s
Pen. = 280m
Gam. = -1.0732
U.Mag. = -0.1613
P.Mag. = 0.9430

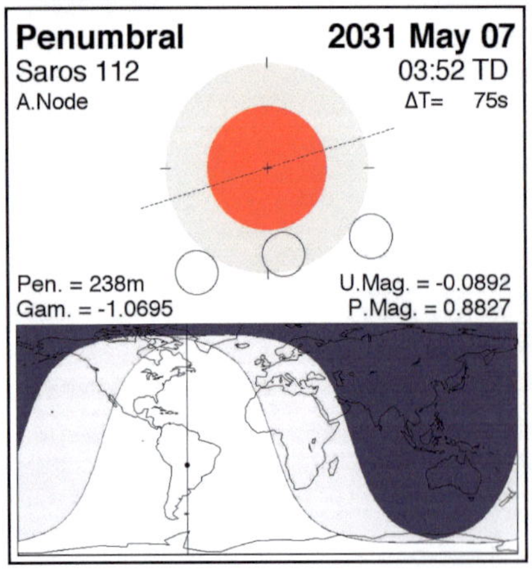
Penumbral
2031 May 07
Saros 112
03:52 TD
A.Node
ΔT= 75s
Pen. = 238m
Gam. = -1.0695
U.Mag. = -0.0892
P.Mag. = 0.8827

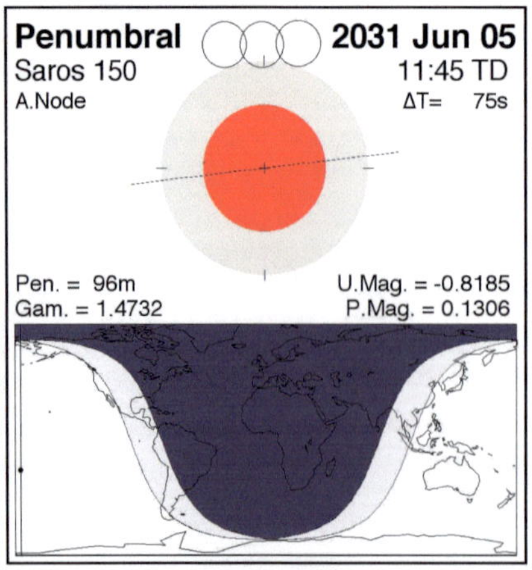
Penumbral
2031 Jun 05
Saros 150
11:45 TD
A.Node
ΔT= 75s
Pen. = 96m
Gam. = 1.4732
U.Mag. = -0.8185
P.Mag. = 0.1306

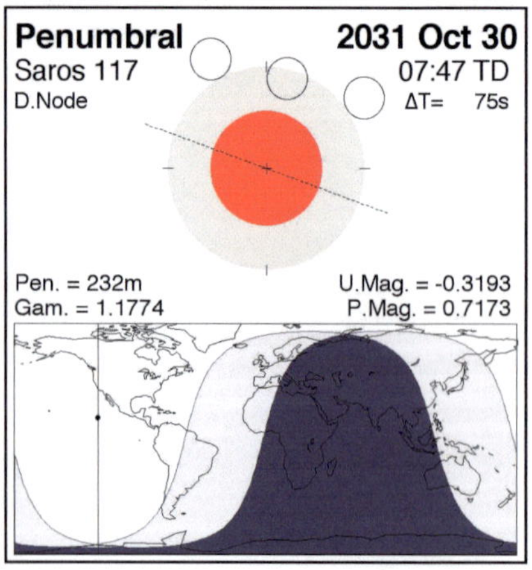
Penumbral
2031 Oct 30
Saros 117
07:47 TD
D.Node
ΔT= 75s
Pen. = 232m
Gam. = 1.1774
U.Mag. = -0.3193
P.Mag. = 0.7173

Plate 006

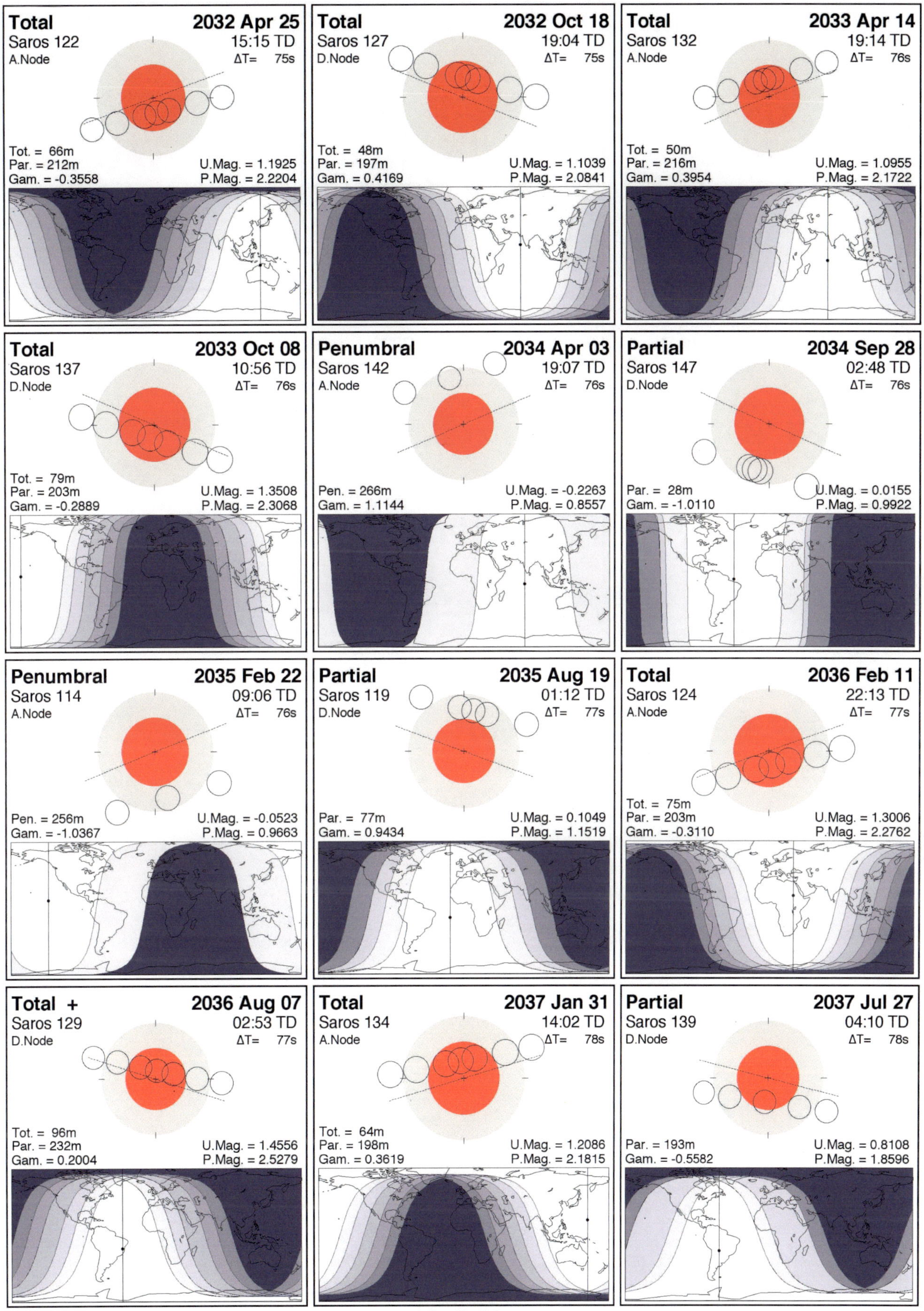
Total
2032 Apr 25
Saros 122
15:15 TD
A.Node
ΔT= 75s
Tot. = 66m
Par. = 212m
Gam. = -0.3558
U.Mag. = 1.1925
P.Mag. = 2.2204
Total
2032 Oct 18
Saros 127
19:04 TD
D.Node
ΔT= 75s
Tot. = 48m
Par. = 197m
Gam. = 0.4169
U.Mag. = 1.1039
P.Mag. = 2.0841
Total
2033 Apr 14
Saros 132
19:14 TD
A.Node
ΔT= 76s
Tot. = 50m
Par. = 216m
Gam. = 0.3954
U.Mag. = 1.0955
P.Mag. = 2.1722
Total
2033 Oct 08
Saros 137
10:56 TD
D.Node
ΔT= 76s
Tot. = 79m
Par. = 203m
Gam. = -0.2889
U.Mag. = 1.3508
P.Mag. = 2.3068
Penumbral
2034 Apr 03
Saros 142
19:07 TD
A.Node
ΔT= 76s
Pen. = 266m
Gam. = 1.1144
U.Mag. = -0.2263
P.Mag. = 0.8557
Partial
2034 Sep 28
Saros 147
02:48 TD
D.Node
ΔT= 76s
Par. = 28m
Gam. = -1.0110
U.Mag. = 0.0155
P.Mag. = 0.9922
Penumbral
2035 Feb 22
Saros 114
09:06 TD
A.Node
ΔT= 76s
Pen. = 256m
Gam. = -1.0367
U.Mag. = -0.0523
P.Mag. = 0.9663
Partial
2035 Aug 19
Saros 119
01:12 TD
D.Node
ΔT= 77s
Par. = 77m
Gam. = 0.9434
U.Mag. = 0.1049
P.Mag. = 1.1519
Total
2036 Feb 11
Saros 124
22:13 TD
A.Node
ΔT= 77s
Tot. = 75m
Par. = 203m
Gam. = -0.3110
U.Mag. = 1.3006
P.Mag. = 2.2762
Total +
2036 Aug 07
Saros 129
02:53 TD
D.Node
ΔT= 77s
Tot. = 96m
Par. = 232m
Gam. = 0.2004
U.Mag. = 1.4556
P.Mag. = 2.5279
Total
2037 Jan 31
Saros 134
14:02 TD
A.Node
ΔT= 78s
Tot. = 64m
Par. = 198m
Gam. = 0.3619
U.Mag. = 1.2086
P.Mag. = 2.1815
Partial
2037 Jul 27
Saros 139
04:10 TD
D.Node
ΔT= 78s
Par. = 193m
Gam. = -0.5582
U.Mag. = 0.8108
P.Mag. = 1.8596

Plate 007

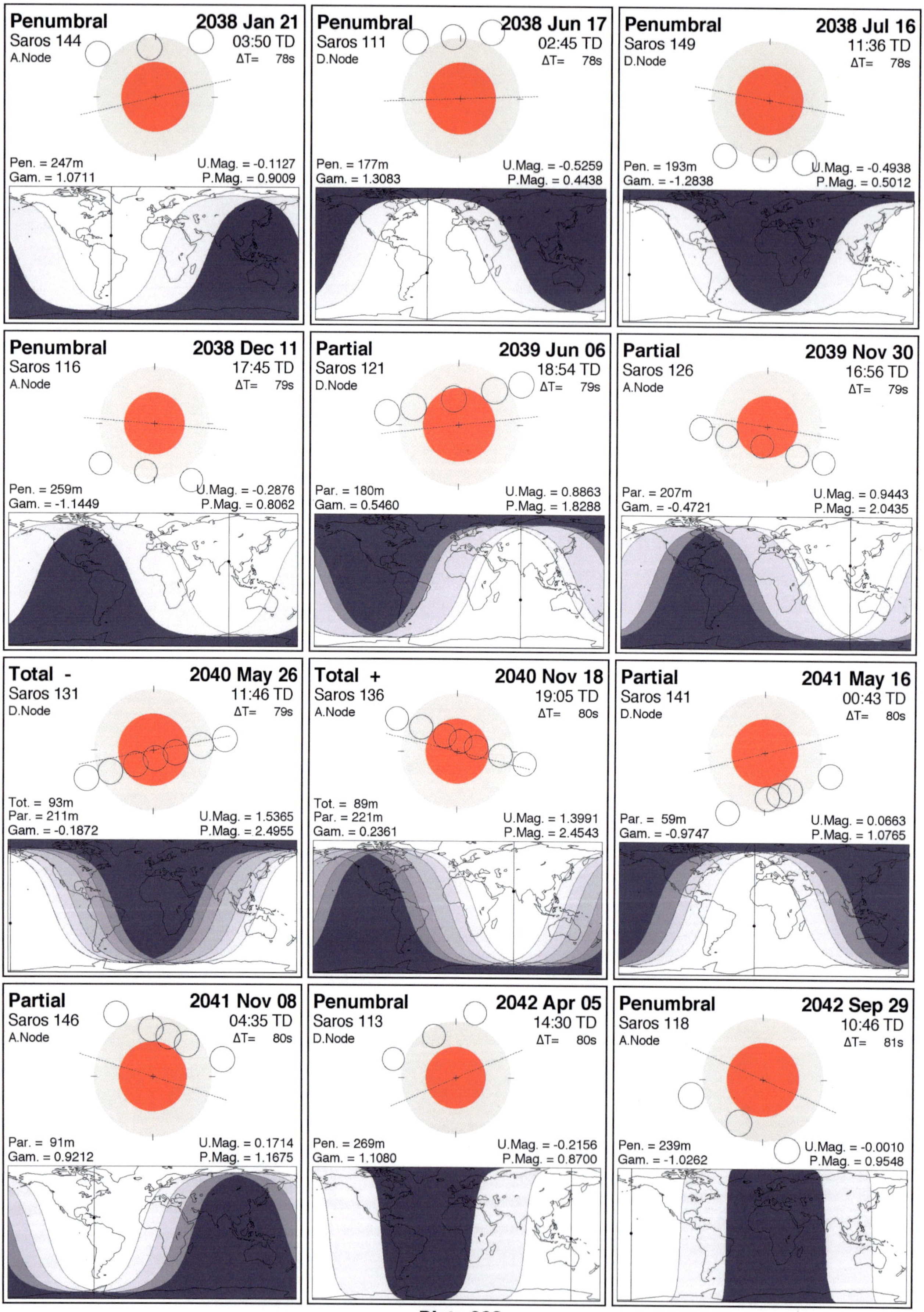
Penumbral
2038 Jan 21
Saros 144
03:50 TD
A.Node
ΔT= 78s
Pen. = 247m
Gam. = 1.0711
U.Mag. = -0.1127
P.Mag. = 0.9009
Penumbral
2038 Jun 17
Saros 111
02:45 TD
D.Node
ΔT= 78s
Pen. = 177m
Gam. = 1.3083
U.Mag. = -0.5259
P.Mag. = 0.4438
Penumbral
2038 Jul 16
Saros 149
11:36 TD
D.Node
ΔT= 78s
Pen. = 193m
Gam. = -1.2838
U.Mag. = -0.4938
P.Mag. = 0.5012
Penumbral
2038 Dec 11
Saros 116
17:45 TD
A.Node
ΔT= 79s
Pen. = 259m
Gam. = -1.1449
U.Mag. = -0.2876
P.Mag. = 0.8062
Partial
2039 Jun 06
Saros 121
18:54 TD
D.Node
ΔT= 79s
Par. = 180m
Gam. = 0.5460
U.Mag. = 0.8863
P.Mag. = 1.8288
Partial
2039 Nov 30
Saros 126
16:56 TD
A.Node
ΔT= 79s
Par. = 207m
Gam. = -0.4721
U.Mag. = 0.9443
P.Mag. = 2.0435
Total -
2040 May 26
Saros 131
11:46 TD
D.Node
ΔT= 79s
Tot. = 93m
Par. = 211m
Gam. = -0.1872
U.Mag. = 1.5365
P.Mag. = 2.4955
Total +
2040 Nov 18
Saros 136
19:05 TD
A.Node
ΔT= 80s
Tot. = 89m
Par. = 221m
Gam. = 0.2361
U.Mag. = 1.3991
P.Mag. = 2.4543
Partial
2041 May 16
Saros 141
00:43 TD
D.Node
ΔT= 80s
Par. = 59m
Gam. = -0.9747
U.Mag. = 0.0663
P.Mag. = 1.0765
Partial
2041 Nov 08
Saros 146
04:35 TD
A.Node
ΔT= 80s
Par. = 91m
Gam. = 0.9212
U.Mag. = 0.1714
P.Mag. = 1.1675
Penumbral
2042 Apr 05
Saros 113
14:30 TD
D.Node
ΔT= 80s
Pen. = 269m
Gam. = 1.1080
U.Mag. = -0.2156
P.Mag. = 0.8700
Penumbral
2042 Sep 29
Saros 118
10:46 TD
A.Node
ΔT= 81s
Pen. = 239m
Gam. = -1.0262
U.Mag. = -0.0010
P.Mag. = 0.9548

Plate 008

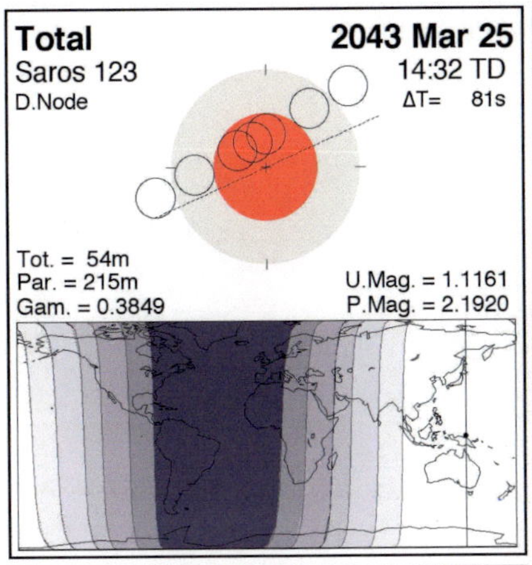
Total
2043 Mar 25
Saros 123
14:32 TD
D.Node
ΔT= 81s
Tot. = 54m
Par. = 215m
Gam. = 0.3849
U.Mag. = 1.1161
P.Mag. = 2.1920

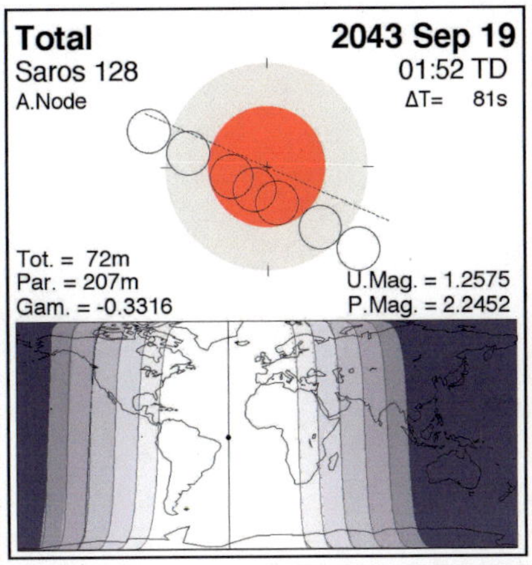
Total
2043 Sep 19
Saros 128
01:52 TD
A.Node
ΔT= 81s
Tot. = 72m
Par. = 207m
Gam. = -0.3316
U.Mag. = 1.2575
P.Mag. = 2.2452

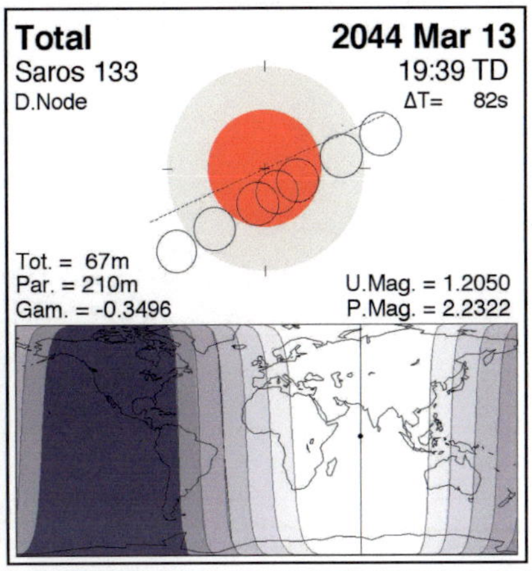
Total
2044 Mar 13
Saros 133
19:39 TD
D.Node
ΔT= 82s
Tot. = 67m
Par. = 210m
Gam. = -0.3496
U.Mag. = 1.2050
P.Mag. = 2.2322

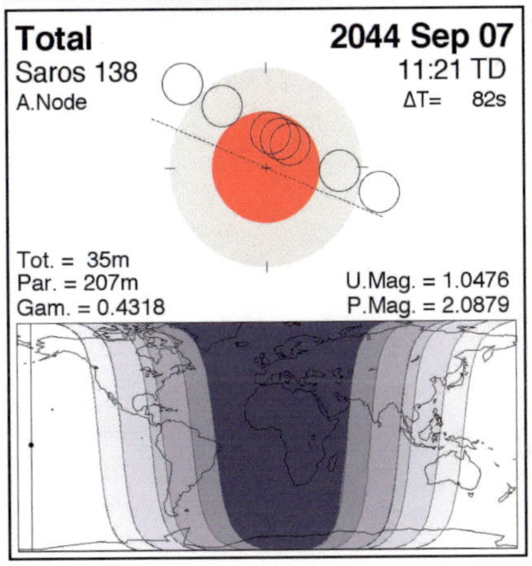
Total
2044 Sep 07
Saros 138
11:21 TD
A.Node
ΔT= 82s
Tot. = 35m
Par. = 207m
Gam. = 0.4318
U.Mag. = 1.0476
P.Mag. = 2.0879

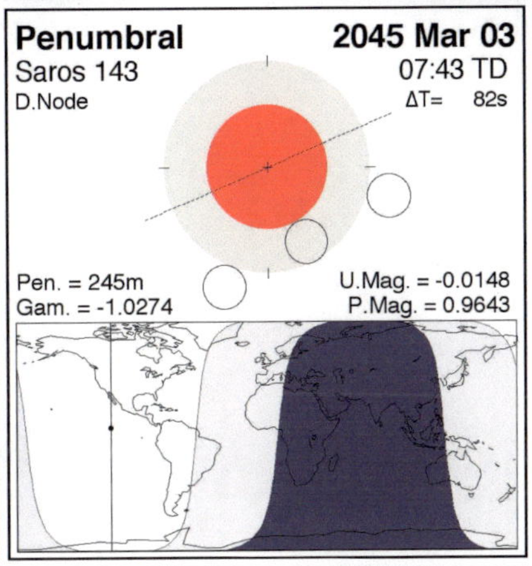
Penumbral
2045 Mar 03
Saros 143
07:43 TD
D.Node
ΔT= 82s
Pen. = 245m
Gam. = -1.0274
U.Mag. = -0.0148
P.Mag. = 0.9643

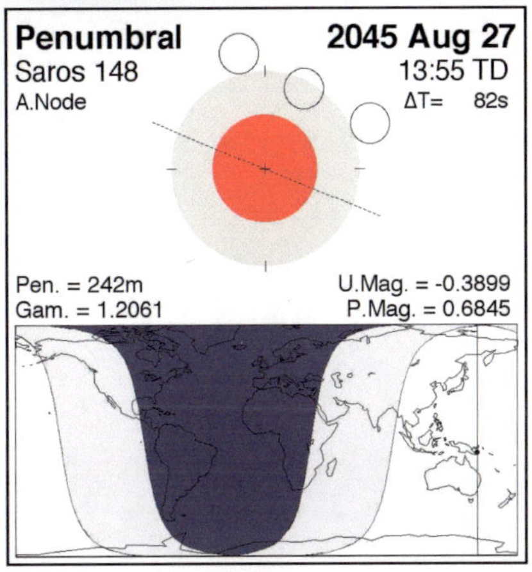
Penumbral
2045 Aug 27
Saros 148
13:55 TD
A.Node
ΔT= 82s
Pen. = 242m
Gam. = 1.2061
U.Mag. = -0.3899
P.Mag. = 0.6845

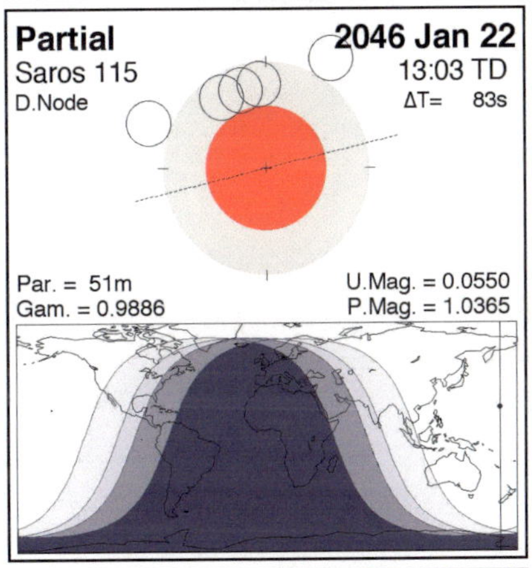
Partial
2046 Jan 22
Saros 115
13:03 TD
D.Node
ΔT= 83s
Par. = 51m
Gam. = 0.9886
U.Mag. = 0.0550
P.Mag. = 1.0365

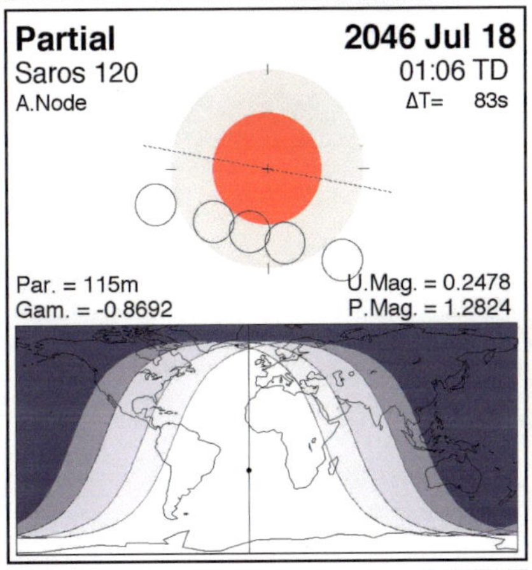
Partial
2046 Jul 18
Saros 120
01:06 TD
A.Node
ΔT= 83s
Par. = 115m
Gam. = -0.8692
U.Mag. = 0.2478
P.Mag. = 1.2824

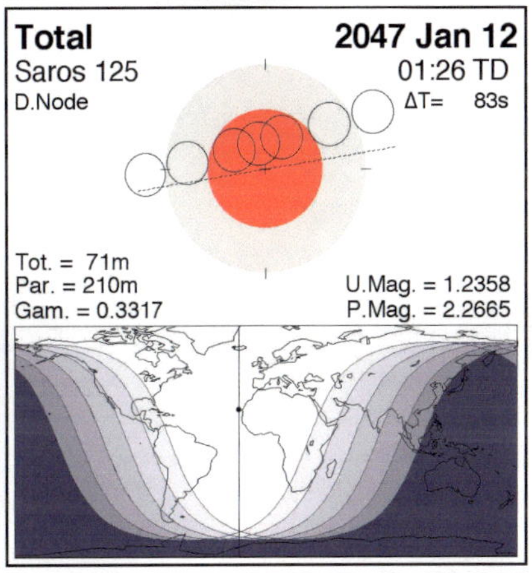
Total
2047 Jan 12
Saros 125
01:26 TD
D.Node
ΔT= 83s
Tot. = 71m
Par. = 210m
Gam. = 0.3317
U.Mag. = 1.2358
P.Mag. = 2.2665

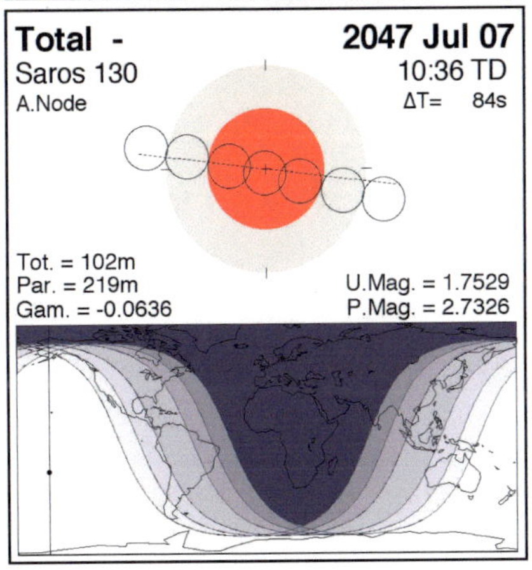
Total -
2047 Jul 07
Saros 130
10:36 TD
A.Node
ΔT= 84s
Tot. = 102m
Par. = 219m
Gam. = -0.0636
U.Mag. = 1.7529
P.Mag. = 2.7326

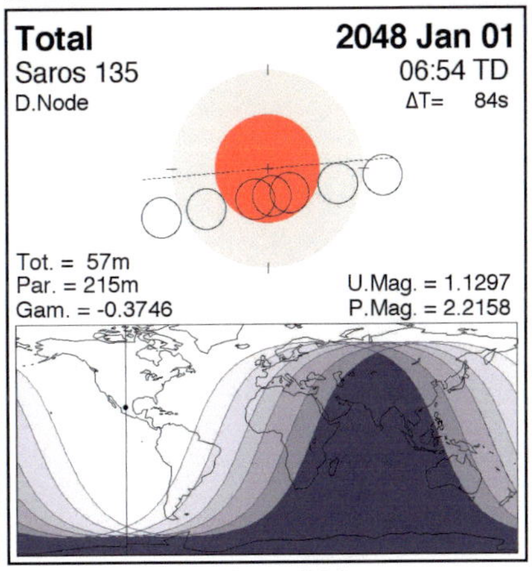
Total
2048 Jan 01
Saros 135
06:54 TD
D.Node
ΔT= 84s
Tot. = 57m
Par. = 215m
Gam. = -0.3746
U.Mag. = 1.1297
P.Mag. = 2.2158

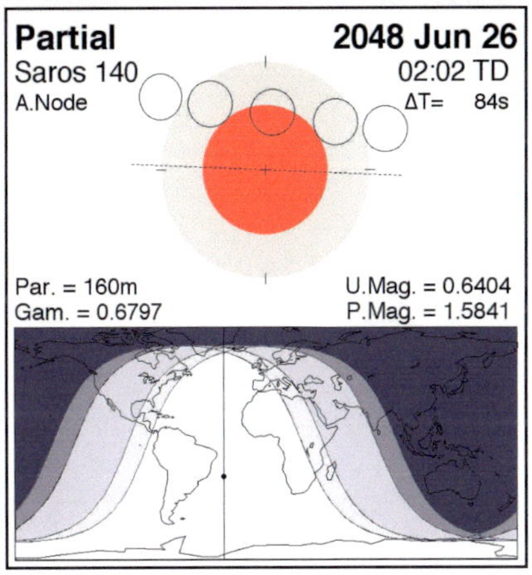
Partial
2048 Jun 26
Saros 140
02:02 TD
A.Node
ΔT= 84s
Par. = 160m
Gam. = 0.6797
U.Mag. = 0.6404
P.Mag. = 1.5841

Plate 009

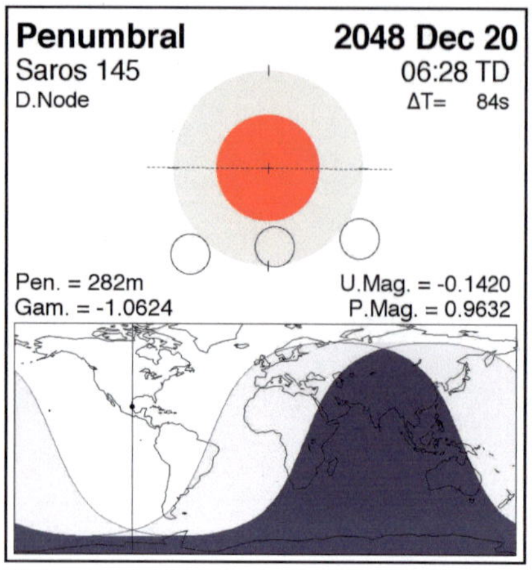
Penumbral
2048 Dec 20
Saros 145
06:28 TD
D.Node
ΔT= 84s
Pen. = 282m
Gam. = -1.0624
U.Mag. = -0.1420
P.Mag. = 0.9632

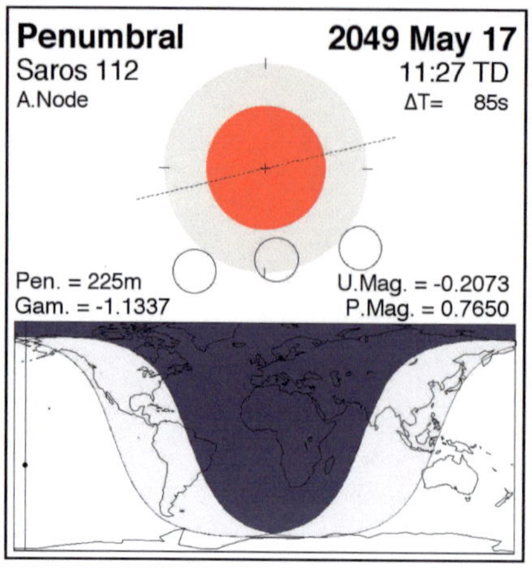
Penumbral
2049 May 17
Saros 112
11:27 TD
A.Node
ΔT= 85s
Pen. = 225m
Gam. = -1.1337
U.Mag. = -0.2073
P.Mag. = 0.7650

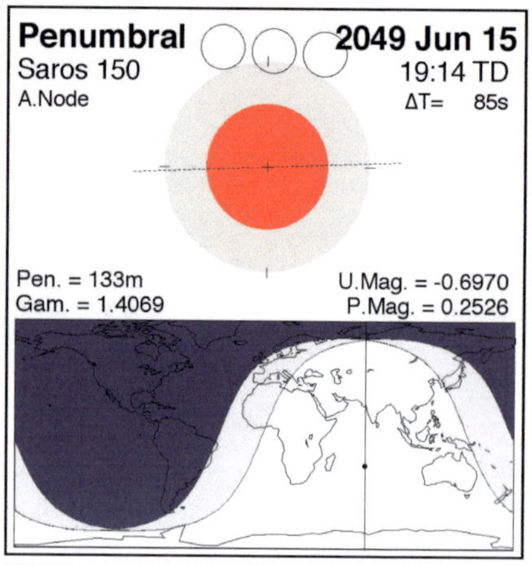
Penumbral
2049 Jun 15
Saros 150
19:14 TD
A.Node
ΔT= 85s
Pen. = 133m
Gam. = 1.4069
U.Mag. = -0.6970
P.Mag. = 0.2526

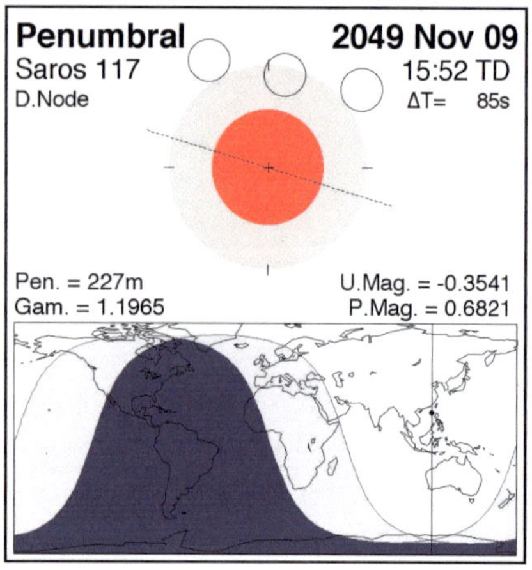
Penumbral
2049 Nov 09
Saros 117
15:52 TD
D.Node
ΔT= 85s
Pen. = 227m
Gam. = 1.1965
U.Mag. = -0.3541
P.Mag. = 0.6821

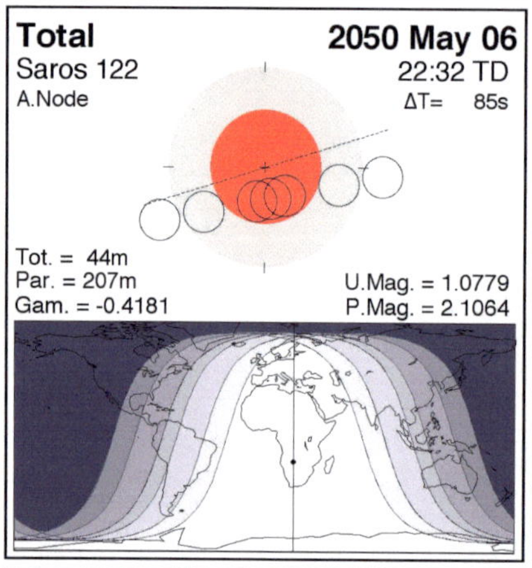
Total
2050 May 06
Saros 122
22:32 TD
A.Node
ΔT= 85s
Tot. = 44m
Par. = 207m
Gam. = -0.4181
U.Mag. = 1.0779
P.Mag. = 2.1064

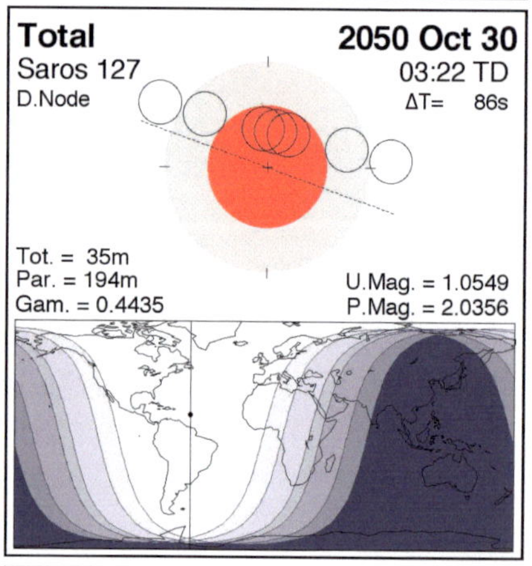
Total
2050 Oct 30
Saros 127
03:22 TD
D.Node
ΔT= 86s
Tot. = 35m
Par. = 194m
Gam. = 0.4435
U.Mag. = 1.0549
P.Mag. = 2.0356

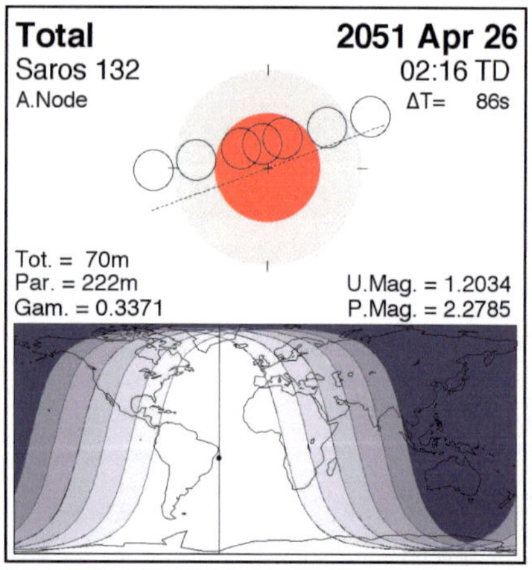
Total
2051 Apr 26
Saros 132
02:16 TD
A.Node
ΔT= 86s
Tot. = 70m
Par. = 222m
Gam. = 0.3371
U.Mag. = 1.2034
P.Mag. = 2.2785

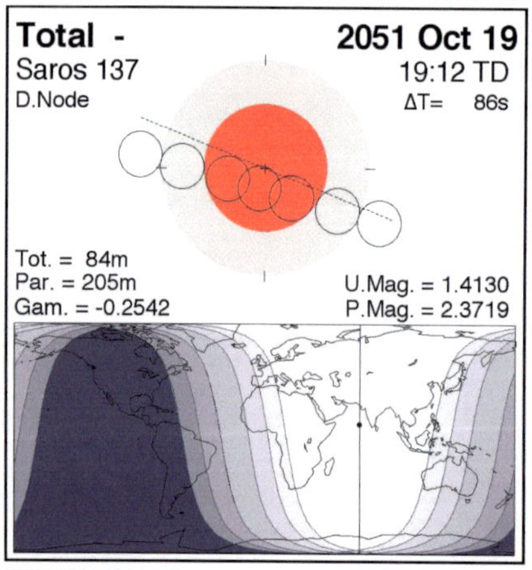
Total -
2051 Oct 19
Saros 137
19:12 TD
D.Node
ΔT= 86s
Tot. = 84m
Par. = 205m
Gam. = -0.2542
U.Mag. = 1.4130
P.Mag. = 2.3719

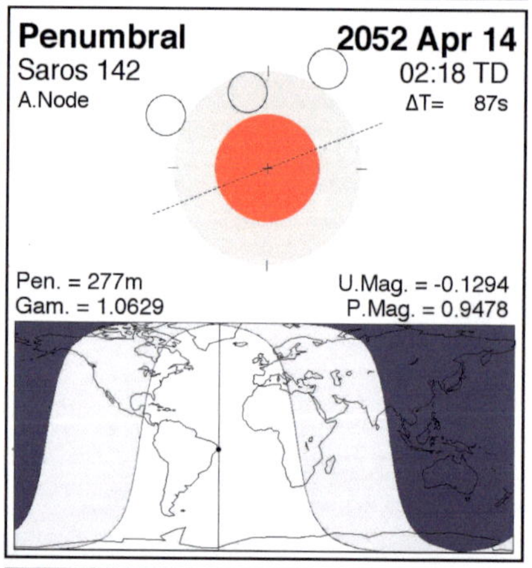
Penumbral
2052 Apr 14
Saros 142
02:18 TD
A.Node
ΔT= 87s
Pen. = 277m
Gam. = 1.0629
U.Mag. = -0.1294
P.Mag. = 0.9478

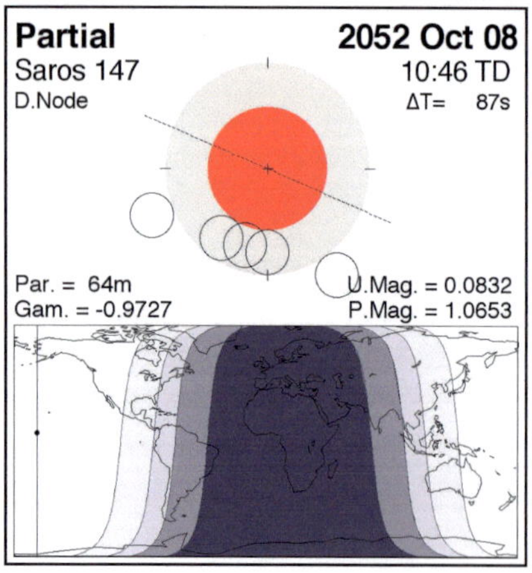
Partial
2052 Oct 08
Saros 147
10:46 TD
D.Node
ΔT= 87s
Par. = 64m
Gam. = -0.9727
U.Mag. = 0.0832
P.Mag. = 1.0653

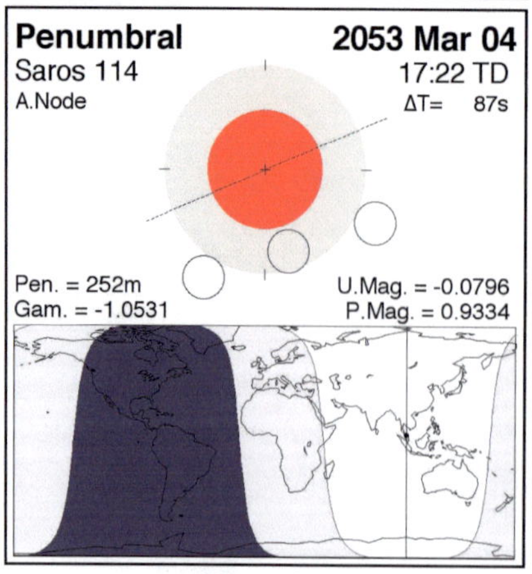
Penumbral
2053 Mar 04
Saros 114
17:22 TD
A.Node
ΔT= 87s
Pen. = 252m
Gam. = -1.0531
U.Mag. = -0.0796
P.Mag. = 0.9334

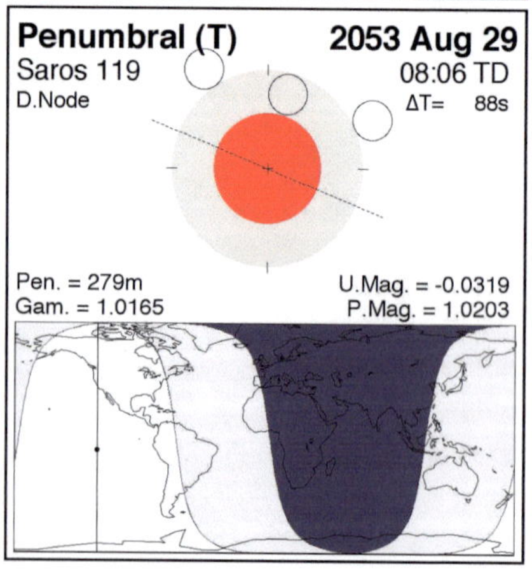
Penumbral (T)
2053 Aug 29
Saros 119
08:06 TD
D.Node
ΔT= 88s
Pen. = 279m
Gam. = 1.0165
U.Mag. = -0.0319
P.Mag. = 1.0203

Plate 010

Total **2054 Feb 22**
Saros 124 06:51 TD
A.Node ΔT= 88s

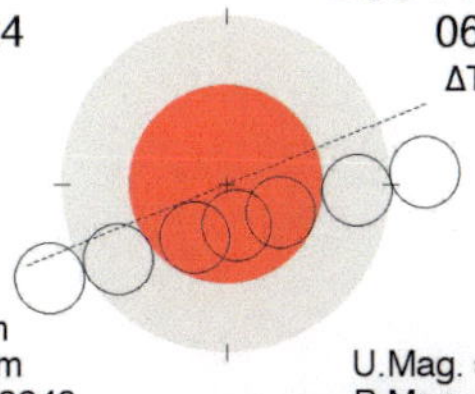

Tot. = 73m
Par. = 202m U.Mag. = 1.2780
Gam. = -0.3242 P.Mag. = 2.2502

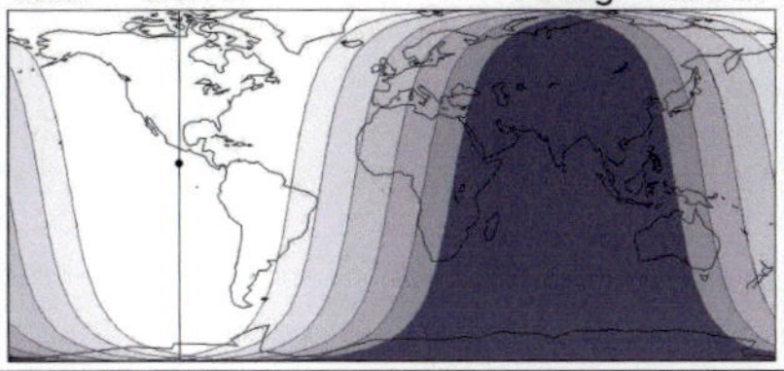

Total **2054 Aug 18**
Saros 129 09:27 TD
D.Node ΔT= 88s

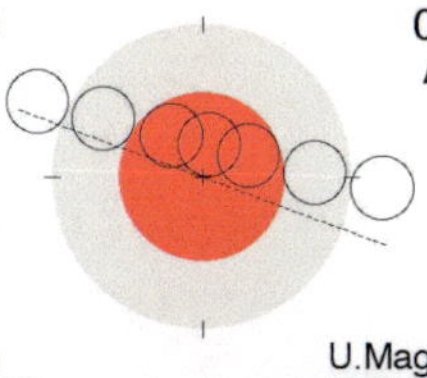

Tot. = 84m
Par. = 227m U.Mag. = 1.3074
Gam. = 0.2806 P.Mag. = 2.3817

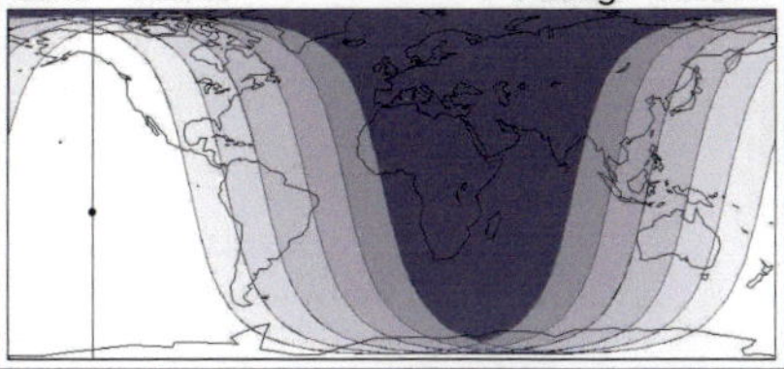

Total **2055 Feb 11**
Saros 134 22:46 TD
A.Node ΔT= 88s

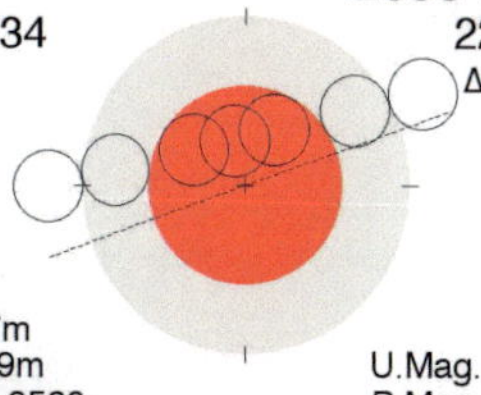

Tot. = 67m
Par. = 199m U.Mag. = 1.2258
Gam. = 0.3526 P.Mag. = 2.1982

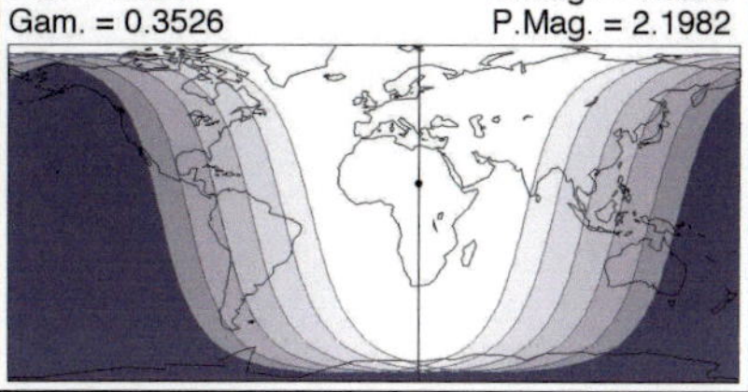

Partial **2055 Aug 07**
Saros 139 10:53 TD
D.Node ΔT= 89s

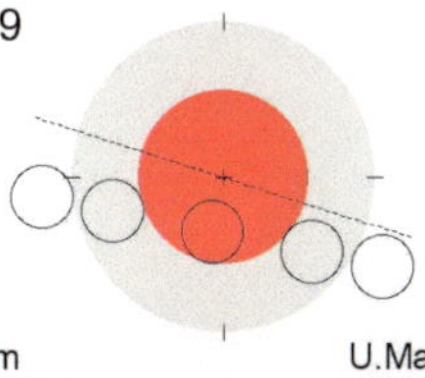

Par. = 204m U.Mag. = 0.9606
Gam. = -0.4769 P.Mag. = 2.0081

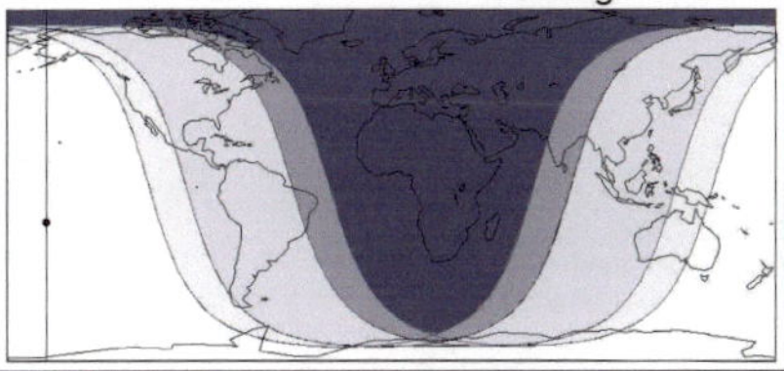

Penumbral **2056 Feb 01**
Saros 144 12:26 TD
A.Node ΔT= 89s

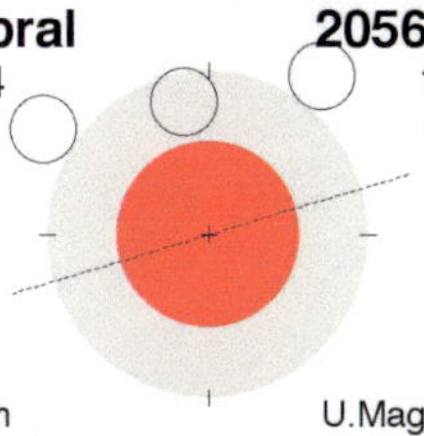

Pen. = 248m U.Mag. = -0.1084
Gam. = 1.0682 P.Mag. = 0.9069

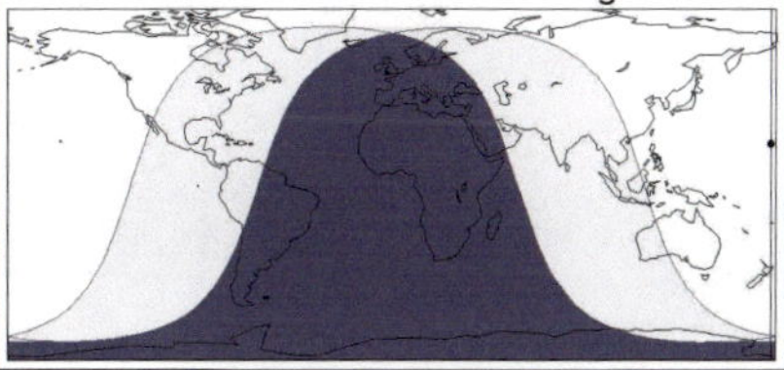

Penumbral **2056 Jun 27**
Saros 111 10:03 TD
D.Node ΔT= 89s

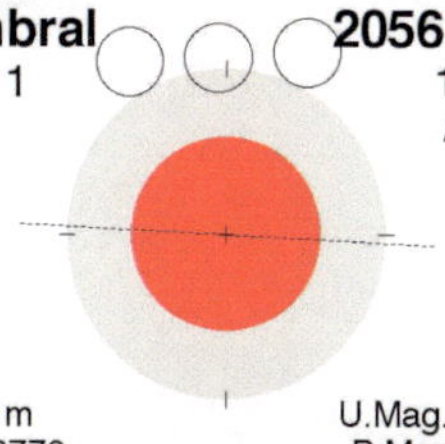

Pen. = 151m U.Mag. = -0.6504
Gam. = 1.3770 P.Mag. = 0.3158

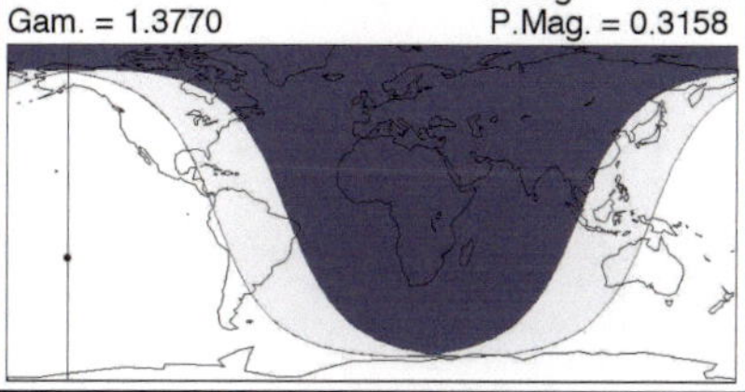

Penumbral **2056 Jul 26**
Saros 149 18:43 TD
D.Node ΔT= 89s

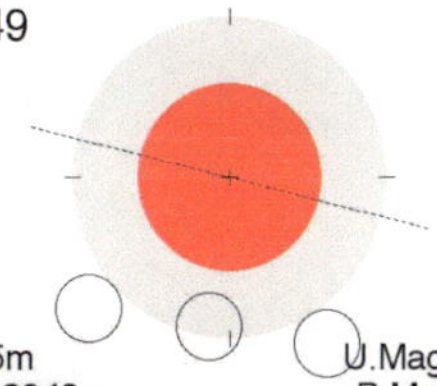

Pen. = 215m U.Mag. = -0.3477
Gam. = -1.2048 P.Mag. = 0.6448

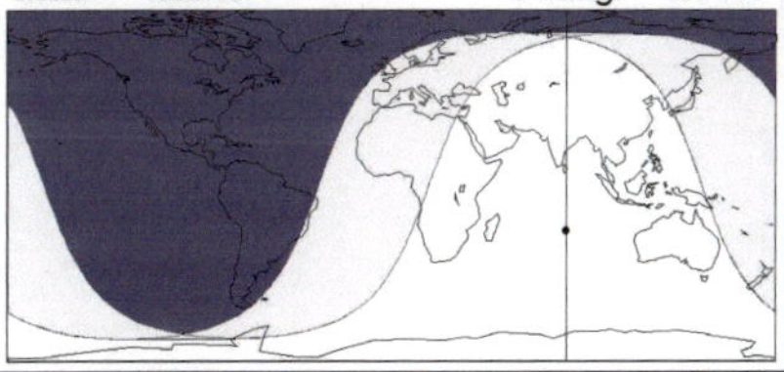

Penumbral **2056 Dec 22**
Saros 116 01:49 TD
A.Node ΔT= 90s

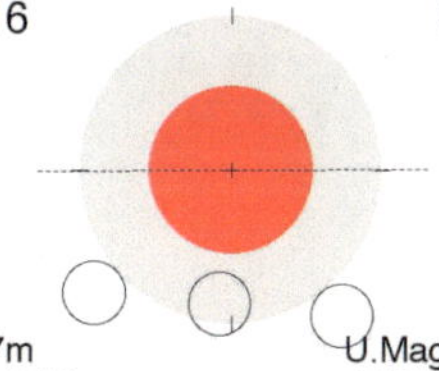

Pen. = 257m U.Mag. = -0.3093
Gam. = -1.1560 P.Mag. = 0.7872

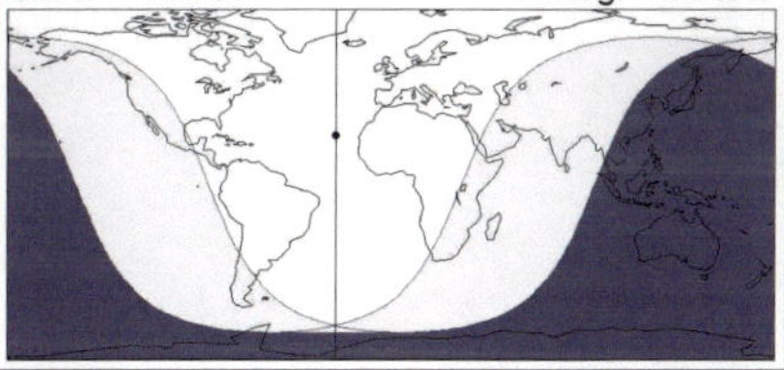

Partial **2057 Jun 17**
Saros 121 02:26 TD
D.Node ΔT= 90s

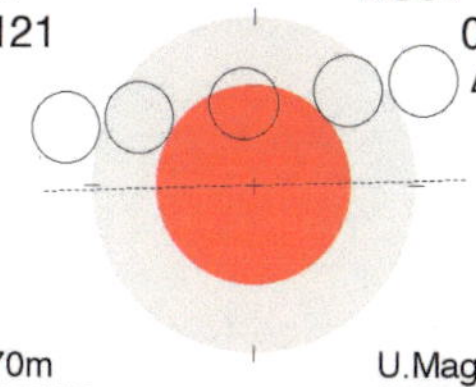

Par. = 170m U.Mag. = 0.7570
Gam. = 0.6168 P.Mag. = 1.6982

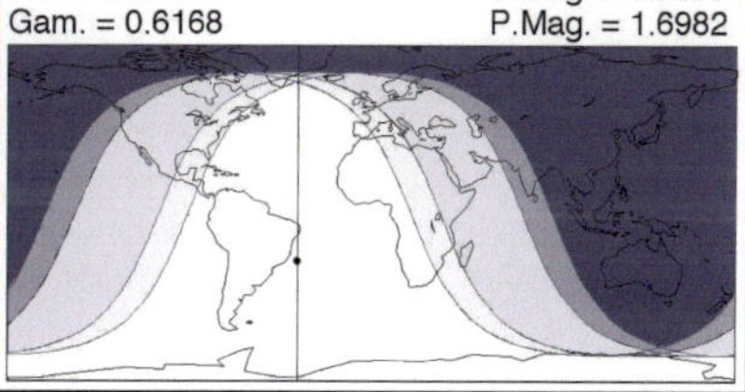

Partial **2057 Dec 11**
Saros 126 00:54 TD
A.Node ΔT= 90s

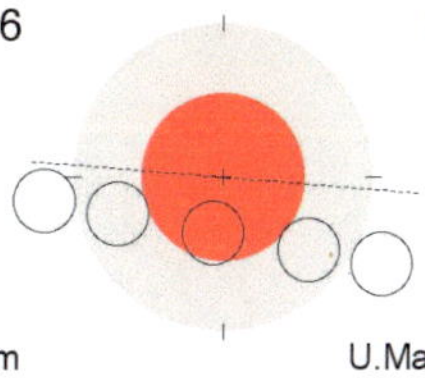

Par. = 205m U.Mag. = 0.9197
Gam. = -0.4853 P.Mag. = 2.0194

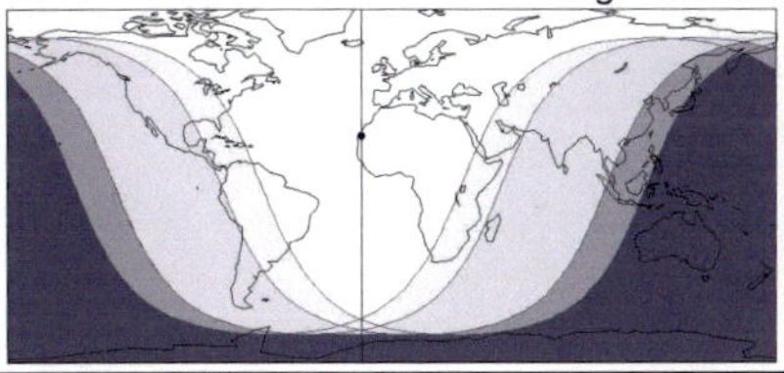

Total - **2058 Jun 06**
Saros 131 19:16 TD
D.Node ΔT= 91s

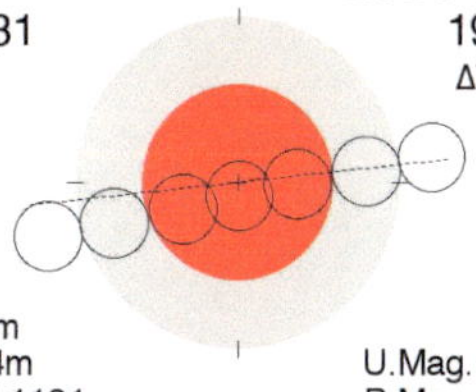

Tot. = 98m
Par. = 214m U.Mag. = 1.6628
Gam. = -0.1181 P.Mag. = 2.6226

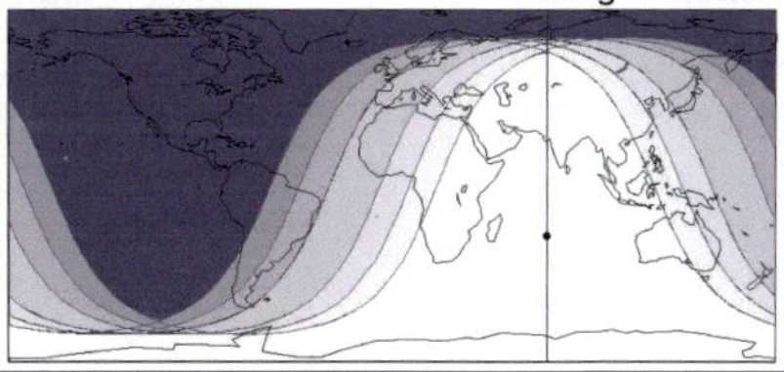

Total + **2058 Nov 30**
Saros 136 03:16 TD
A.Node ΔT= 91s

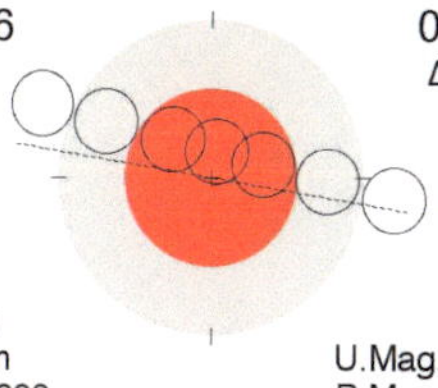

Tot. = 90m
Par. = 221m U.Mag. = 1.4277
Gam. = 0.2208 P.Mag. = 2.4819

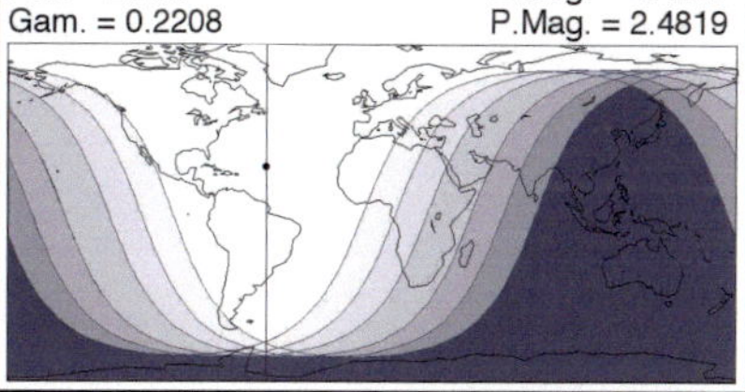

Plate 011

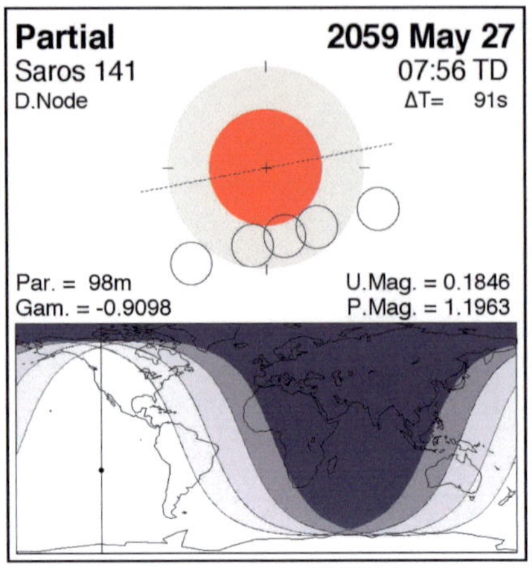
Partial
2059 May 27
Saros 141
07:56 TD
D.Node
ΔT= 91s
Par. = 98m
Gam. = -0.9098
U.Mag. = 0.1846
P.Mag. = 1.1963

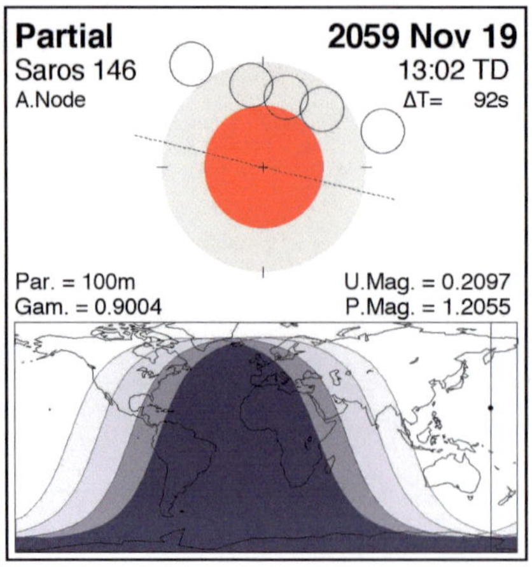
Partial
2059 Nov 19
Saros 146
13:02 TD
A.Node
ΔT= 92s
Par. = 100m
Gam. = 0.9004
U.Mag. = 0.2097
P.Mag. = 1.2055

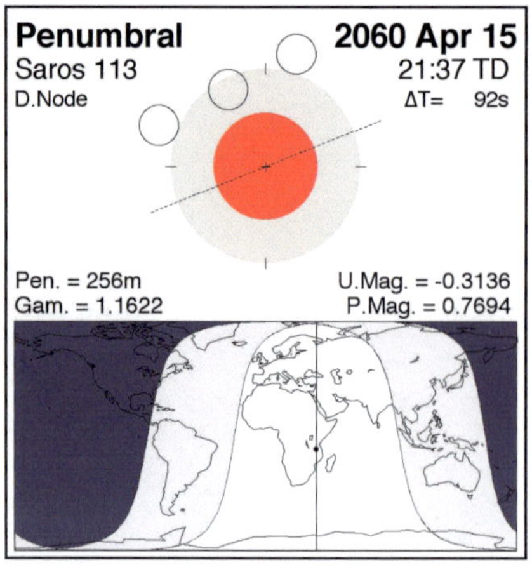
Penumbral
2060 Apr 15
Saros 113
21:37 TD
D.Node
ΔT= 92s
Pen. = 256m
Gam. = 1.1622
U.Mag. = -0.3136
P.Mag. = 0.7694

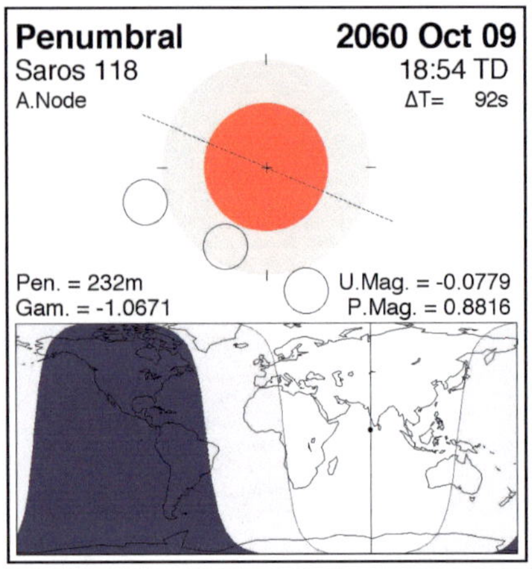
Penumbral
2060 Oct 09
Saros 118
18:54 TD
A.Node
ΔT= 92s
Pen. = 232m
Gam. = -1.0671
U.Mag. = -0.0779
P.Mag. = 0.8816

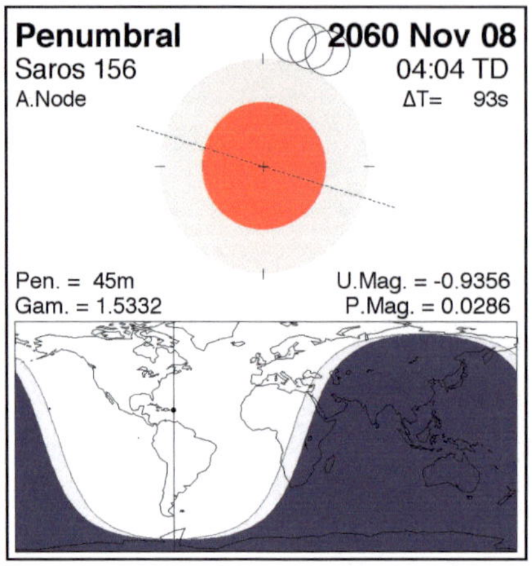
Penumbral
2060 Nov 08
Saros 156
04:04 TD
A.Node
ΔT= 93s
Pen. = 45m
Gam. = 1.5332
U.Mag. = -0.9356
P.Mag. = 0.0286

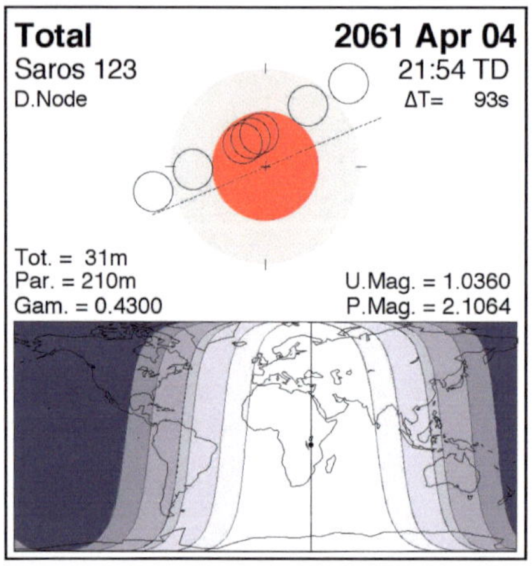
Total
2061 Apr 04
Saros 123
21:54 TD
D.Node
ΔT= 93s
Tot. = 31m
Par. = 210m
Gam. = 0.4300
U.Mag. = 1.0360
P.Mag. = 2.1064

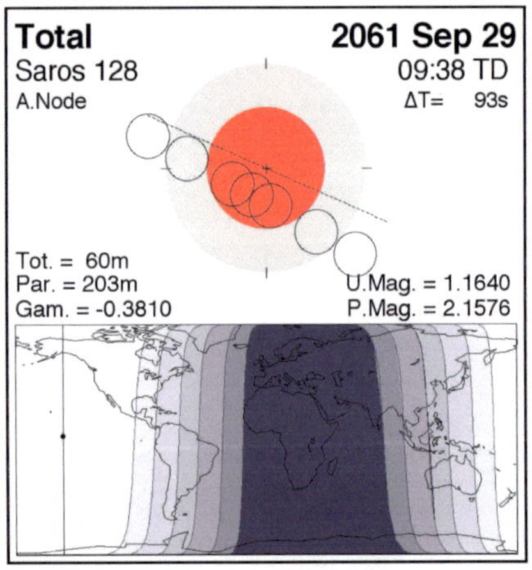
Total
2061 Sep 29
Saros 128
09:38 TD
A.Node
ΔT= 93s
Tot. = 60m
Par. = 203m
Gam. = -0.3810
U.Mag. = 1.1640
P.Mag. = 2.1576

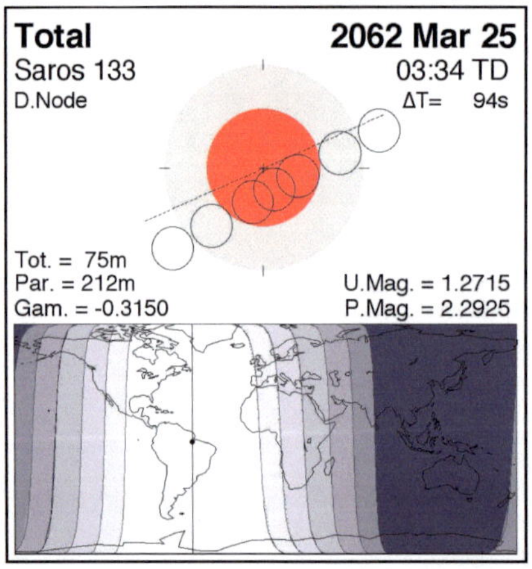
Total
2062 Mar 25
Saros 133
03:34 TD
D.Node
ΔT= 94s
Tot. = 75m
Par. = 212m
Gam. = -0.3150
U.Mag. = 1.2715
P.Mag. = 2.2925

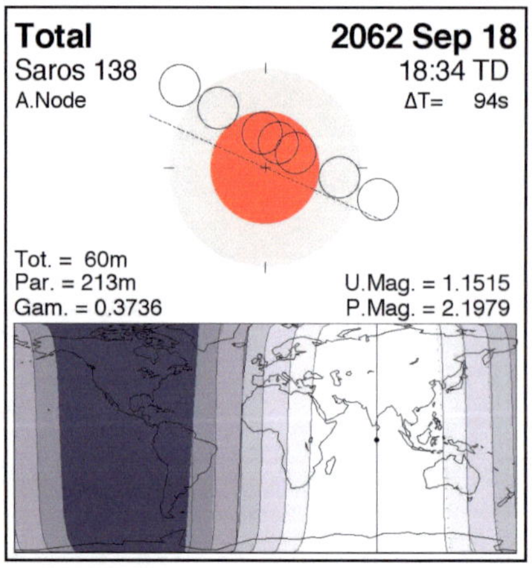
Total
2062 Sep 18
Saros 138
18:34 TD
A.Node
ΔT= 94s
Tot. = 60m
Par. = 213m
Gam. = 0.3736
U.Mag. = 1.1515
P.Mag. = 2.1979

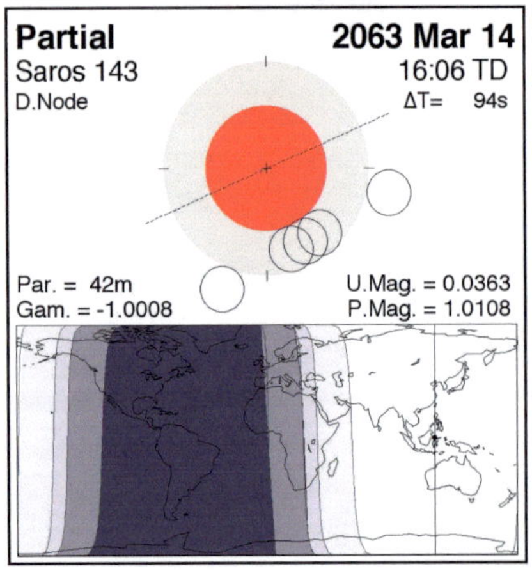
Partial
2063 Mar 14
Saros 143
16:06 TD
D.Node
ΔT= 94s
Par. = 42m
Gam. = -1.0008
U.Mag. = 0.0363
P.Mag. = 1.0108

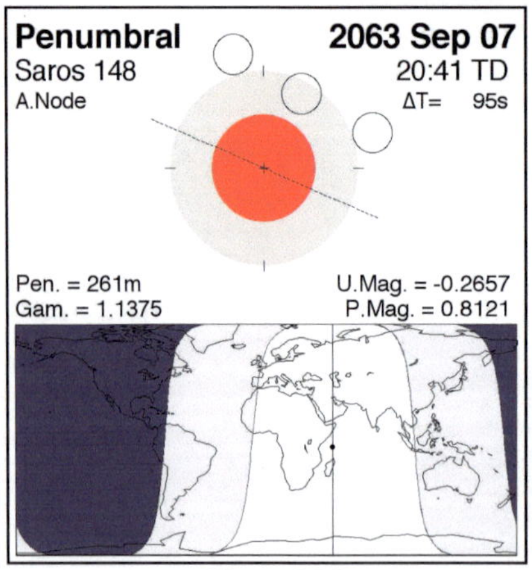
Penumbral
2063 Sep 07
Saros 148
20:41 TD
A.Node
ΔT= 95s
Pen. = 261m
Gam. = 1.1375
U.Mag. = -0.2657
P.Mag. = 0.8121

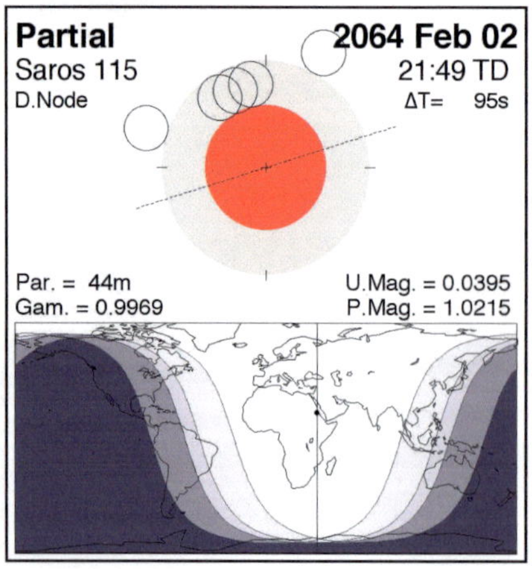
Partial
2064 Feb 02
Saros 115
21:49 TD
D.Node
ΔT= 95s
Par. = 44m
Gam. = 0.9969
U.Mag. = 0.0395
P.Mag. = 1.0215

Plate 012

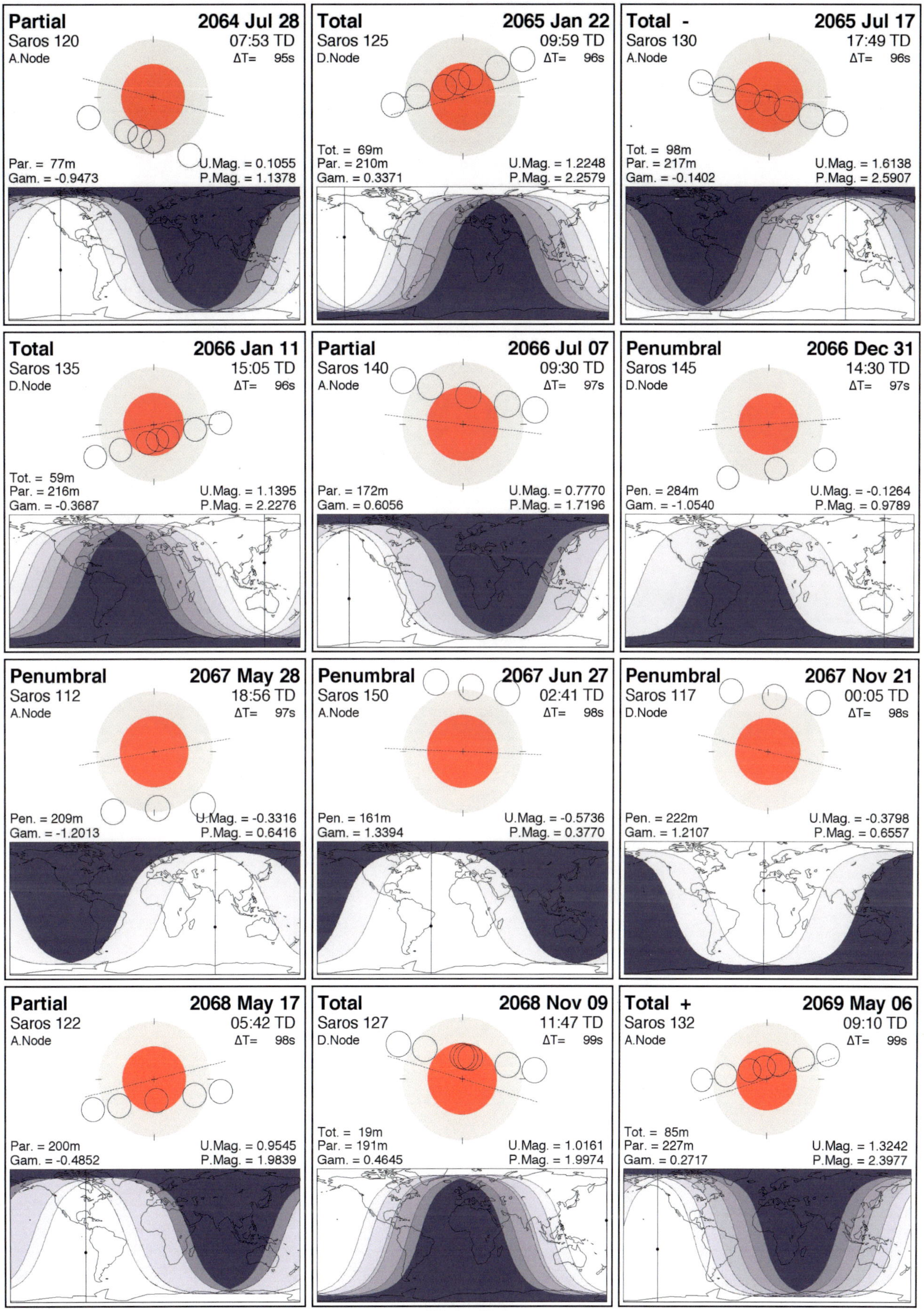
Partial
2064 Jul 28
Saros 120
07:53 TD
A.Node
ΔT= 95s
Par. = 77m
Gam. = -0.9473
U.Mag. = 0.1055
P.Mag. = 1.1378
Total
2065 Jan 22
Saros 125
09:59 TD
D.Node
ΔT= 96s
Tot. = 69m
Par. = 210m
Gam. = 0.3371
U.Mag. = 1.2248
P.Mag. = 2.2579
Total -
2065 Jul 17
Saros 130
17:49 TD
A.Node
ΔT= 96s
Tot. = 98m
Par. = 217m
Gam. = -0.1402
U.Mag. = 1.6138
P.Mag. = 2.5907
Total
2066 Jan 11
Saros 135
15:05 TD
D.Node
ΔT= 96s
Tot. = 59m
Par. = 216m
Gam. = -0.3687
U.Mag. = 1.1395
P.Mag. = 2.2276
Partial
2066 Jul 07
Saros 140
09:30 TD
A.Node
ΔT= 97s
Par. = 172m
Gam. = 0.6056
U.Mag. = 0.7770
P.Mag. = 1.7196
Penumbral
2066 Dec 31
Saros 145
14:30 TD
D.Node
ΔT= 97s
Pen. = 284m
Gam. = -1.0540
U.Mag. = -0.1264
P.Mag. = 0.9789
Penumbral
2067 May 28
Saros 112
18:56 TD
A.Node
ΔT= 97s
Pen. = 209m
Gam. = -1.2013
U.Mag. = -0.3316
P.Mag. = 0.6416
Penumbral
2067 Jun 27
Saros 150
02:41 TD
A.Node
ΔT= 98s
Pen. = 161m
Gam. = 1.3394
U.Mag. = -0.5736
P.Mag. = 0.3770
Penumbral
2067 Nov 21
Saros 117
00:05 TD
D.Node
ΔT= 98s
Pen. = 222m
Gam. = 1.2107
U.Mag. = -0.3798
P.Mag. = 0.6557
Partial
2068 May 17
Saros 122
05:42 TD
A.Node
ΔT= 98s
Par. = 200m
Gam. = -0.4852
U.Mag. = 0.9545
P.Mag. = 1.9839
Total
2068 Nov 09
Saros 127
11:47 TD
D.Node
ΔT= 99s
Tot. = 19m
Par. = 191m
Gam. = 0.4645
U.Mag. = 1.0161
P.Mag. = 1.9974
Total +
2069 May 06
Saros 132
09:10 TD
A.Node
ΔT= 99s
Tot. = 85m
Par. = 227m
Gam. = 0.2717
U.Mag. = 1.3242
P.Mag. = 2.3977

Plate 013

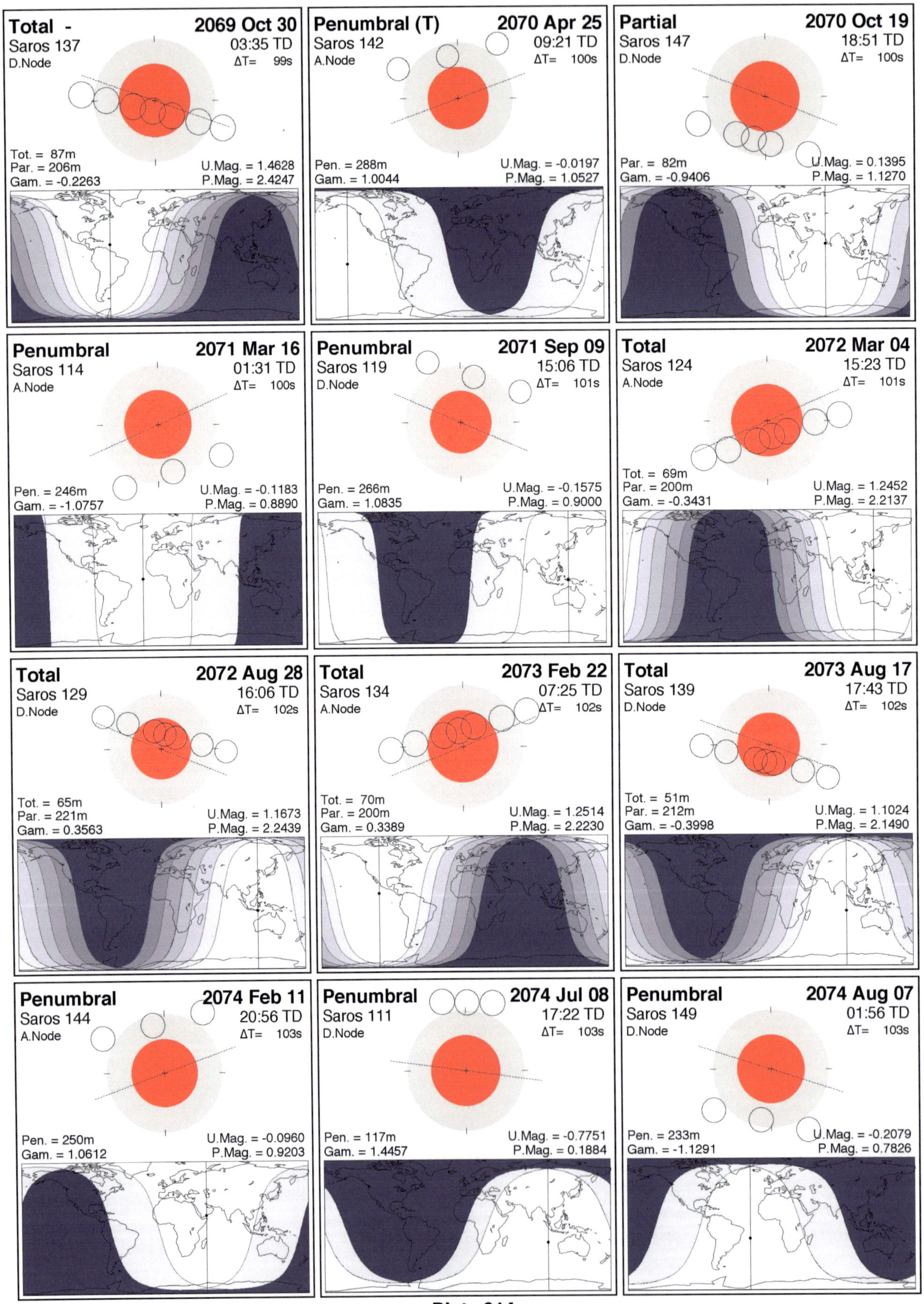
Total - 2069 Oct 30
Saros 137 03:35 TD
D.Node ΔT= 99s
Tot. = 87m
Par. = 206m
Gam. = -0.2263
U.Mag. = 1.4628
P.Mag. = 2.4247
Penumbral (T) 2070 Apr 25
Saros 142 09:21 TD
A.Node ΔT= 100s
Pen. = 288m
Gam. = 1.0044
U.Mag. = -0.0197
P.Mag. = 1.0527
Partial 2070 Oct 19
Saros 147 18:51 TD
D.Node ΔT= 100s
Par. = 82m
Gam. = -0.9406
U.Mag. = 0.1395
P.Mag. = 1.1270
Penumbral 2071 Mar 16
Saros 114 01:31 TD
A.Node ΔT= 100s
Pen. = 246m
Gam. = -1.0757
U.Mag. = -0.1183
P.Mag. = 0.8890
Penumbral 2071 Sep 09
Saros 119 15:06 TD
D.Node ΔT= 101s
Pen. = 266m
Gam. = 1.0835
U.Mag. = -0.1575
P.Mag. = 0.9000
Total 2072 Mar 04
Saros 124 15:23 TD
A.Node ΔT= 101s
Tot. = 69m
Par. = 200m
Gam. = -0.3431
U.Mag. = 1.2452
P.Mag. = 2.2137
Total 2072 Aug 28
Saros 129 16:06 TD
D.Node ΔT= 102s
Tot. = 65m
Par. = 221m
Gam. = 0.3563
U.Mag. = 1.1673
P.Mag. = 2.2439
Total 2073 Feb 22
Saros 134 07:25 TD
A.Node ΔT= 102s
Tot. = 70m
Par. = 200m
Gam. = 0.3389
U.Mag. = 1.2514
P.Mag. = 2.2230
Total 2073 Aug 17
Saros 139 17:43 TD
D.Node ΔT= 102s
Tot. = 51m
Par. = 212m
Gam. = -0.3998
U.Mag. = 1.1024
P.Mag. = 2.1490
Penumbral 2074 Feb 11
Saros 144 20:56 TD
A.Node ΔT= 103s
Pen. = 250m
Gam. = 1.0612
U.Mag. = -0.0960
P.Mag. = 0.9203
Penumbral 2074 Jul 08
Saros 111 17:22 TD
D.Node ΔT= 103s
Pen. = 117m
Gam. = 1.4457
U.Mag. = -0.7751
P.Mag. = 0.1884
Penumbral 2074 Aug 07
Saros 149 01:56 TD
D.Node ΔT= 103s
Pen. = 233m
Gam. = -1.1291
U.Mag. = -0.2079
P.Mag. = 0.7826

Plate 014

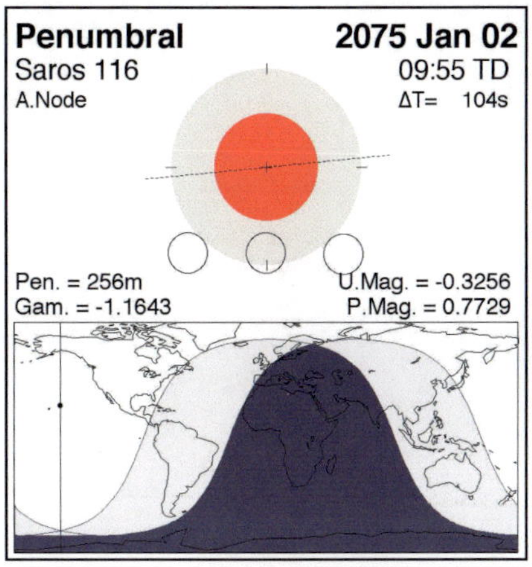
Penumbral
2075 Jan 02
Saros 116
09:55 TD
A.Node
ΔT= 104s
Pen. = 256m
Gam. = -1.1643
U.Mag. = -0.3256
P.Mag. = 0.7729

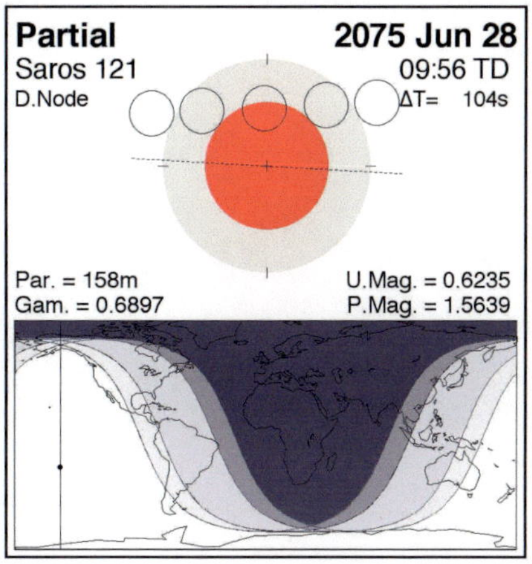
Partial
2075 Jun 28
Saros 121
09:56 TD
D.Node
ΔT= 104s
Par. = 158m
Gam. = 0.6897
U.Mag. = 0.6235
P.Mag. = 1.5639

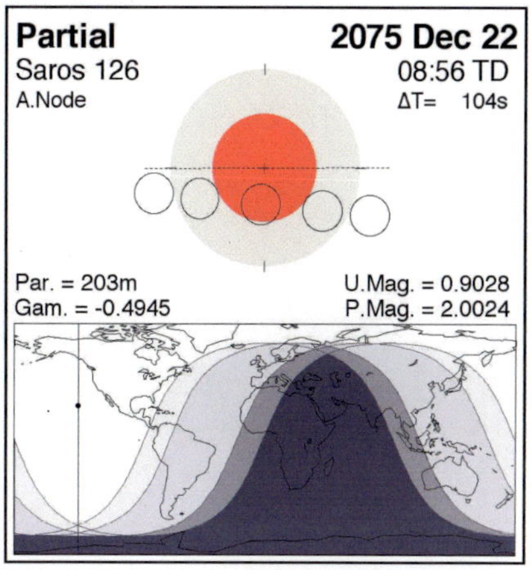
Partial
2075 Dec 22
Saros 126
08:56 TD
A.Node
ΔT= 104s
Par. = 203m
Gam. = -0.4945
U.Mag. = 0.9028
P.Mag. = 2.0024

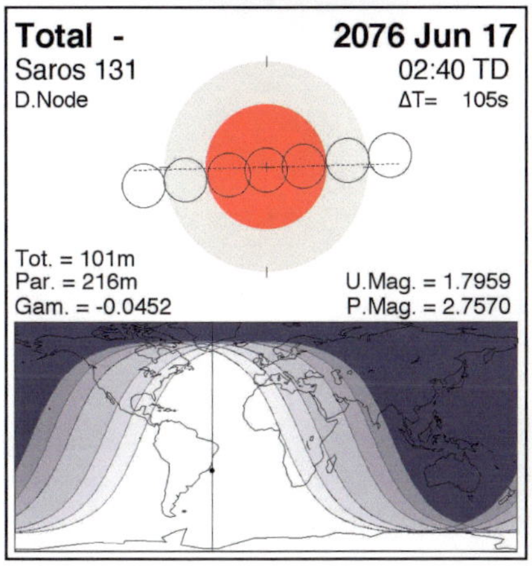
Total -
2076 Jun 17
Saros 131
02:40 TD
D.Node
ΔT= 105s
Tot. = 101m
Par. = 216m
Gam. = -0.0452
U.Mag. = 1.7959
P.Mag. = 2.7570

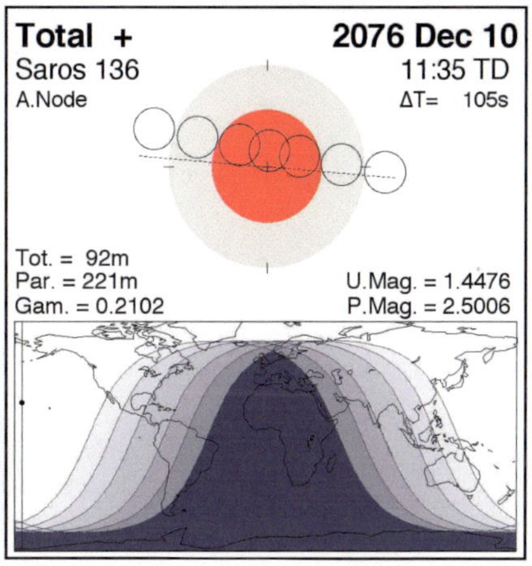
Total +
2076 Dec 10
Saros 136
11:35 TD
A.Node
ΔT= 105s
Tot. = 92m
Par. = 221m
Gam. = 0.2102
U.Mag. = 1.4476
P.Mag. = 2.5006

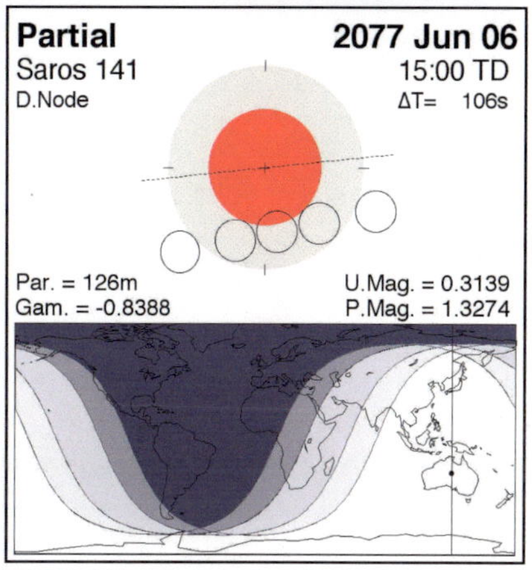
Partial
2077 Jun 06
Saros 141
15:00 TD
D.Node
ΔT= 106s
Par. = 126m
Gam. = -0.8388
U.Mag. = 0.3139
P.Mag. = 1.3274

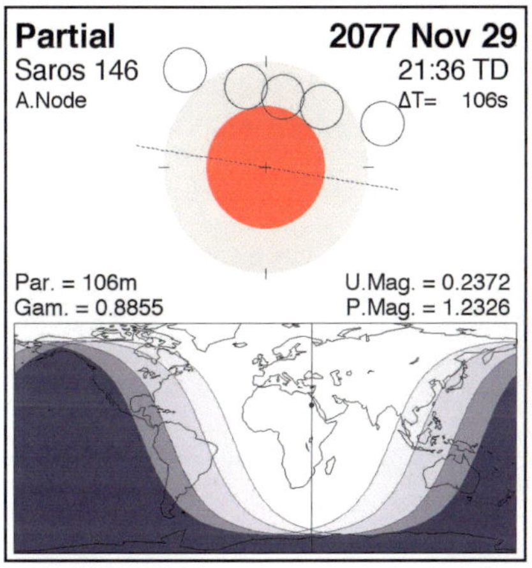
Partial
2077 Nov 29
Saros 146
21:36 TD
A.Node
ΔT= 106s
Par. = 106m
Gam. = 0.8855
U.Mag. = 0.2372
P.Mag. = 1.2326

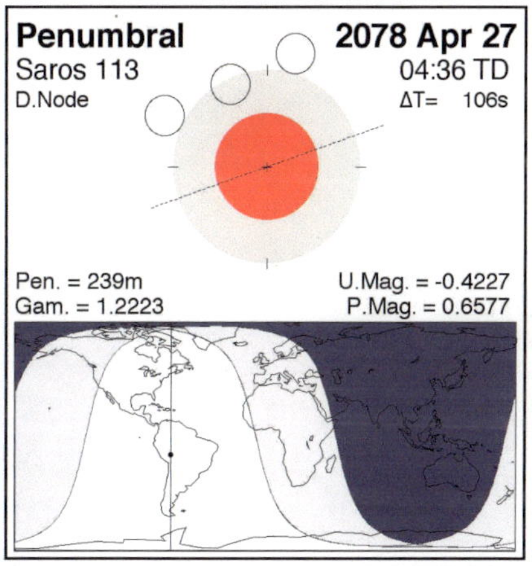
Penumbral
2078 Apr 27
Saros 113
04:36 TD
D.Node
ΔT= 106s
Pen. = 239m
Gam. = 1.2223
U.Mag. = -0.4227
P.Mag. = 0.6577

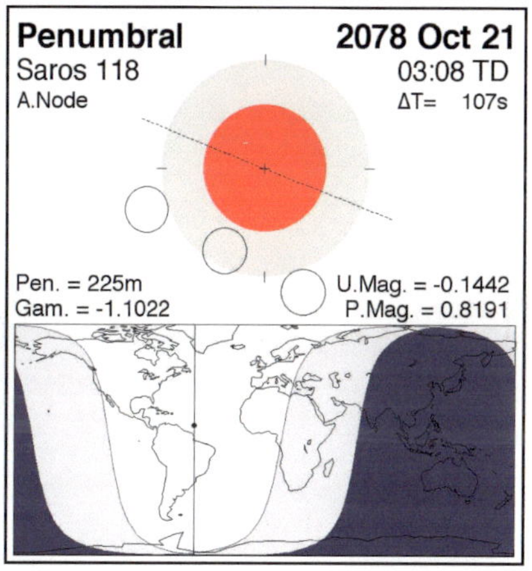
Penumbral
2078 Oct 21
Saros 118
03:08 TD
A.Node
ΔT= 107s
Pen. = 225m
Gam. = -1.1022
U.Mag. = -0.1442
P.Mag. = 0.8191

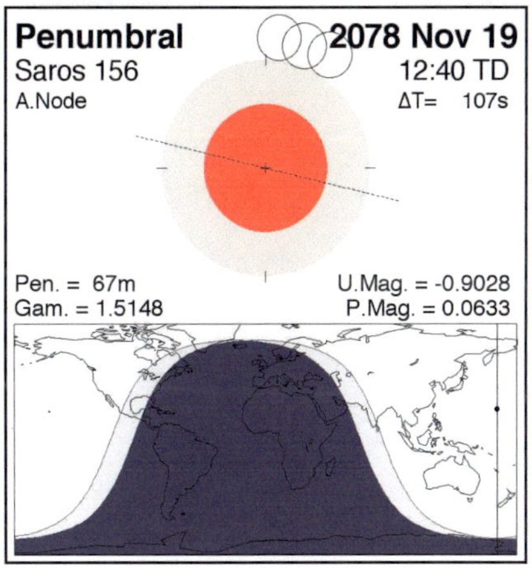
Penumbral
2078 Nov 19
Saros 156
12:40 TD
A.Node
ΔT= 107s
Pen. = 67m
Gam. = 1.5148
U.Mag. = -0.9028
P.Mag. = 0.0633

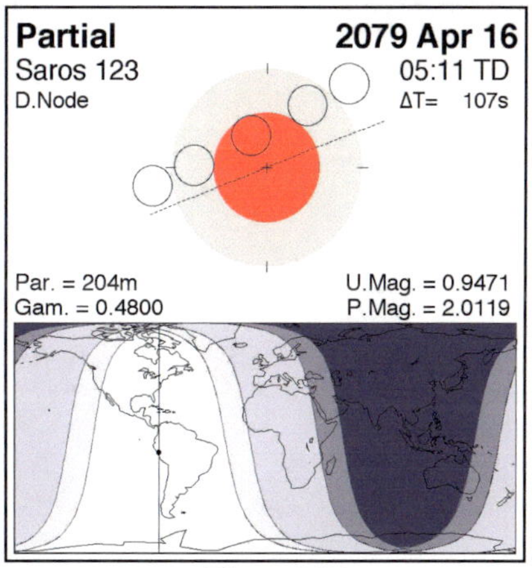
Partial
2079 Apr 16
Saros 123
05:11 TD
D.Node
ΔT= 107s
Par. = 204m
Gam. = 0.4800
U.Mag. = 0.9471
P.Mag. = 2.0119

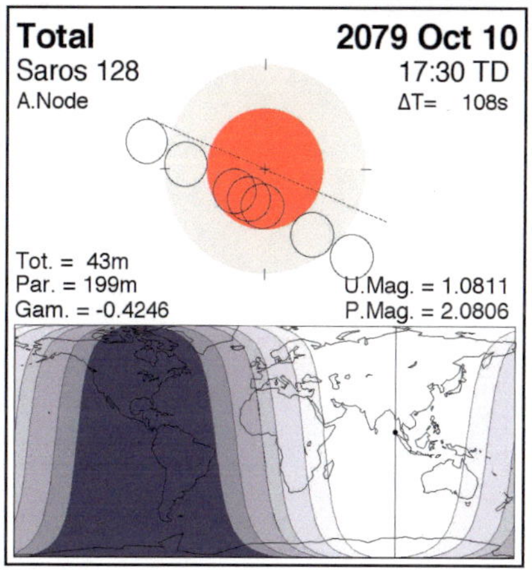
Total
2079 Oct 10
Saros 128
17:30 TD
A.Node
ΔT= 108s
Tot. = 43m
Par. = 199m
Gam. = -0.4246
U.Mag. = 1.0811
P.Mag. = 2.0806

Plate 015

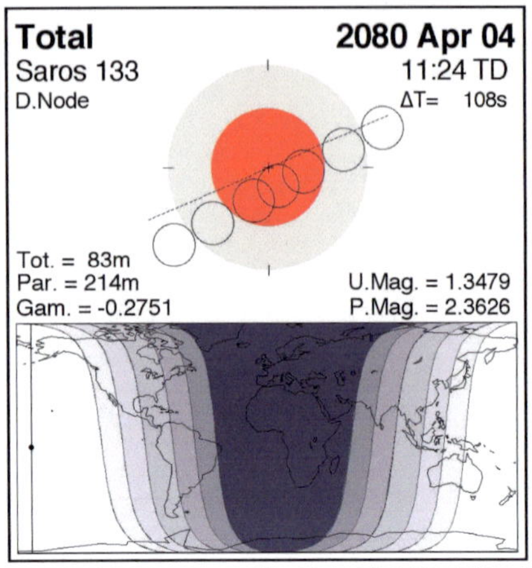
Total
2080 Apr 04
Saros 133
11:24 TD
D.Node
ΔT= 108s
Tot. = 83m
Par. = 214m
Gam. = -0.2751
U.Mag. = 1.3479
P.Mag. = 2.3626

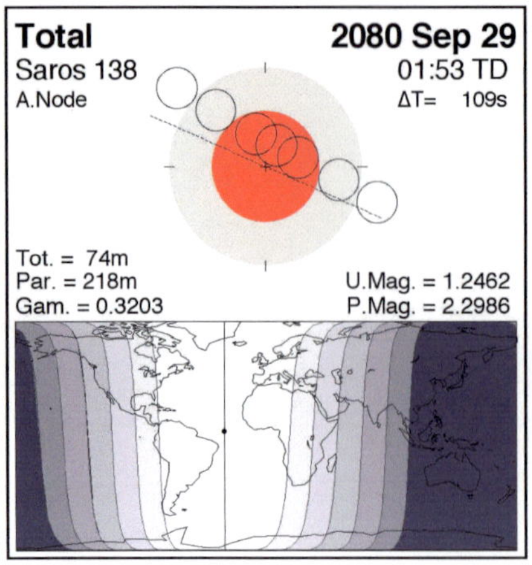
Total
2080 Sep 29
Saros 138
01:53 TD
A.Node
ΔT= 109s
Tot. = 74m
Par. = 218m
Gam. = 0.3203
U.Mag. = 1.2462
P.Mag. = 2.2986

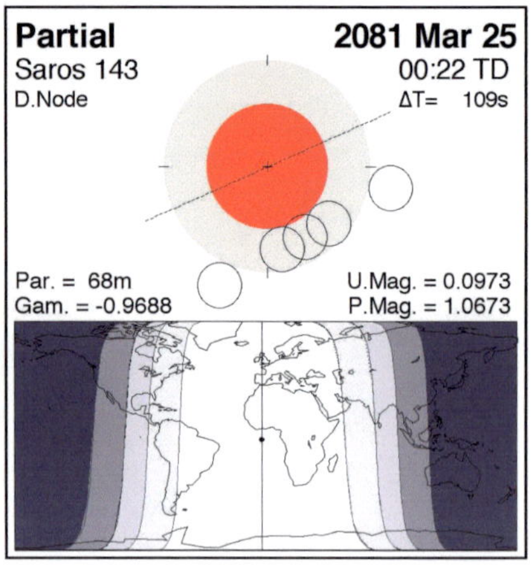
Partial
2081 Mar 25
Saros 143
00:22 TD
D.Node
ΔT= 109s
Par. = 68m
Gam. = -0.9688
U.Mag. = 0.0973
P.Mag. = 1.0673

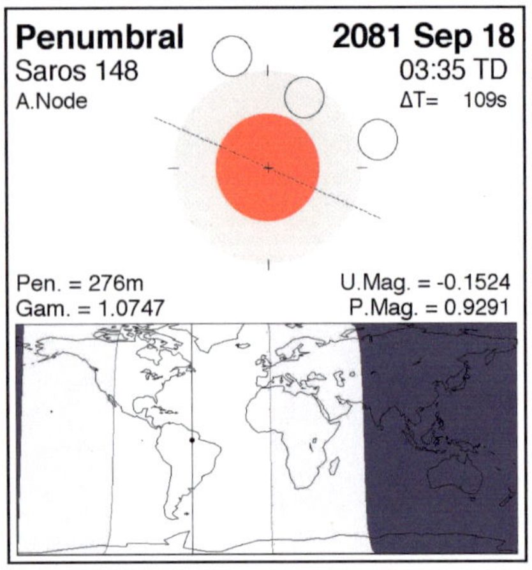
Penumbral
2081 Sep 18
Saros 148
03:35 TD
A.Node
ΔT= 109s
Pen. = 276m
Gam. = 1.0747
U.Mag. = -0.1524
P.Mag. = 0.9291

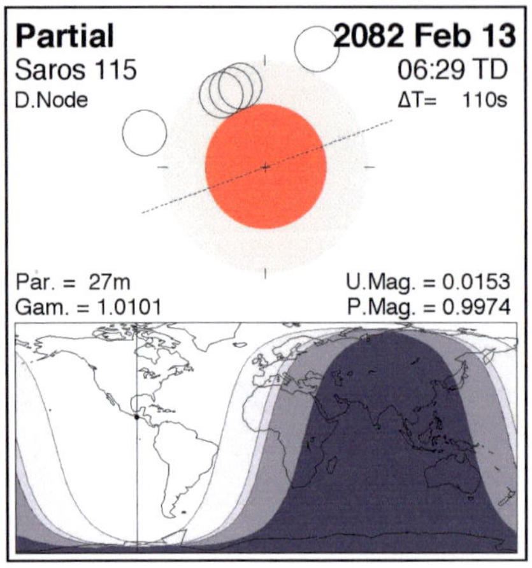
Partial
2082 Feb 13
Saros 115
06:29 TD
D.Node
ΔT= 110s
Par. = 27m
Gam. = 1.0101
U.Mag. = 0.0153
P.Mag. = 0.9974

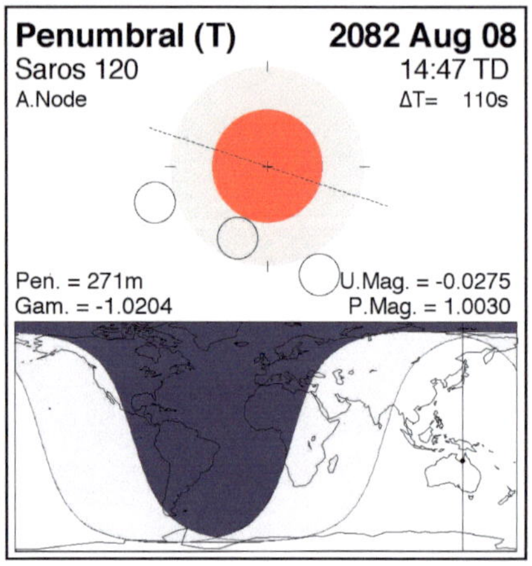
Penumbral (T)
2082 Aug 08
Saros 120
14:47 TD
A.Node
ΔT= 110s
Pen. = 271m
Gam. = -1.0204
U.Mag. = -0.0275
P.Mag. = 1.0030

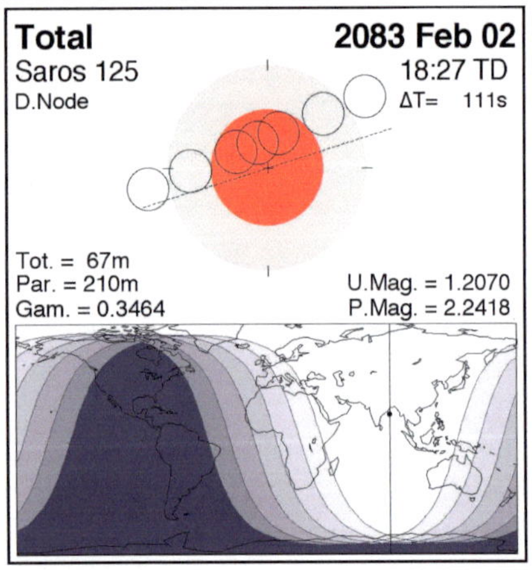
Total
2083 Feb 02
Saros 125
18:27 TD
D.Node
ΔT= 111s
Tot. = 67m
Par. = 210m
Gam. = 0.3464
U.Mag. = 1.2070
P.Mag. = 2.2418

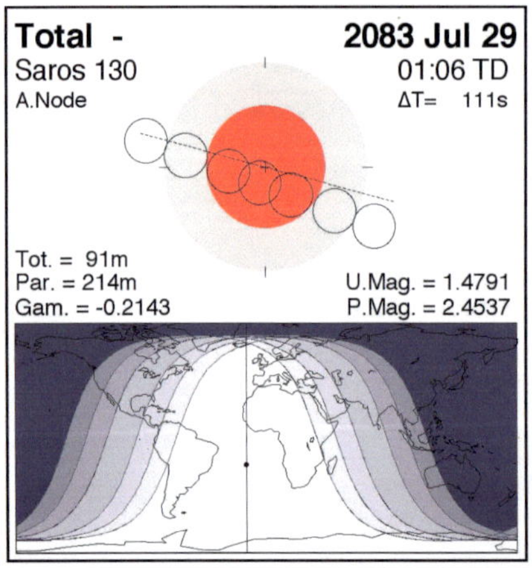
Total -
2083 Jul 29
Saros 130
01:06 TD
A.Node
ΔT= 111s
Tot. = 91m
Par. = 214m
Gam. = -0.2143
U.Mag. = 1.4791
P.Mag. = 2.4537

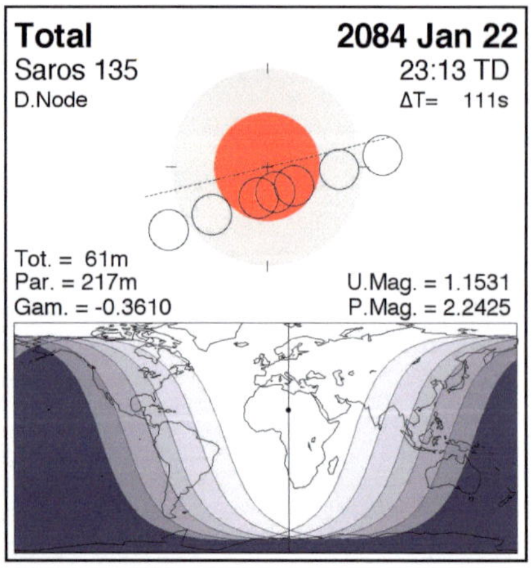
Total
2084 Jan 22
Saros 135
23:13 TD
D.Node
ΔT= 111s
Tot. = 61m
Par. = 217m
Gam. = -0.3610
U.Mag. = 1.1531
P.Mag. = 2.2425

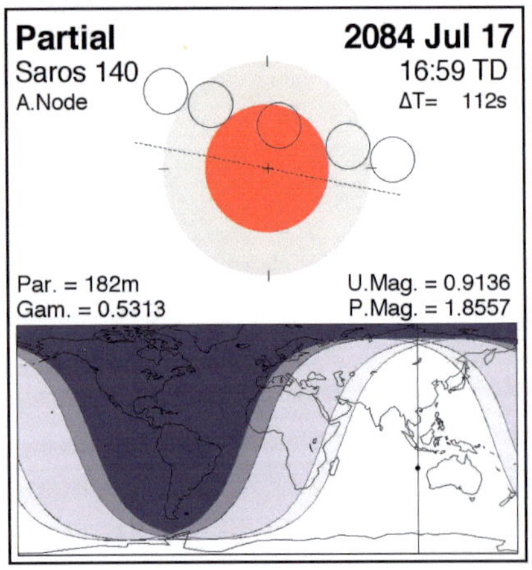
Partial
2084 Jul 17
Saros 140
16:59 TD
A.Node
ΔT= 112s
Par. = 182m
Gam. = 0.5313
U.Mag. = 0.9136
P.Mag. = 1.8557

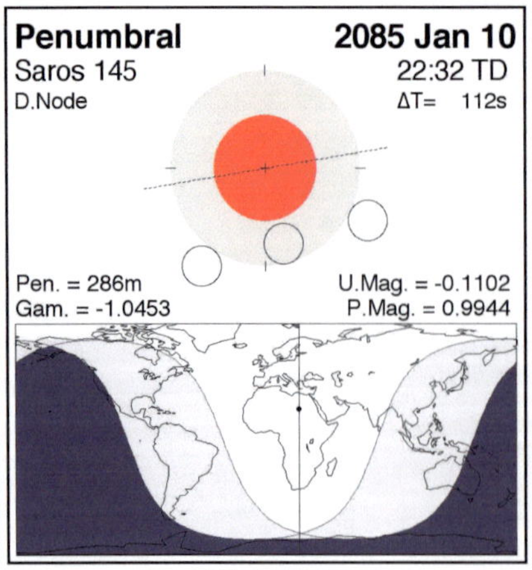
Penumbral
2085 Jan 10
Saros 145
22:32 TD
D.Node
ΔT= 112s
Pen. = 286m
Gam. = -1.0453
U.Mag. = -0.1102
P.Mag. = 0.9944

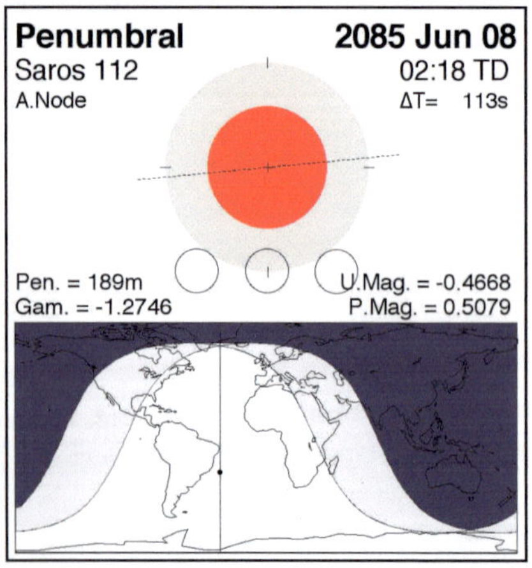
Penumbral
2085 Jun 08
Saros 112
02:18 TD
A.Node
ΔT= 113s
Pen. = 189m
Gam. = -1.2746
U.Mag. = -0.4668
P.Mag. = 0.5079

Plate 016

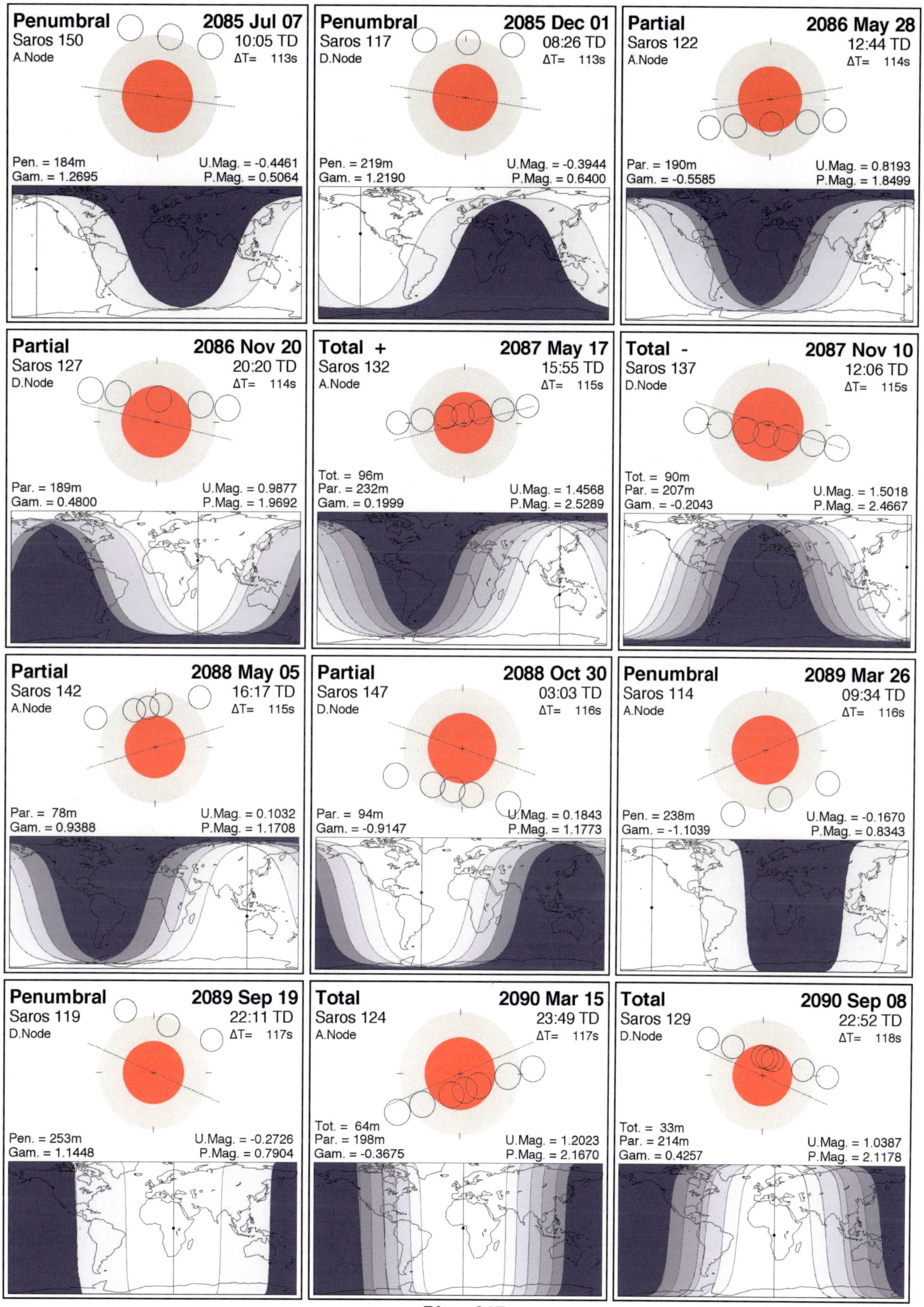
Penumbral
2085 Jul 07
Saros 150
10:05 TD
A.Node
ΔT= 113s
Pen. = 184m
Gam. = 1.2695
U.Mag. = -0.4461
P.Mag. = 0.5064
Penumbral
2085 Dec 01
Saros 117
08:26 TD
D.Node
ΔT= 113s
Pen. = 219m
Gam. = 1.2190
U.Mag. = -0.3944
P.Mag. = 0.6400
Partial
2086 May 28
Saros 122
12:44 TD
A.Node
ΔT= 114s
Par. = 190m
Gam. = -0.5585
U.Mag. = 0.8193
P.Mag. = 1.8499
Partial
2086 Nov 20
Saros 127
20:20 TD
D.Node
ΔT= 114s
Par. = 189m
Gam. = 0.4800
U.Mag. = 0.9877
P.Mag. = 1.9692
Total +
2087 May 17
Saros 132
15:55 TD
A.Node
ΔT= 115s
Tot. = 96m
Par. = 232m
Gam. = 0.1999
U.Mag. = 1.4568
P.Mag. = 2.5289
Total -
2087 Nov 10
Saros 137
12:06 TD
D.Node
ΔT= 115s
Tot. = 90m
Par. = 207m
Gam. = -0.2043
U.Mag. = 1.5018
P.Mag. = 2.4667
Partial
2088 May 05
Saros 142
16:17 TD
A.Node
ΔT= 115s
Par. = 78m
Gam. = 0.9388
U.Mag. = 0.1032
P.Mag. = 1.1708
Partial
2088 Oct 30
Saros 147
03:03 TD
D.Node
ΔT= 116s
Par. = 94m
Gam. = -0.9147
U.Mag. = 0.1843
P.Mag. = 1.1773
Penumbral
2089 Mar 26
Saros 114
09:34 TD
A.Node
ΔT= 116s
Pen. = 238m
Gam. = -1.1039
U.Mag. = -0.1670
P.Mag. = 0.8343
Penumbral
2089 Sep 19
Saros 119
22:11 TD
D.Node
ΔT= 117s
Pen. = 253m
Gam. = 1.1448
U.Mag. = -0.2726
P.Mag. = 0.7904
Total
2090 Mar 15
Saros 124
23:49 TD
A.Node
ΔT= 117s
Tot. = 64m
Par. = 198m
Gam. = -0.3675
U.Mag. = 1.2023
P.Mag. = 2.1670
Total
2090 Sep 08
Saros 129
22:52 TD
D.Node
ΔT= 118s
Tot. = 33m
Par. = 214m
Gam. = 0.4257
U.Mag. = 1.0387
P.Mag. = 2.1178

Plate 017

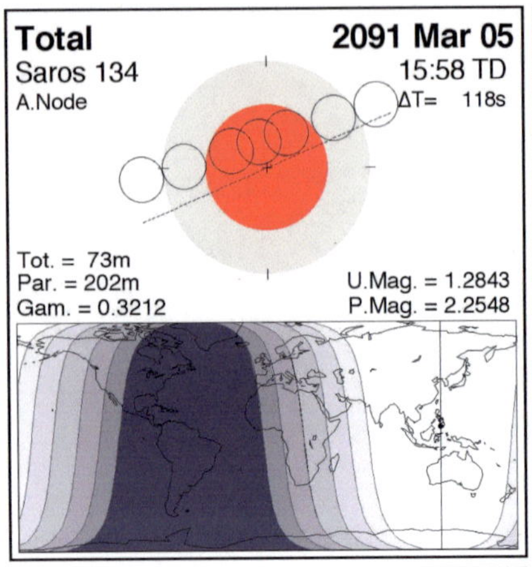
Total
2091 Mar 05
Saros 134
15:58 TD
A.Node
ΔT= 118s
Tot. = 73m
Par. = 202m
Gam. = 0.3212
U.Mag. = 1.2843
P.Mag. = 2.2548

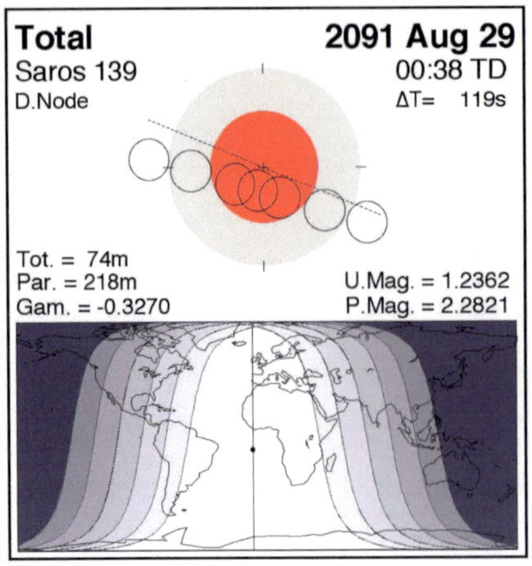
Total
2091 Aug 29
Saros 139
00:38 TD
D.Node
ΔT= 119s
Tot. = 74m
Par. = 218m
Gam. = -0.3270
U.Mag. = 1.2362
P.Mag. = 2.2821

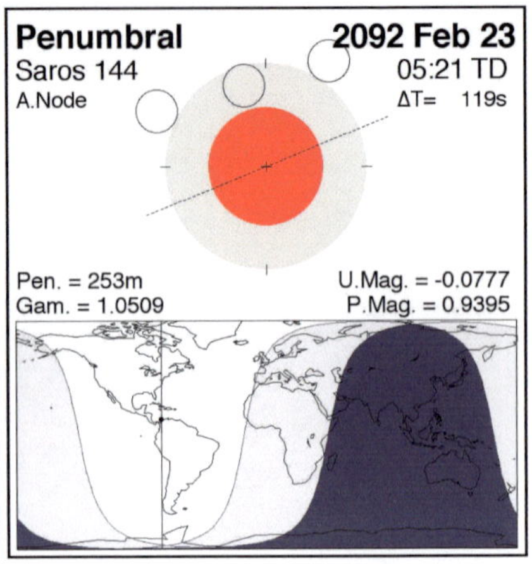
Penumbral
2092 Feb 23
Saros 144
05:21 TD
A.Node
ΔT= 119s
Pen. = 253m
Gam. = 1.0509
U.Mag. = -0.0777
P.Mag. = 0.9395

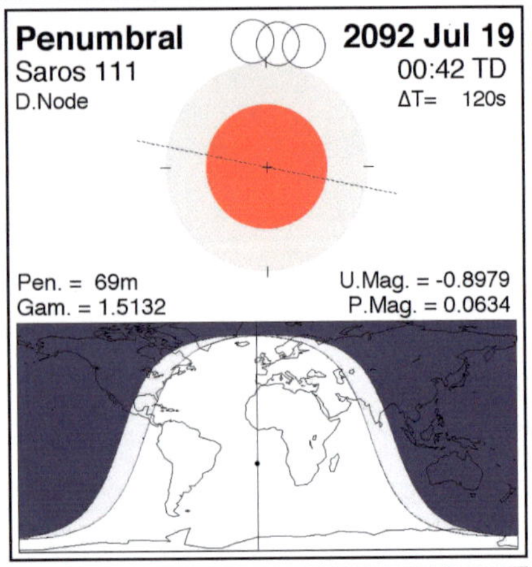
Penumbral
2092 Jul 19
Saros 111
00:42 TD
D.Node
ΔT= 120s
Pen. = 69m
Gam. = 1.5132
U.Mag. = -0.8979
P.Mag. = 0.0634

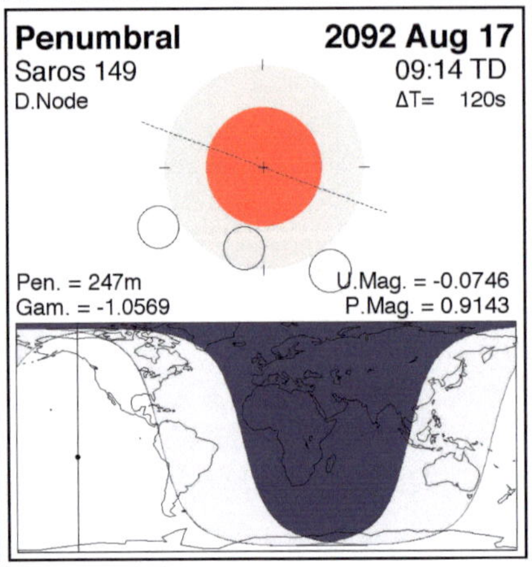
Penumbral
2092 Aug 17
Saros 149
09:14 TD
D.Node
ΔT= 120s
Pen. = 247m
Gam. = -1.0569
U.Mag. = -0.0746
P.Mag. = 0.9143

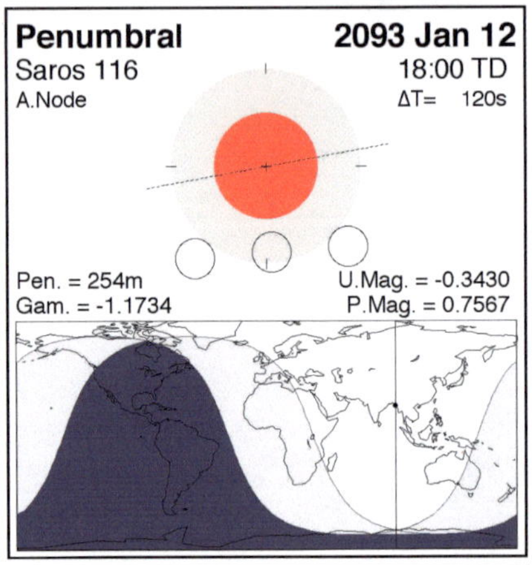
Penumbral
2093 Jan 12
Saros 116
18:00 TD
A.Node
ΔT= 120s
Pen. = 254m
Gam. = -1.1734
U.Mag. = -0.3430
P.Mag. = 0.7567

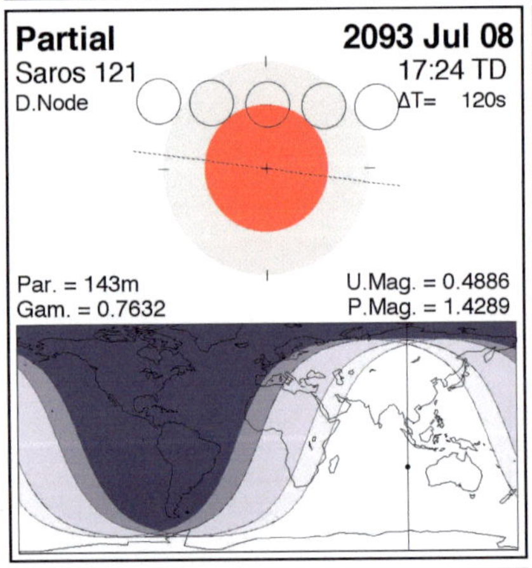
Partial
2093 Jul 08
Saros 121
17:24 TD
D.Node
ΔT= 120s
Par. = 143m
Gam. = 0.7632
U.Mag. = 0.4886
P.Mag. = 1.4289

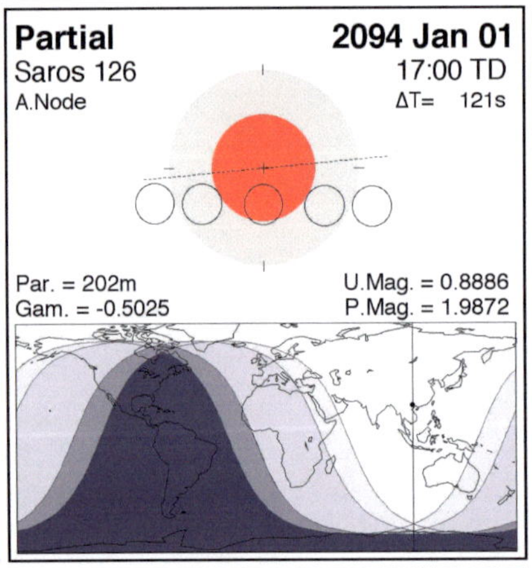
Partial
2094 Jan 01
Saros 126
17:00 TD
A.Node
ΔT= 121s
Par. = 202m
Gam. = -0.5025
U.Mag. = 0.8886
P.Mag. = 1.9872

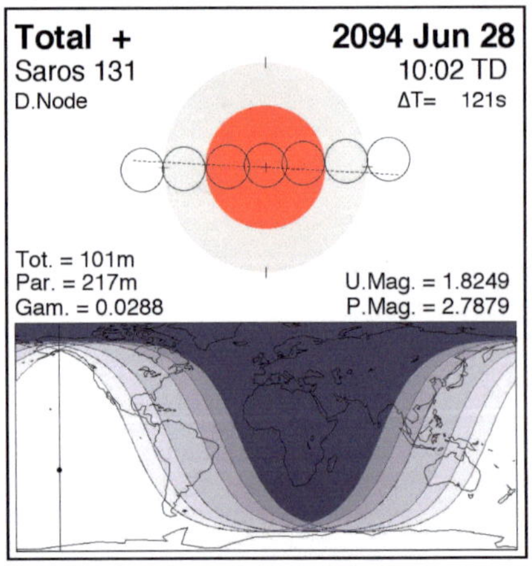
Total +
2094 Jun 28
Saros 131
10:02 TD
D.Node
ΔT= 121s
Tot. = 101m
Par. = 217m
Gam. = 0.0288
U.Mag. = 1.8249
P.Mag. = 2.7879

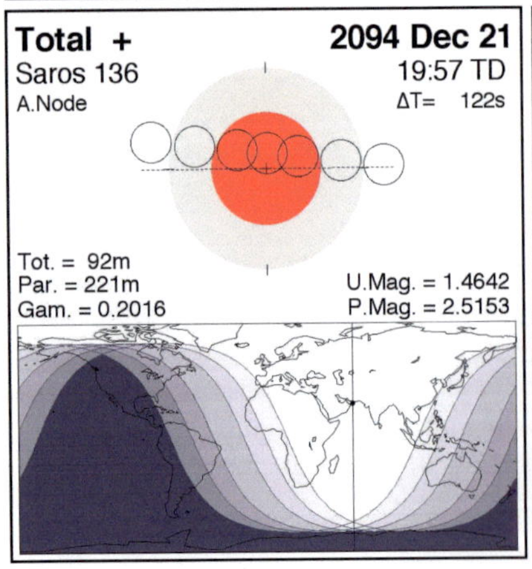
Total +
2094 Dec 21
Saros 136
19:57 TD
A.Node
ΔT= 122s
Tot. = 92m
Par. = 221m
Gam. = 0.2016
U.Mag. = 1.4642
P.Mag. = 2.5153

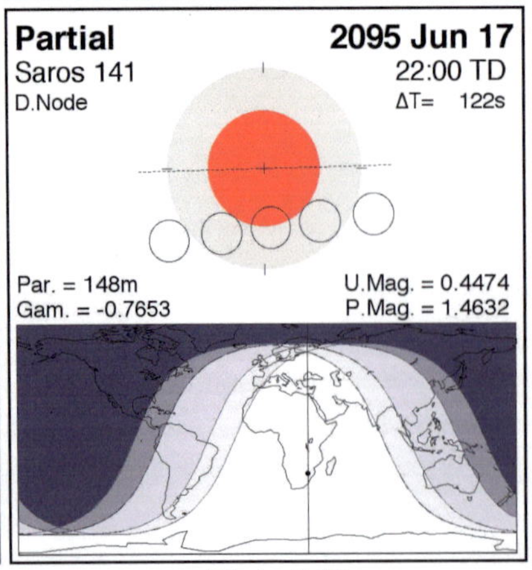
Partial
2095 Jun 17
Saros 141
22:00 TD
D.Node
ΔT= 122s
Par. = 148m
Gam. = -0.7653
U.Mag. = 0.4474
P.Mag. = 1.4632

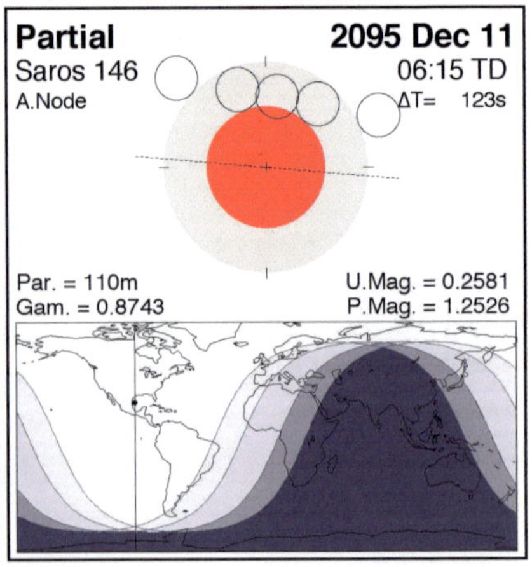
Partial
2095 Dec 11
Saros 146
06:15 TD
A.Node
ΔT= 123s
Par. = 110m
Gam. = 0.8743
U.Mag. = 0.2581
P.Mag. = 1.2526

Plate 018

Penumbral **2096 May 07**
Saros 113 11:25 TD
D.Node ΔT= 123s

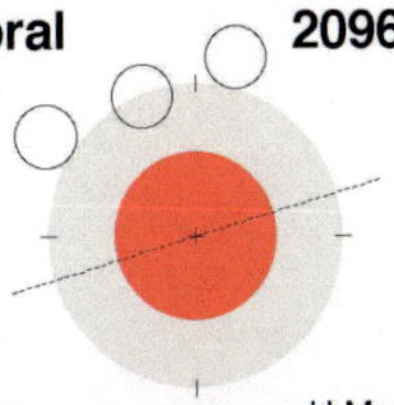

Pen. = 218m U.Mag. = -0.5451
Gam. = 1.2897 P.Mag. = 0.5327

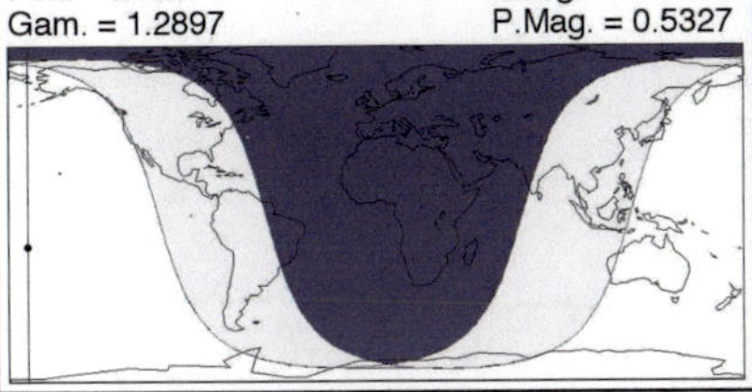

Penumbral **2096 Jun 06**
Saros 151 02:44 TD
D.Node ΔT= 123s

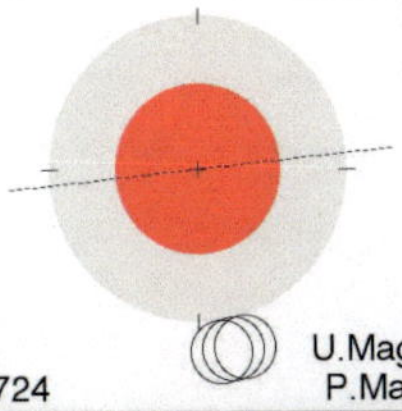

Pen. = 25m U.Mag. = -1.0567
Gam. = -1.5724 P.Mag. = 0.0064

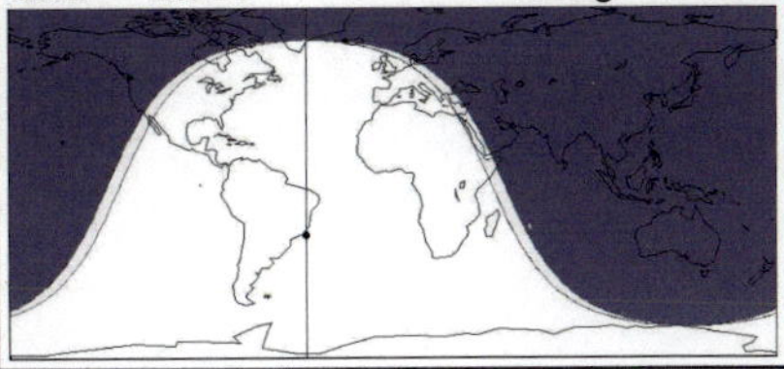

Penumbral **2096 Oct 31**
Saros 118 11:30 TD
A.Node ΔT= 124s

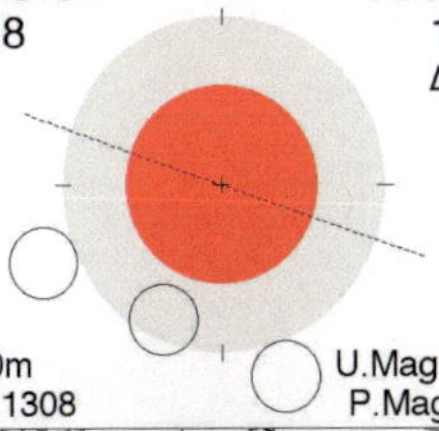

Pen. = 220m U.Mag. = -0.1987
Gam. = -1.1308 P.Mag. = 0.7685

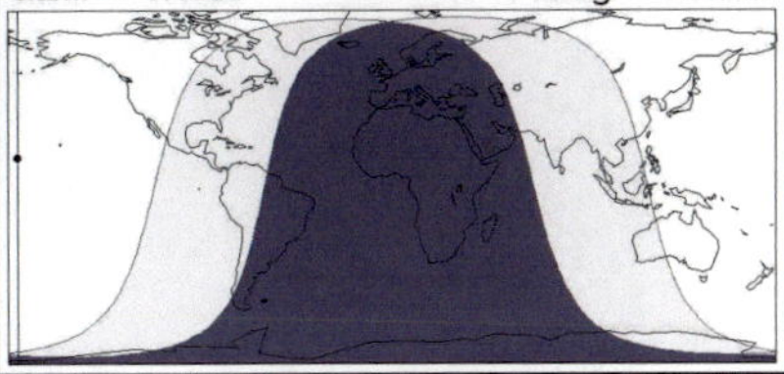

Penumbral **2096 Nov 29**
Saros 156 21:22 TD
A.Node ΔT= 124s

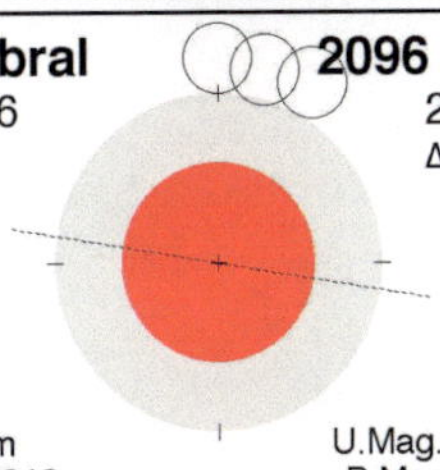

Pen. = 79m U.Mag. = -0.8799
Gam. = 1.5018 P.Mag. = 0.0879

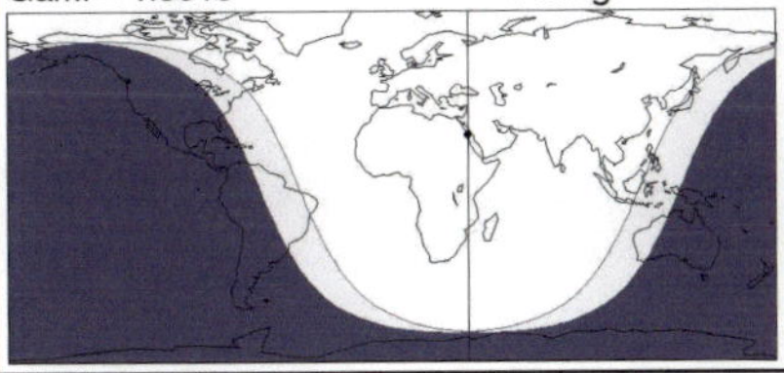

Partial **2097 Apr 26**
Saros 123 12:18 TD
D.Node ΔT= 124s

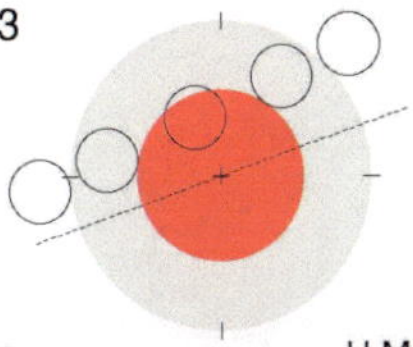

Par. = 196m U.Mag. = 0.8439
Gam. = 0.5377 P.Mag. = 1.9032

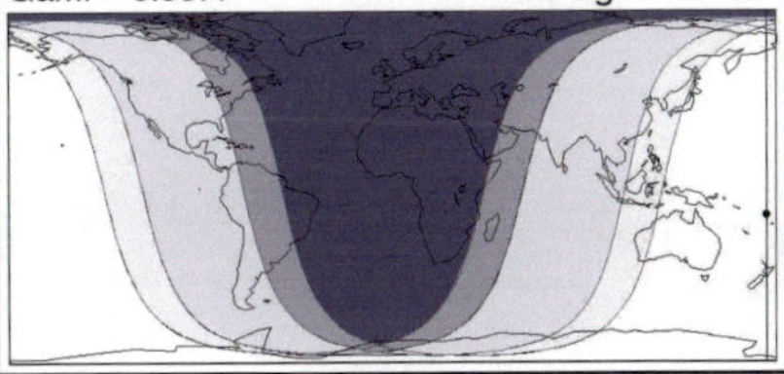

Total **2097 Oct 21**
Saros 128 01:31 TD
A.Node ΔT= 125s

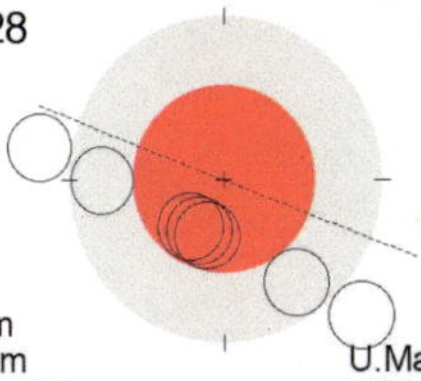

Tot. = 17m
Par. = 196m U.Mag. = 1.0116
Gam. = -0.4608 P.Mag. = 2.0171

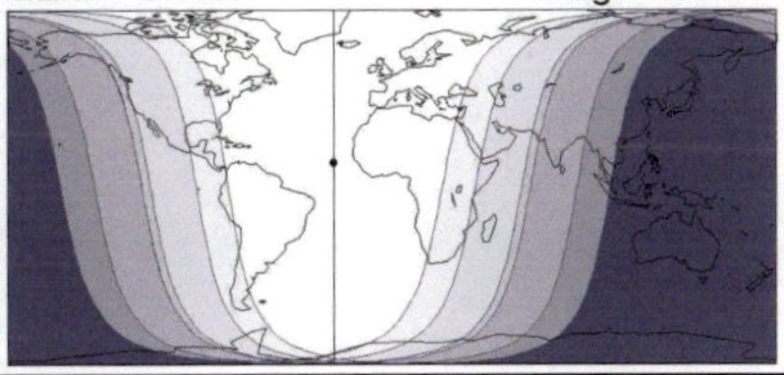

Total - **2098 Apr 15**
Saros 133 19:05 TD
D.Node ΔT= 125s

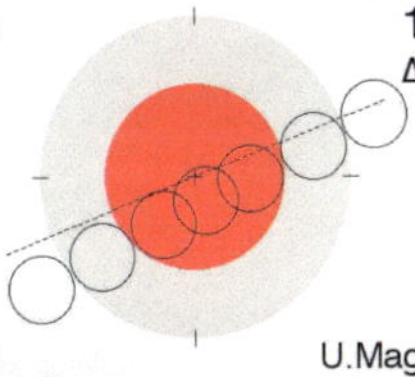

Tot. = 90m
Par. = 216m U.Mag. = 1.4389
Gam. = -0.2272 P.Mag. = 2.4474

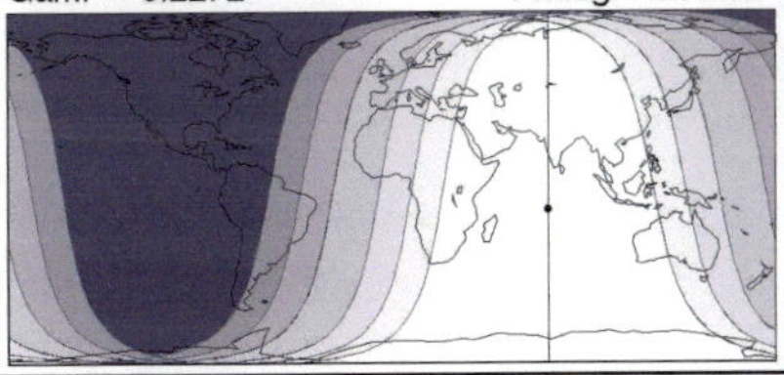

Total **2098 Oct 10**
Saros 138 09:20 TD
A.Node ΔT= 126s

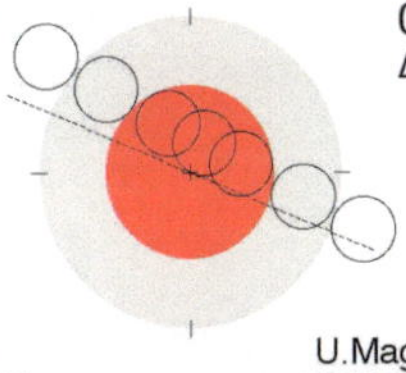

Tot. = 83m
Par. = 222m U.Mag. = 1.3266
Gam. = 0.2749 P.Mag. = 2.3850

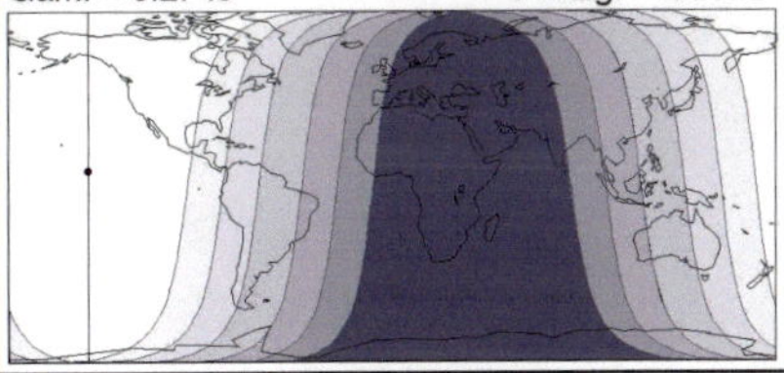

Partial **2099 Apr 05**
Saros 143 08:31 TD
D.Node ΔT= 126s

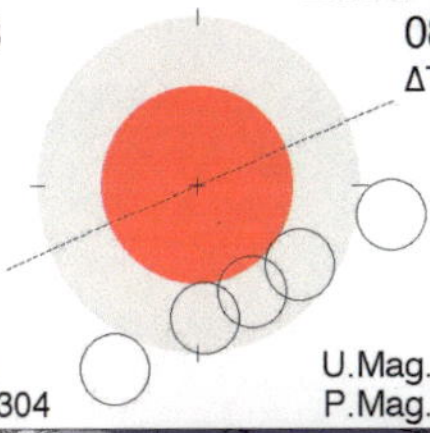

Par. = 89m U.Mag. = 0.1700
Gam. = -0.9304 P.Mag. = 1.1353

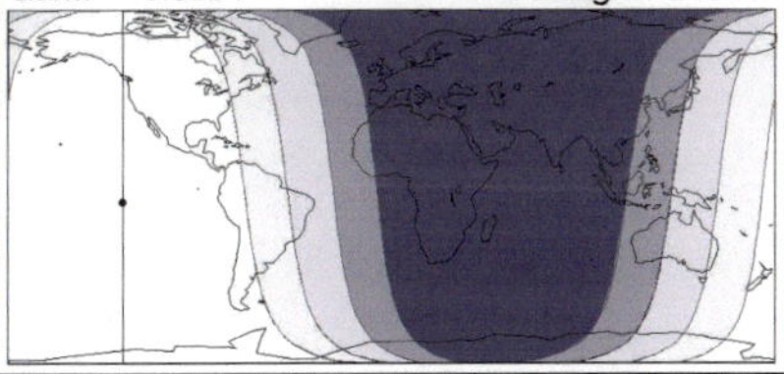

Penumbral (T) **2099 Sep 29**
Saros 148 10:37 TD
A.Node ΔT= 127s

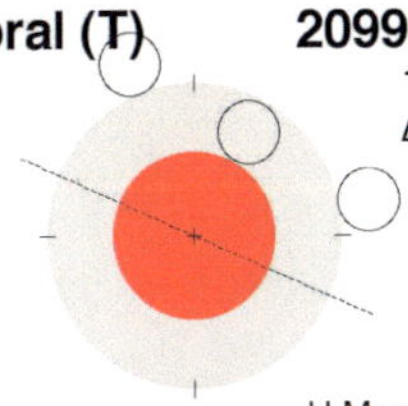

Pen. = 289m U.Mag. = -0.0491
Gam. = 1.0175 P.Mag. = 1.0360

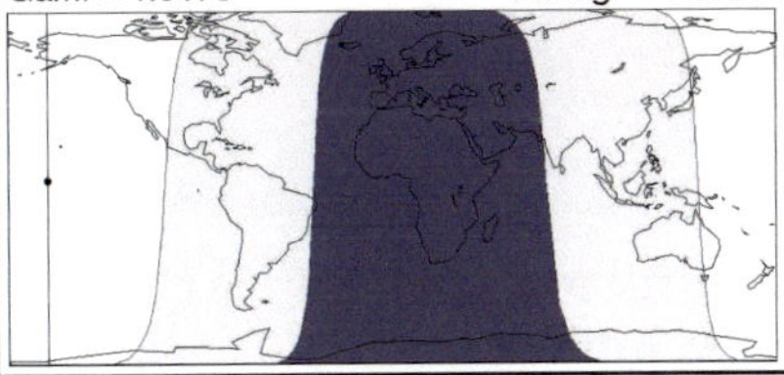

Penumbral **2100 Feb 24**
Saros 115 15:05 TD
D.Node ΔT= 127s

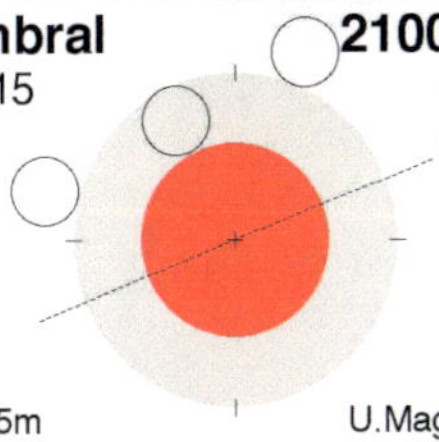

Pen. = 245m U.Mag. = -0.0151
Gam. = 1.0267 P.Mag. = 0.9669

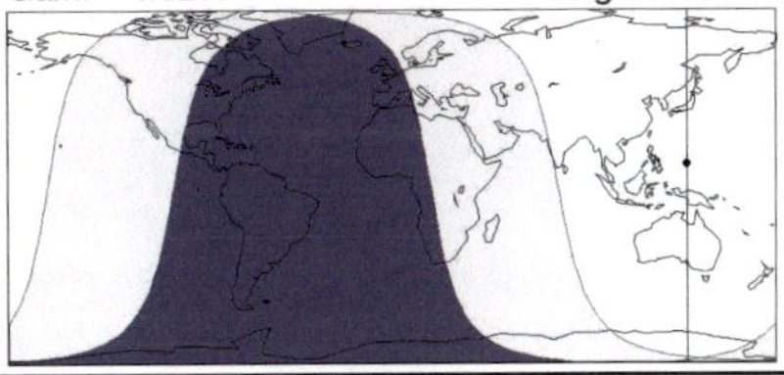

Penumbral **2100 Aug 19**
Saros 120 21:45 TD
A.Node ΔT= 128s

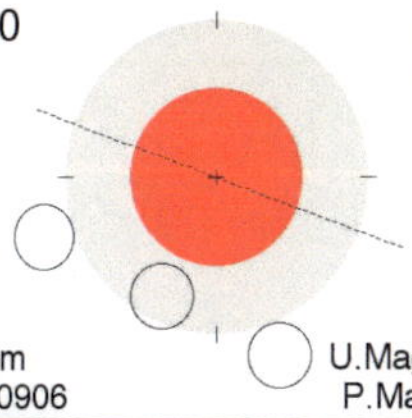

Pen. = 255m U.Mag. = -0.1556
Gam. = -1.0906 P.Mag. = 0.8735

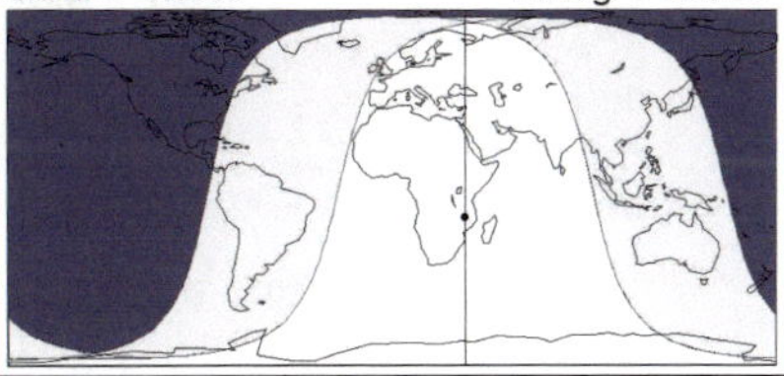

Plate 019

Photo 10 – A time sequence shows the partial phases and totality during the total lunar eclipse of 2000 July 16 from Maui, Hawaii (Copyright ©2000 by Fred Espenak).

Appendix C

Full Page Lunar Eclipse Figures: 2021 to 2070

Key to Full Page Lunar Eclipse Figures

Figure 5–2: Key to Full Page Lunar Eclipse Figures (Appendix C)

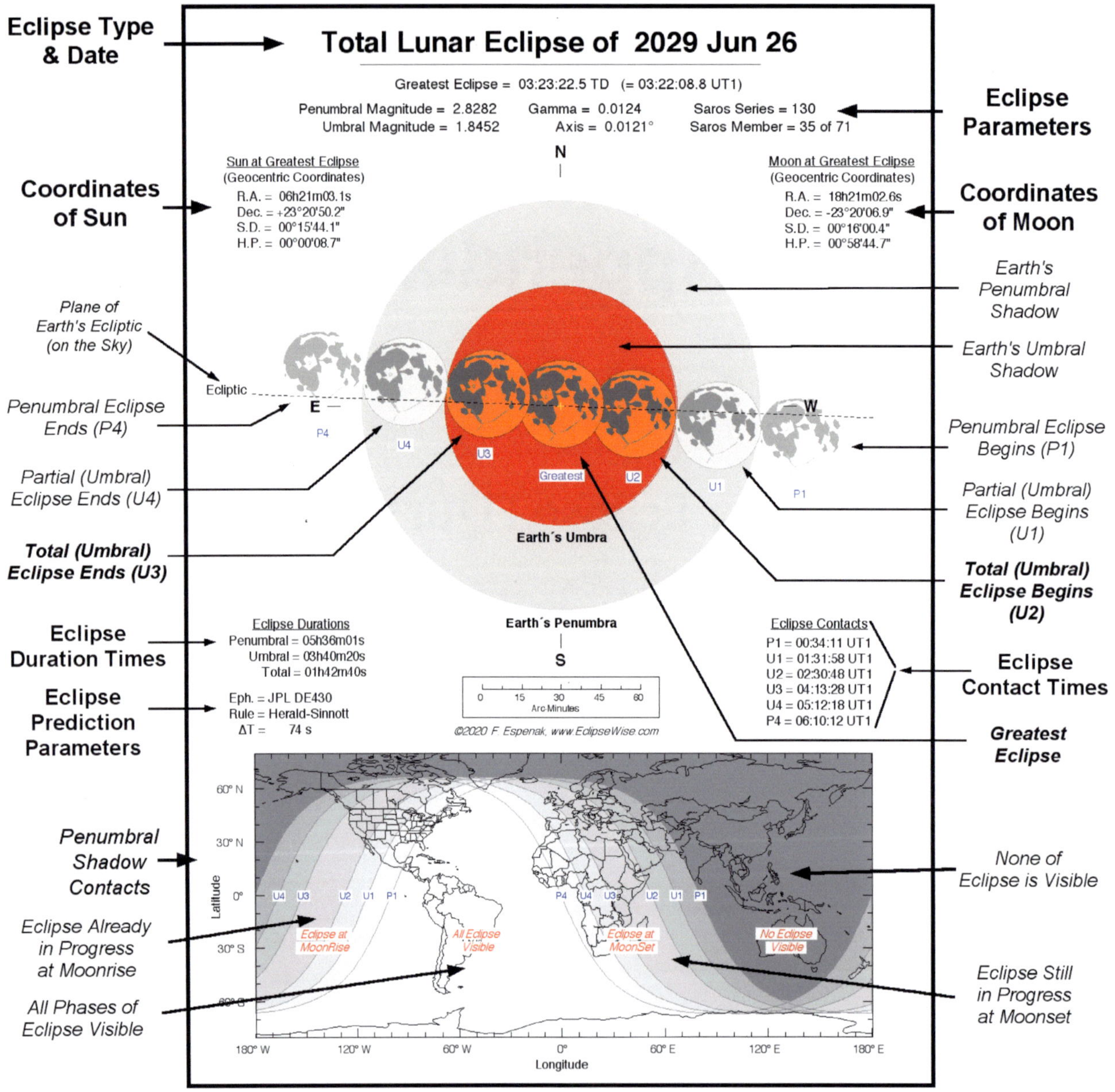

Total Lunar Eclipse of 2021 May 26

Greatest Eclipse = 11:19:52.7 TD (= 11:18:42.7 UT1)

Penumbral Magnitude = 1.9558 | Gamma = 0.4774 | Saros Series = 121
Umbral Magnitude = 1.0112 | Axis = 0.4880° | Saros Member = 55 of 82

Sun at Greatest Eclipse
(Geocentric Coordinates)
R.A. = 04h14m03.6s
Dec. = +21°12'25.4"
S.D. = 00°15'47.3"
H.P. = 00°00'08.7"

Moon at Greatest Eclipse
(Geocentric Coordinates)
R.A. = 16h14m37.8s
Dec. = -20°44'15.0"
S.D. = 00°16'42.9"
H.P. = 01°01'20.5"

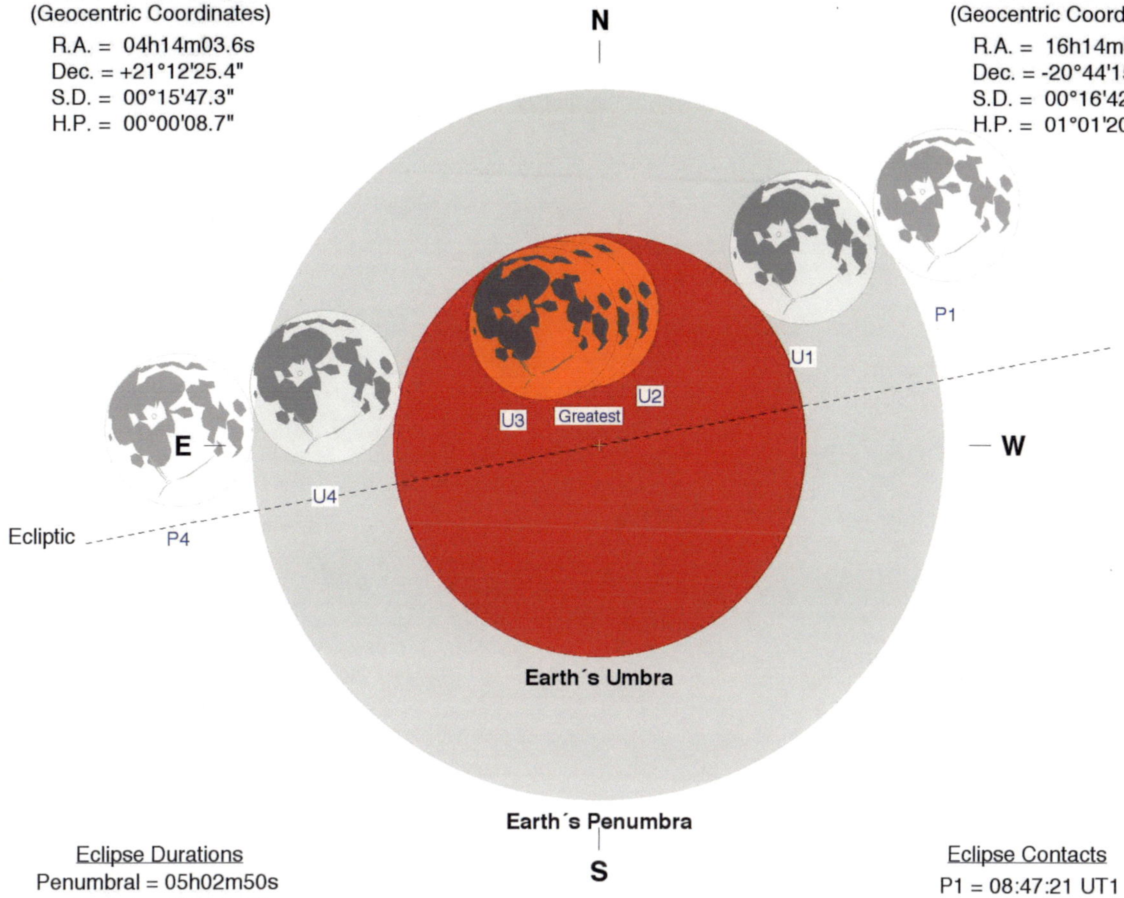

Eclipse Durations
Penumbral = 05h02m50s
Umbral = 03h08m07s
Total = 00h15m52s

Eph. = JPL DE430
Rule = Herald-Sinnott
ΔT = 70 s

0 15 30 45 60
Arc-Minutes

©2020 F. Espenak, www.EclipseWise.com

Eclipse Contacts
P1 = 08:47:21 UT1
U1 = 09:44:43 UT1
U2 = 11:10:56 UT1
U3 = 11:26:48 UT1
U4 = 12:52:50 UT1
P4 = 13:50:11 UT1

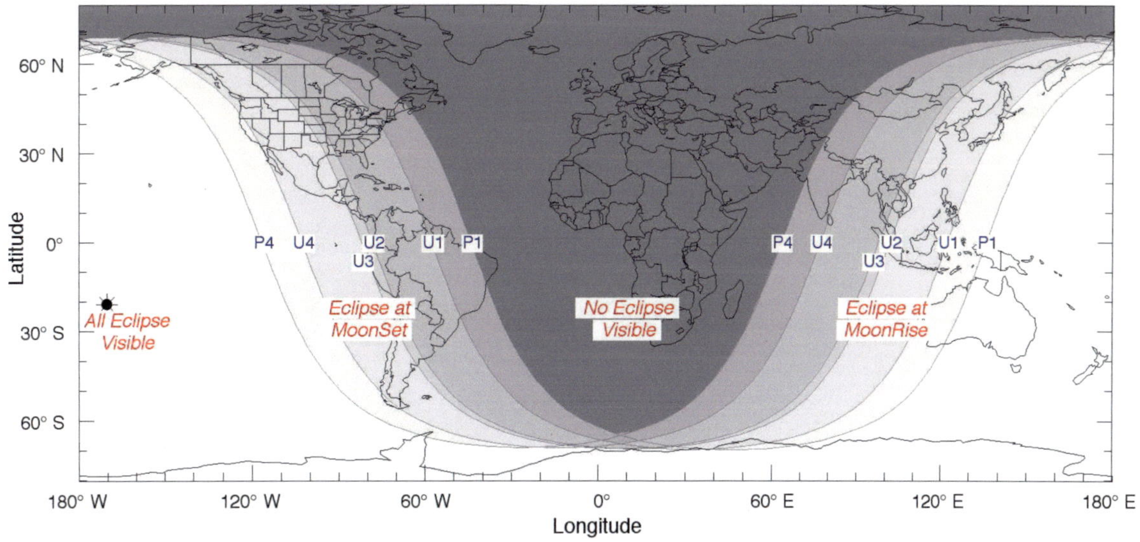

Partial Lunar Eclipse of 2021 Nov 19

Greatest Eclipse = 09:04:05.8 TD (= 09:02:55.5 UT1)

Penumbral Magnitude = 2.0738 Gamma = -0.4552 Saros Series = 126
Umbral Magnitude = 0.9760 Axis = 0.4104° Saros Member = 45 of 70

Sun at Greatest Eclipse
(Geocentric Coordinates)
R.A. = 15h39m50.9s
Dec. = -19°32'33.1"
S.D. = 00°16'11.0"
H.P. = 00°00'08.9"

Moon at Greatest Eclipse
(Geocentric Coordinates)
R.A. = 03h40m24.8s
Dec. = +19°09'15.5"
S.D. = 00°14'44.5"
H.P. = 00°54'06.1"

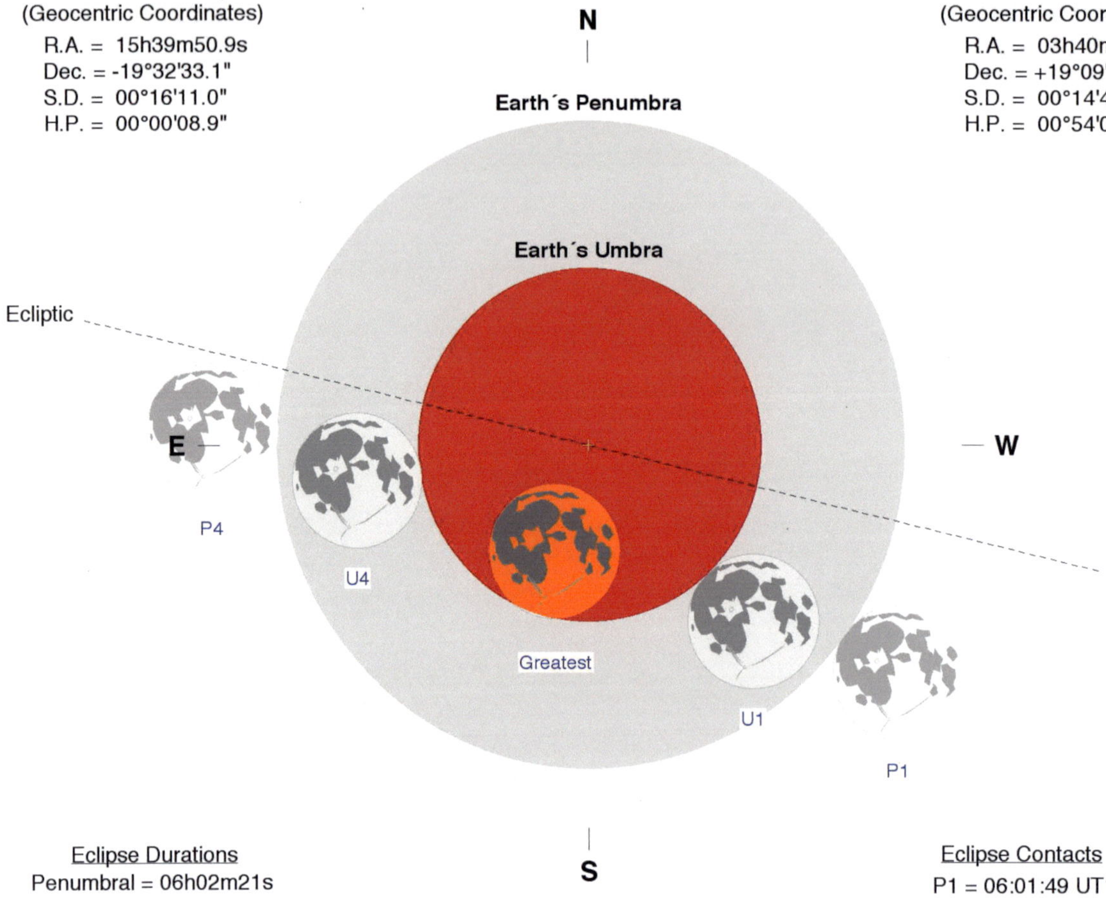

Eclipse Durations
Penumbral = 06h02m21s
Umbral = 03h29m09s

Eclipse Contacts
P1 = 06:01:49 UT1
U1 = 07:18:26 UT1
U4 = 10:47:36 UT1
P4 = 12:04:11 UT1

Eph. = JPL DE430
Rule = Herald-Sinnott
ΔT = 70 s

0 15 30 45 60
Arc-Minutes

©2020 F. Espenak, www.EclipseWise.com

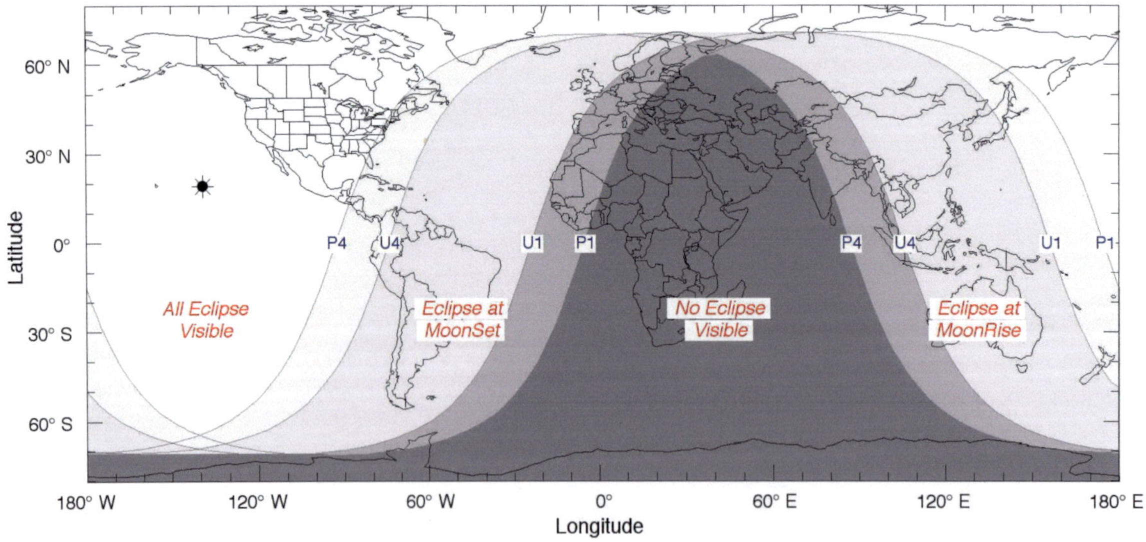

Total Lunar Eclipse of 2022 May 16

Greatest Eclipse = 04:12:41.7 TD (= 04:11:31.2 UT1)

Penumbral Magnitude = 2.3743
Umbral Magnitude = 1.4154

Gamma = -0.2532
Axis = 0.2555°

Saros Series = 131
Saros Member = 34 of 72

Sun at Greatest Eclipse
(Geocentric Coordinates)
R.A. = 03h31m49.5s
Dec. = +19°05'13.4"
S.D. = 00°15'49.2"
H.P. = 00°00'08.7"

Moon at Greatest Eclipse
(Geocentric Coordinates)
R.A. = 15h31m27.8s
Dec. = -19°19'40.4"
S.D. = 00°16'29.9"
H.P. = 01°00'33.1"

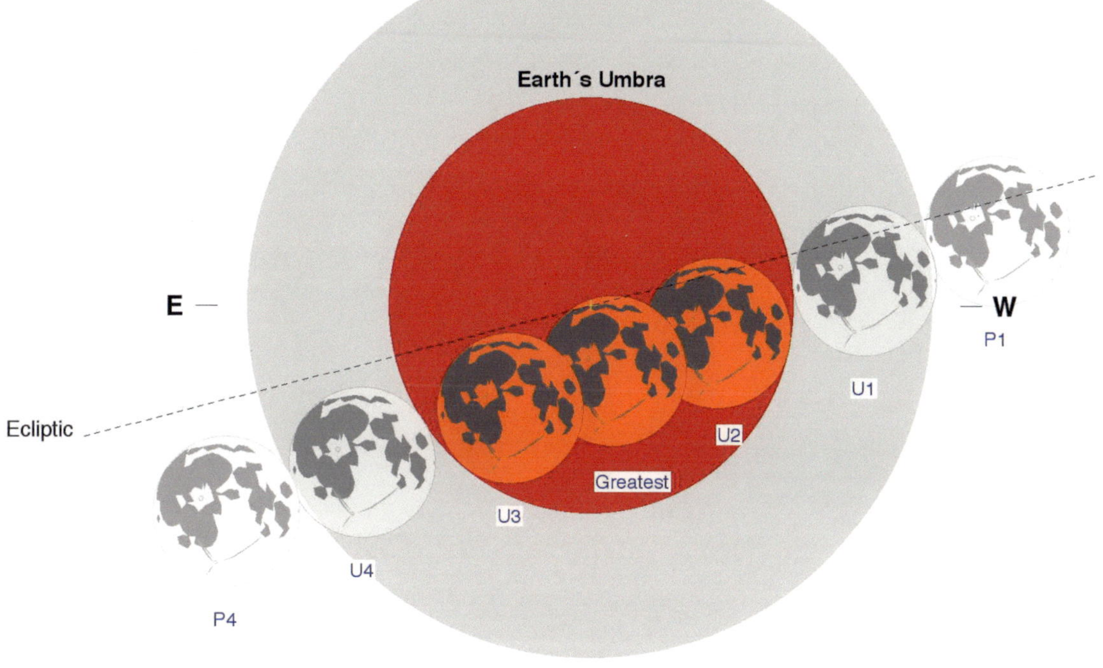

Eclipse Durations
Penumbral = 05h19m28s
Umbral = 03h27m57s
Total = 01h25m32s

Eph. = JPL DE430
Rule = Herald-Sinnott
ΔT = 70 s

0 15 30 45 60
Arc-Minutes

©2020 F. Espenak, www.EclipseWise.com

Eclipse Contacts
P1 = 01:31:44 UT1
U1 = 02:27:31 UT1
U2 = 03:28:40 UT1
U3 = 04:54:11 UT1
U4 = 05:55:27 UT1
P4 = 06:51:11 UT1

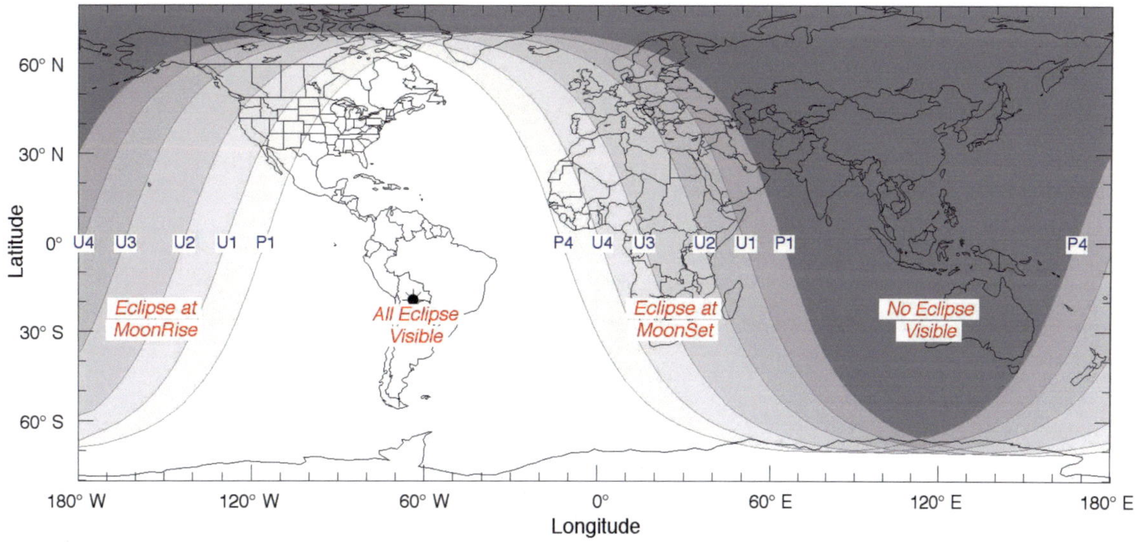

Total Lunar Eclipse of 2022 Nov 08

Greatest Eclipse = 11:00:22.0 TD (= 10:59:11.3 UT1)

Penumbral Magnitude = 2.4161   Gamma = 0.2570   Saros Series = 136
Umbral Magnitude = 1.3607   Axis = 0.2404°   Saros Member = 20 of 72

Sun at Greatest Eclipse
(Geocentric Coordinates)
R.A. = 14h54m11.2s
Dec. = -16°37'47.0"
S.D. = 00°16'08.5"
H.P. = 00°00'08.9"

Moon at Greatest Eclipse
(Geocentric Coordinates)
R.A. = 02h53m48.1s
Dec. = +16°51'06.7"
S.D. = 00°15'17.7"
H.P. = 00°56'07.8"

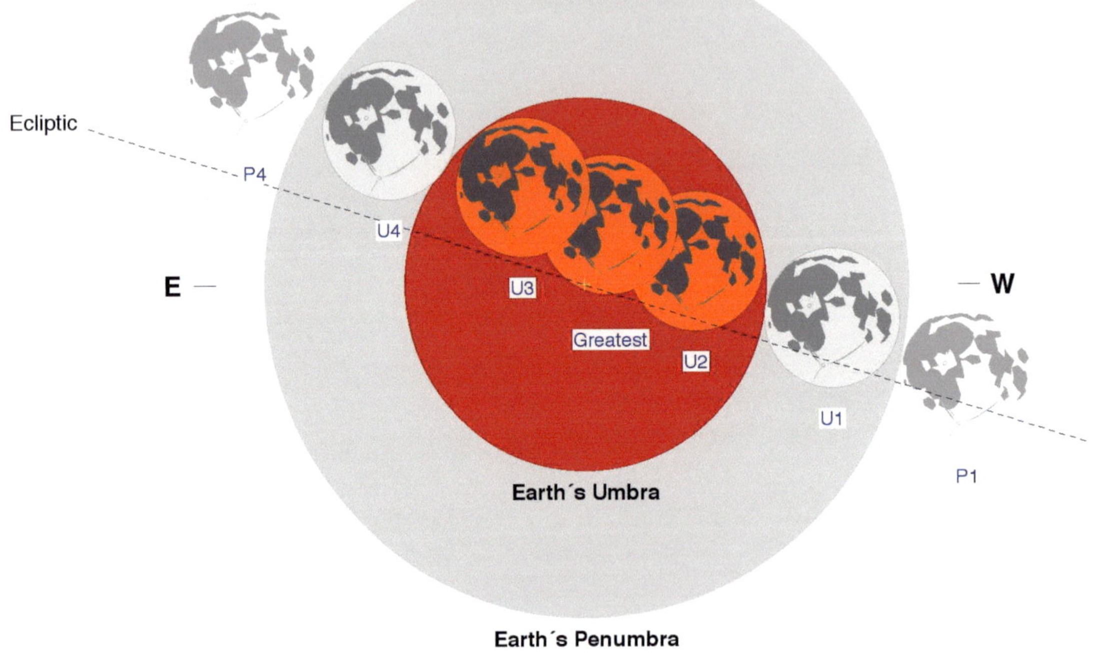

Eclipse Durations
Penumbral = 05h54m41s
Umbral = 03h40m35s
Total = 01h25m40s

Eph. = JPL DE430
Rule = Herald-Sinnott
ΔT = 71 s

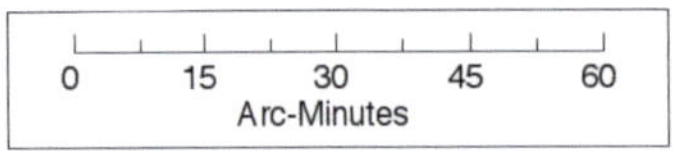

©2020 F. Espenak, www.EclipseWise.com

Eclipse Contacts
P1 = 08:01:52 UT1
U1 = 09:08:49 UT1
U2 = 10:16:12 UT1
U3 = 11:41:52 UT1
U4 = 12:49:24 UT1
P4 = 13:56:32 UT1

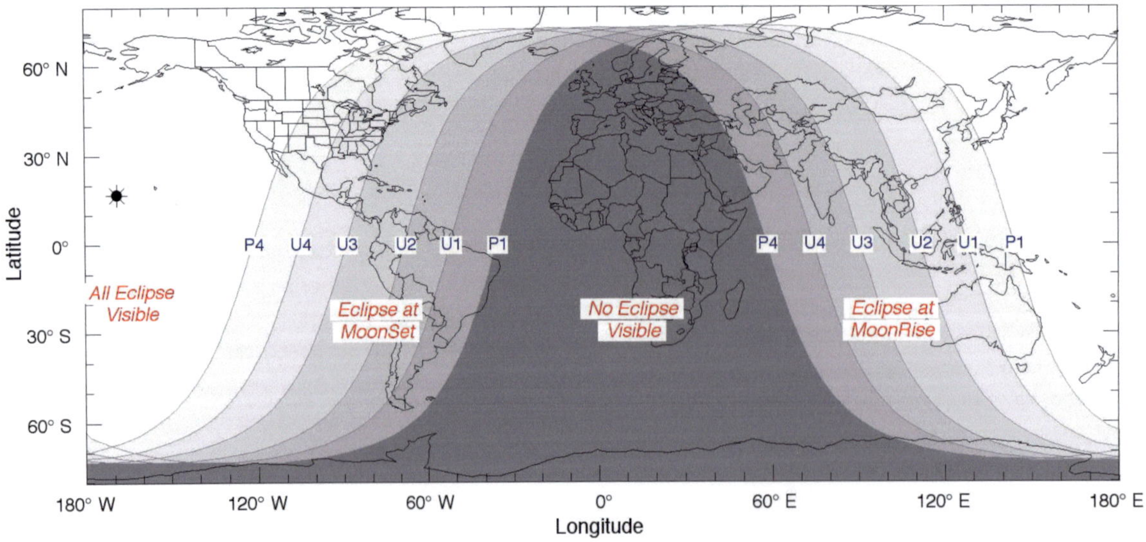

Penumbral Lunar Eclipse of 2023 May 05

Greatest Eclipse = 17:24:04.3 TD (= 17:22:53.5 UT1)

Penumbral Magnitude = 0.9655
Umbral Magnitude = -0.0438

Gamma = -1.0350
Axis = 0.9946°

Saros Series = 141
Saros Member = 24 of 72

Sun at Greatest Eclipse
(Geocentric Coordinates)
R.A. = 02h49m59.7s
Dec. = +16°19'27.9"
S.D. = 00°15'51.6"
H.P. = 00°00'08.7"

Moon at Greatest Eclipse
(Geocentric Coordinates)
R.A. = 14h48m23.5s
Dec. = -17°14'31.7"
S.D. = 00°15'42.8"
H.P. = 00°57'40.1"

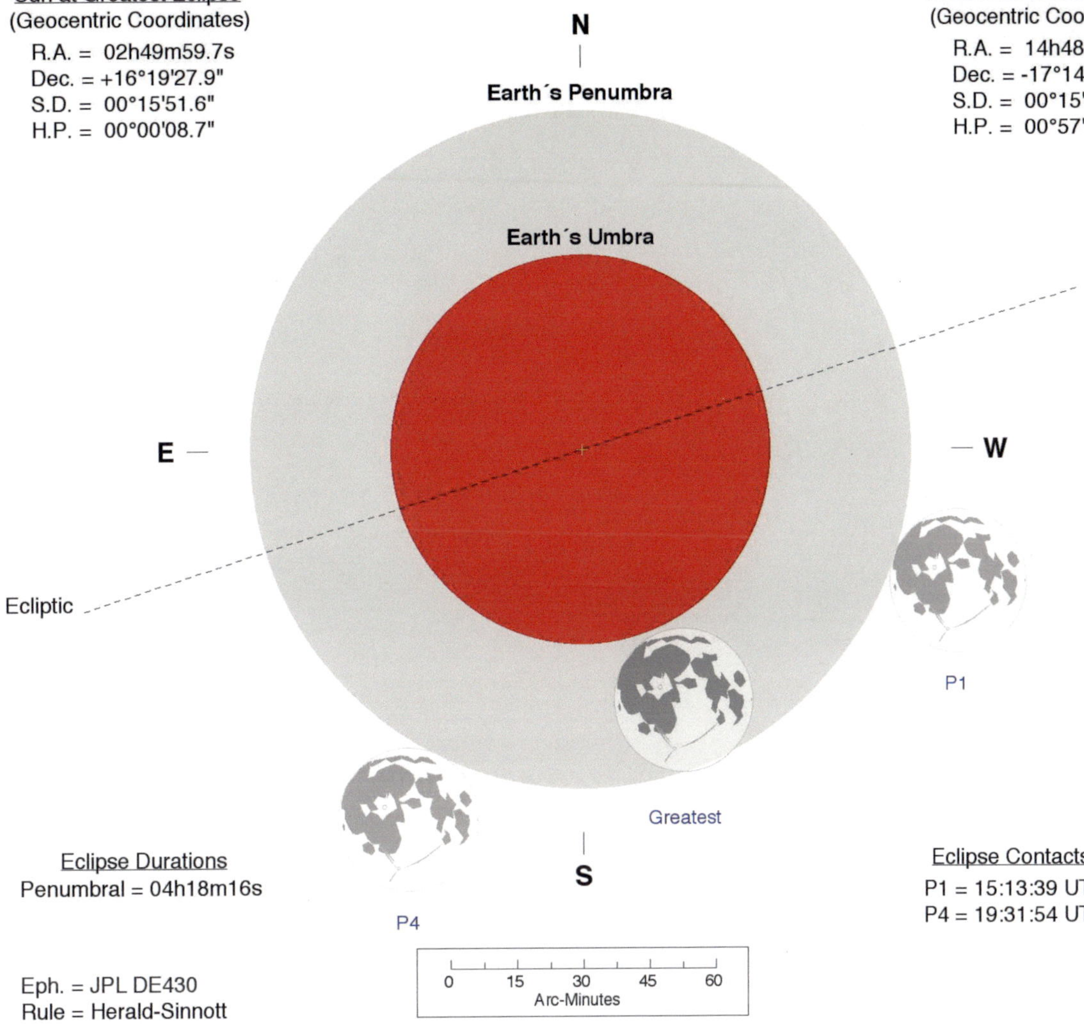

Eclipse Durations
Penumbral = 04h18m16s

Eclipse Contacts
P1 = 15:13:39 UT1
P4 = 19:31:54 UT1

Eph. = JPL DE430
Rule = Herald-Sinnott
ΔT = 71 s

©2020 F. Espenak, www.EclipseWise.com

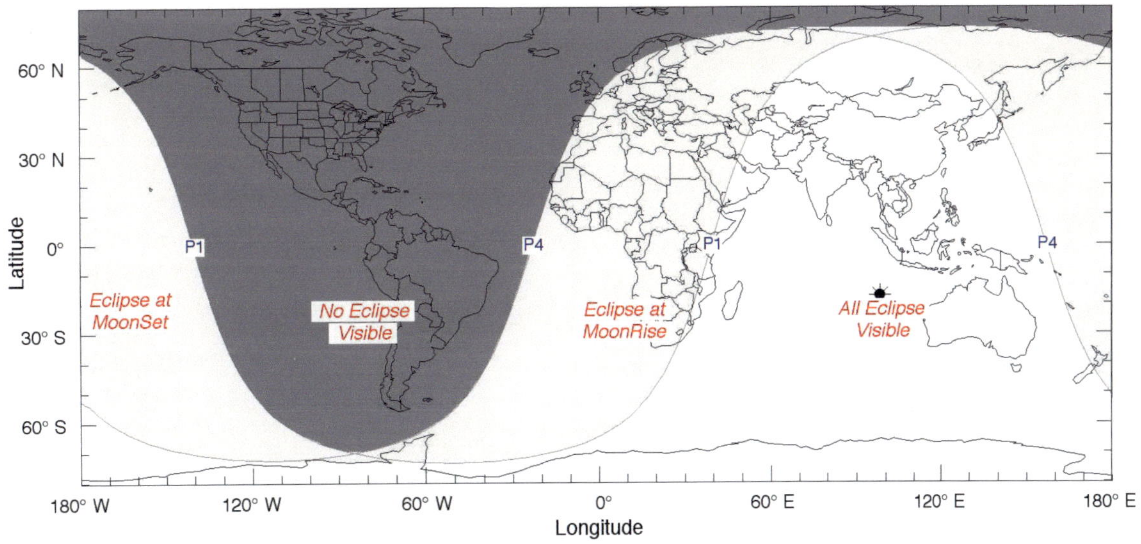

Partial Lunar Eclipse of 2023 Oct 28

Greatest Eclipse = 20:15:17.0 TD (= 20:14:05.9 UT1)

Penumbral Magnitude = 1.1200   Gamma = 0.9472   Saros Series = 146
Umbral Magnitude = 0.1239   Axis = 0.9362°   Saros Member = 11 of 72

<u>Sun at Greatest Eclipse</u>
(Geocentric Coordinates)
R.A. = 14h11m25.9s
Dec. = -13°14'10.5"
S.D. = 00°16'05.9"
H.P. = 00°00'08.9"

<u>Moon at Greatest Eclipse</u>
(Geocentric Coordinates)
R.A. = 02h09m47.6s
Dec. = +14°05'01.6"
S.D. = 00°16'09.7"
H.P. = 00°59'18.9"

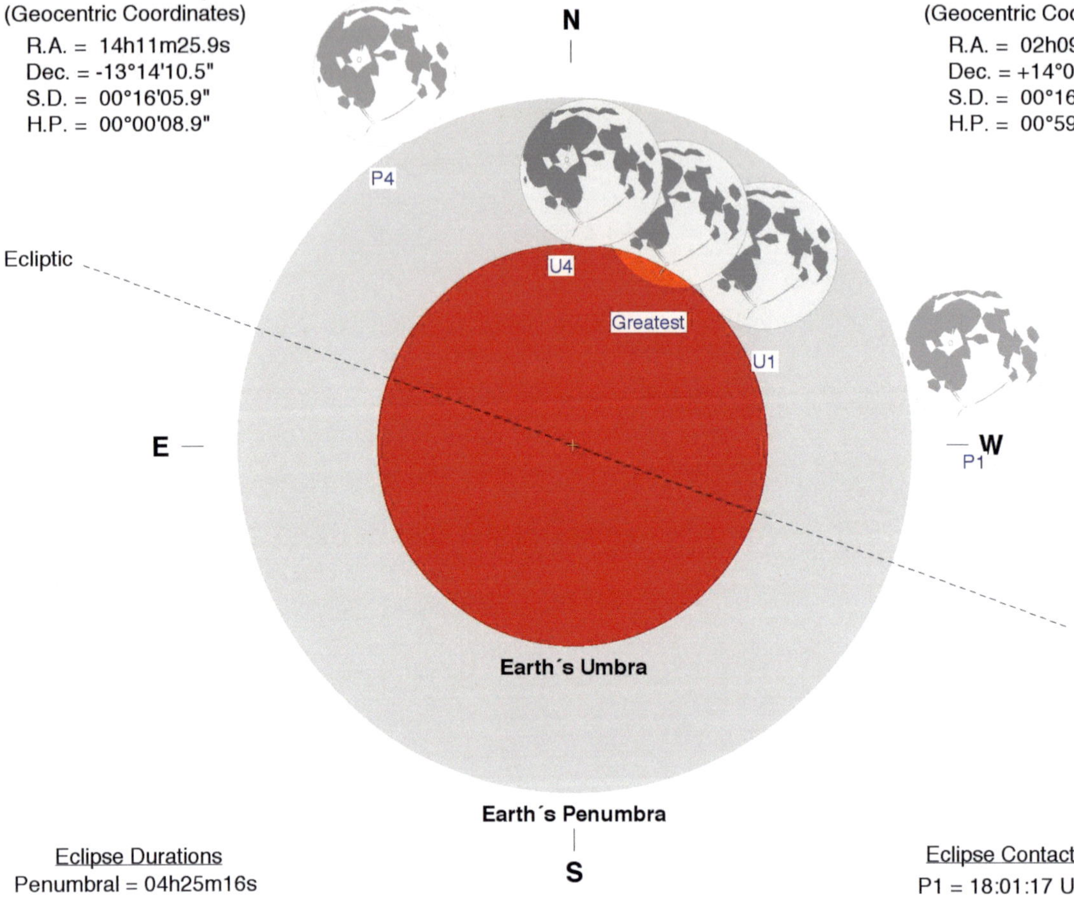

<u>Eclipse Durations</u>
Penumbral = 04h25m16s
Umbral = 01h18m09s

<u>Eclipse Contacts</u>
P1 = 18:01:17 UT1
U1 = 19:34:41 UT1
U4 = 20:52:50 UT1
P4 = 22:26:33 UT1

Eph. = JPL DE430
Rule = Herald-Sinnott
ΔT = 71 s

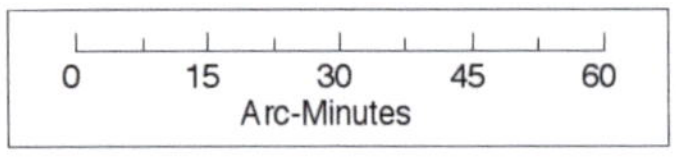

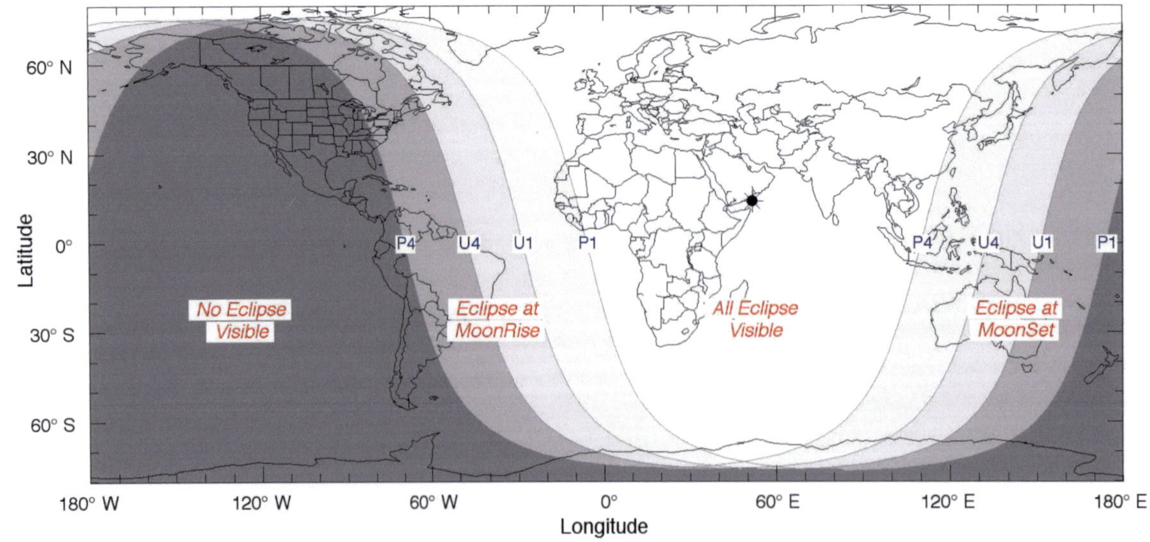

Penumbral Lunar Eclipse of 2024 Mar 25

Greatest Eclipse = 07:14:00.1 TD (= 07:12:48.8 UT1)

Penumbral Magnitude = 0.9577 Gamma = 1.0610 Saros Series = 113
Umbral Magnitude = -0.1304 Axis = 0.9564° Saros Member = 64 of 71

Sun at Greatest Eclipse
(Geocentric Coordinates)
R.A. = 00h18m49.9s
Dec. = +02°02'16.6"
S.D. = 00°16'02.2"
H.P. = 00°00'08.8"

Moon at Greatest Eclipse
(Geocentric Coordinates)
R.A. = 12h20m41.3s
Dec. = -01°12'05.6"
S.D. = 00°14'44.3"
H.P. = 00°54'05.4"

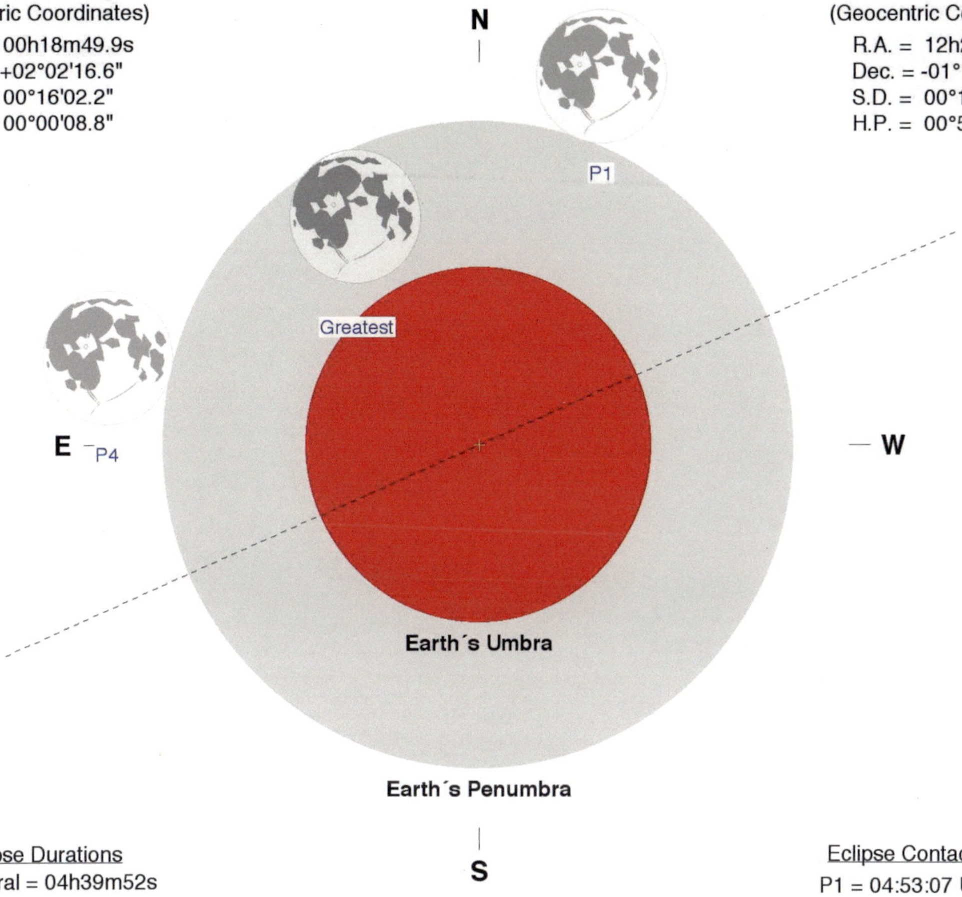

Eclipse Durations
Penumbral = 04h39m52s

Eclipse Contacts
P1 = 04:53:07 UT1
P4 = 09:32:59 UT1

Eph. = JPL DE430
Rule = Herald-Sinnott
ΔT = 71 s

0 15 30 45 60
Arc-Minutes

©2020 F. Espenak, www.EclipseWise.com

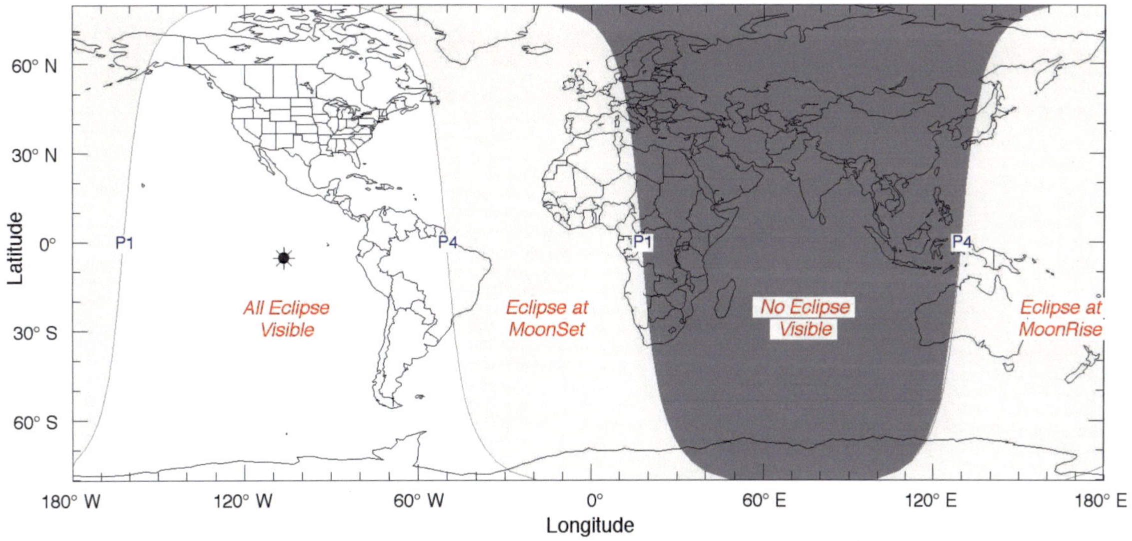

Partial Lunar Eclipse of 2024 Sep 18

Greatest Eclipse = 02:45:25.5 TD (= 02:44:14.1 UT1)

Penumbral Magnitude = 1.0392 Gamma = -0.9792 Saros Series = 118
Umbral Magnitude = 0.0869 Axis = 1.0009° Saros Member = 52 of 73

Sun at Greatest Eclipse
(Geocentric Coordinates)
R.A. = 11h44m09.7s
Dec. = +01°42'52.9"
S.D. = 00°15'55.1"
H.P. = 00°00'08.8"

Moon at Greatest Eclipse
(Geocentric Coordinates)
R.A. = 23h46m06.1s
Dec. = -02°35'26.7"
S.D. = 00°16'42.8"
H.P. = 01°01'20.4"

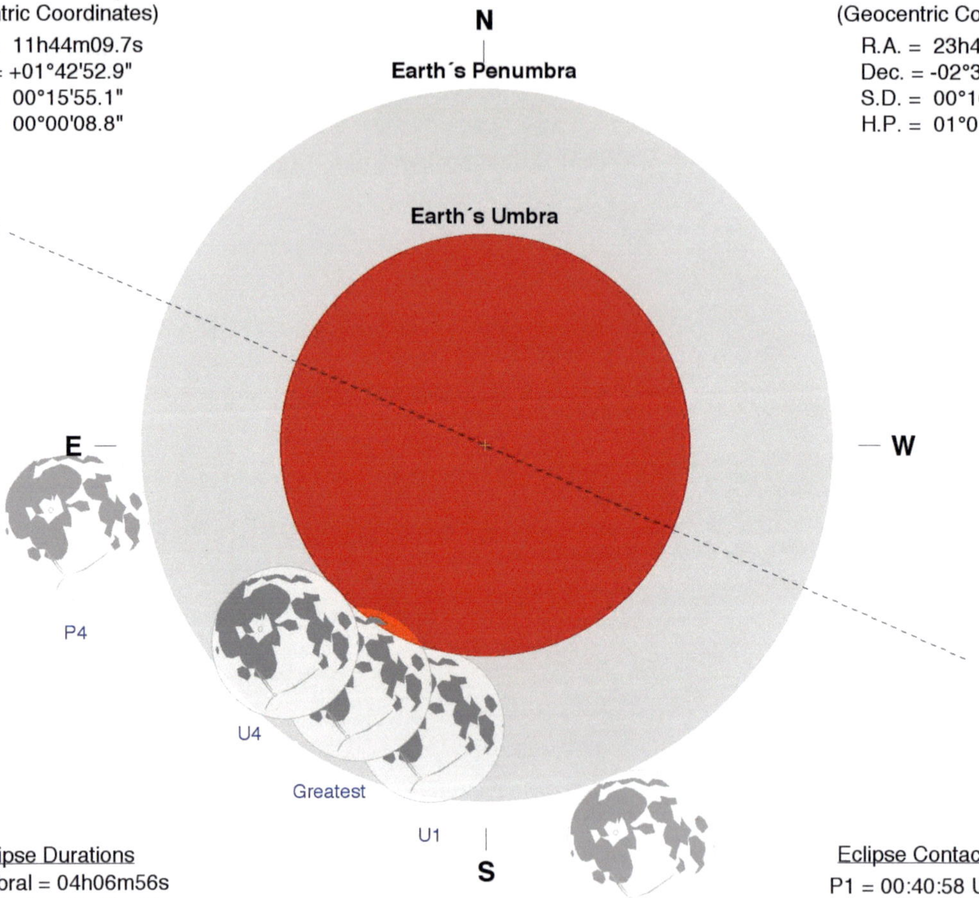

Eclipse Durations
Penumbral = 04h06m56s
Umbral = 01h03m40s

Eclipse Contacts
P1 = 00:40:58 UT1
U1 = 02:12:42 UT1
U4 = 03:16:22 UT1
P4 = 04:47:54 UT1

Eph. = JPL DE430
Rule = Herald-Sinnott
ΔT = 71 s

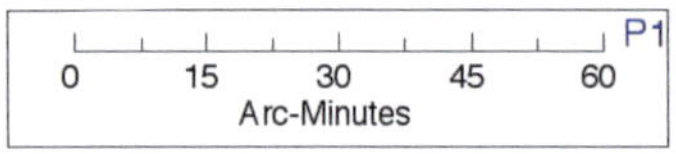

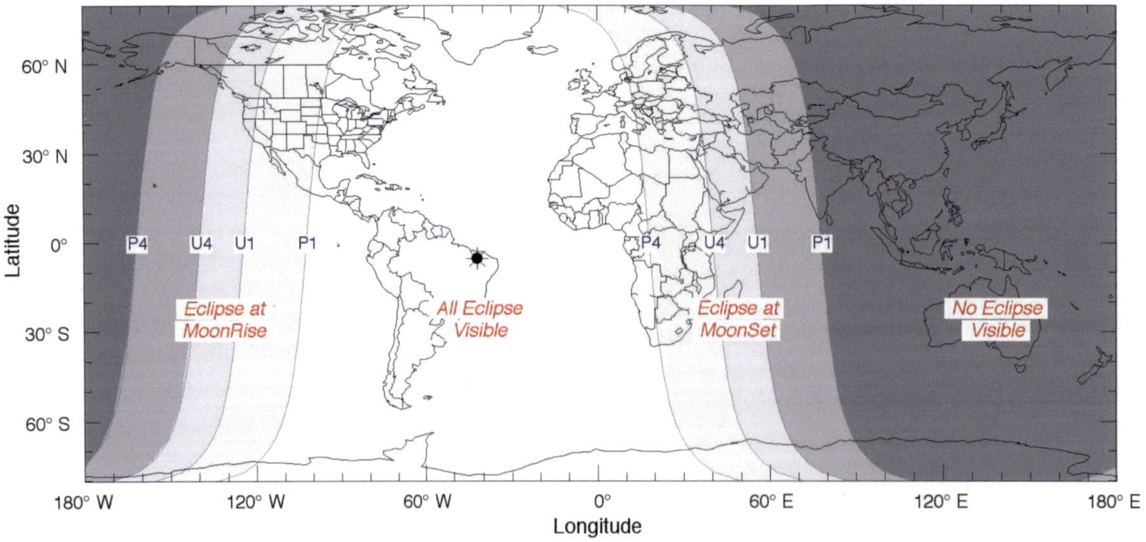

Total Lunar Eclipse of 2025 Mar 14

Greatest Eclipse = 06:59:56.2 TD (= 06:58:44.5 UT1)

Penumbral Magnitude = 2.2615	Gamma = 0.3485	Saros Series = 123
Umbral Magnitude = 1.1804	Axis = 0.3171°	Saros Member = 53 of 72

Sun at Greatest Eclipse
(Geocentric Coordinates)

R.A. = 23h37m46.0s
Dec. = -02°24'16.8"
S.D. = 00°16'05.2"
H.P. = 00°00'08.8"

Moon at Greatest Eclipse
(Geocentric Coordinates)

R.A. = 11h38m23.0s
Dec. = +02°40'54.6"
S.D. = 00°14'52.8"
H.P. = 00°54'36.8"

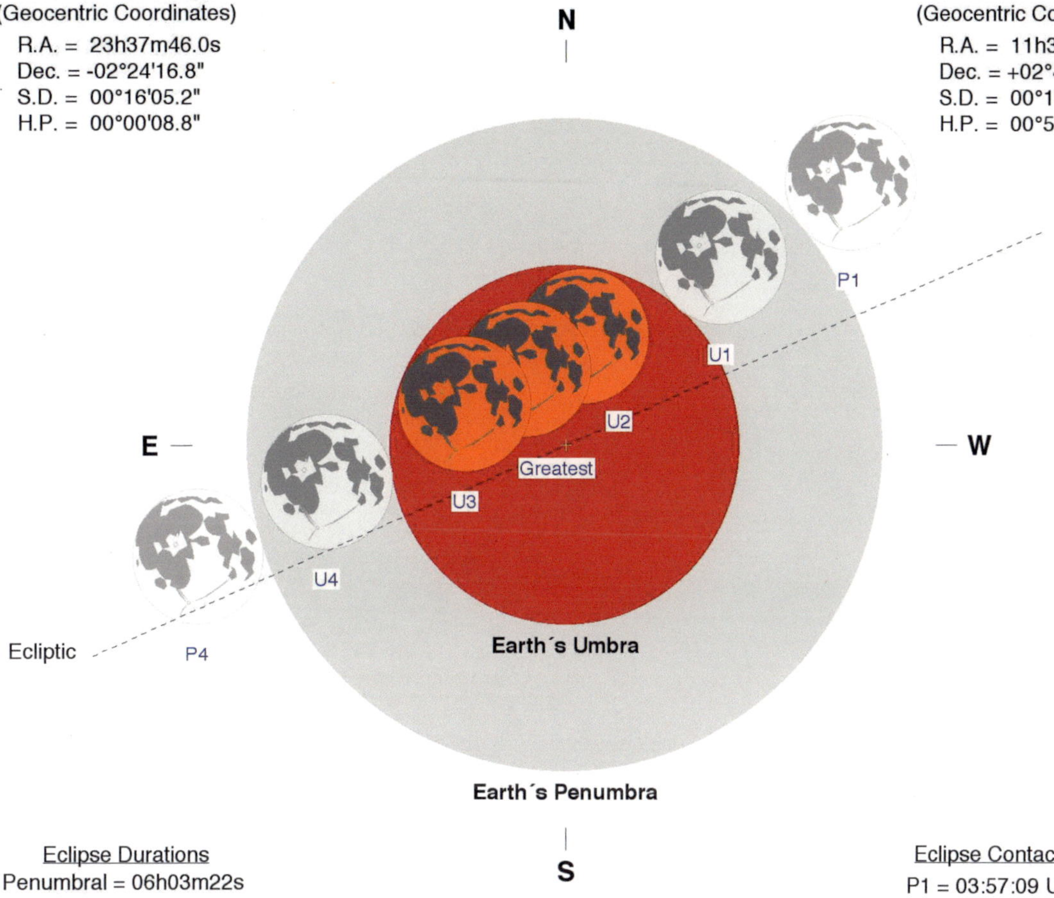

Eclipse Durations

Penumbral = 06h03m22s
Umbral = 03h38m56s
Total = 01h06m04s

Eph. = JPL DE430
Rule = Herald-Sinnott
ΔT = 72 s

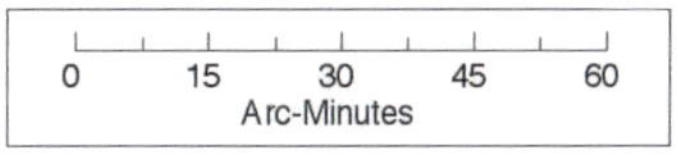

Eclipse Contacts

P1 = 03:57:09 UT1
U1 = 05:09:23 UT1
U2 = 06:25:58 UT1
U3 = 07:32:01 UT1
U4 = 08:48:18 UT1
P4 = 10:00:32 UT1

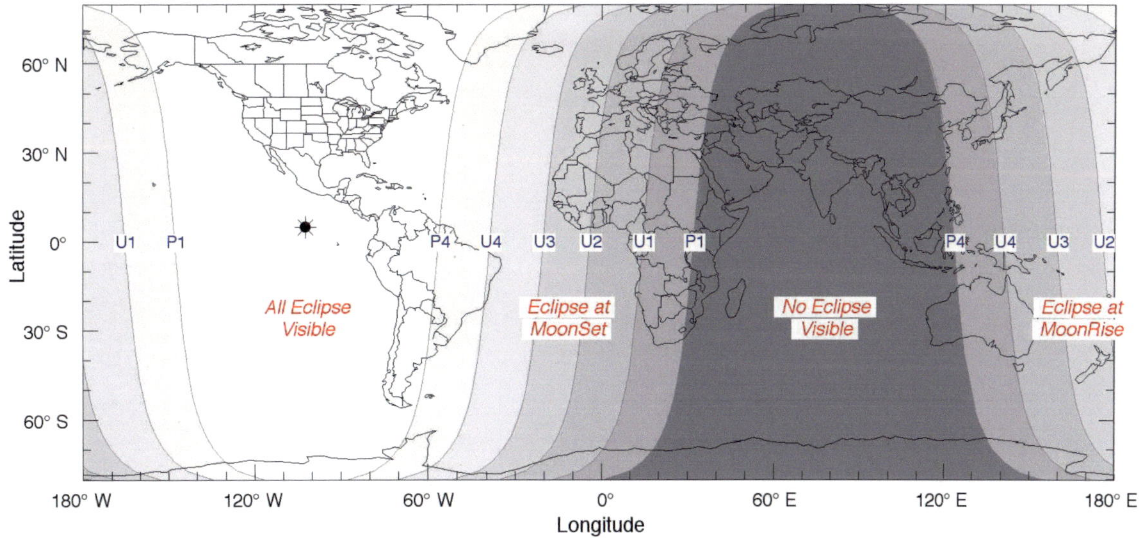

Total Lunar Eclipse of 2025 Sep 07

Greatest Eclipse = 18:12:58.0 TD (= 18:11:46.1 UT1)

Penumbral Magnitude = 2.3459 Gamma = -0.2752 Saros Series = 128
Umbral Magnitude = 1.3638 Axis = 0.2721° Saros Member = 41 of 71

Sun at Greatest Eclipse
(Geocentric Coordinates)
R.A. = 11h06m09.1s
Dec. = +05°45'47.6"
S.D. = 00°15'52.4"
H.P. = 00°00'08.7"

Moon at Greatest Eclipse
(Geocentric Coordinates)
R.A. = 23h06m40.4s
Dec. = -06°00'08.9"
S.D. = 00°16'09.8"
H.P. = 00°59'19.1"

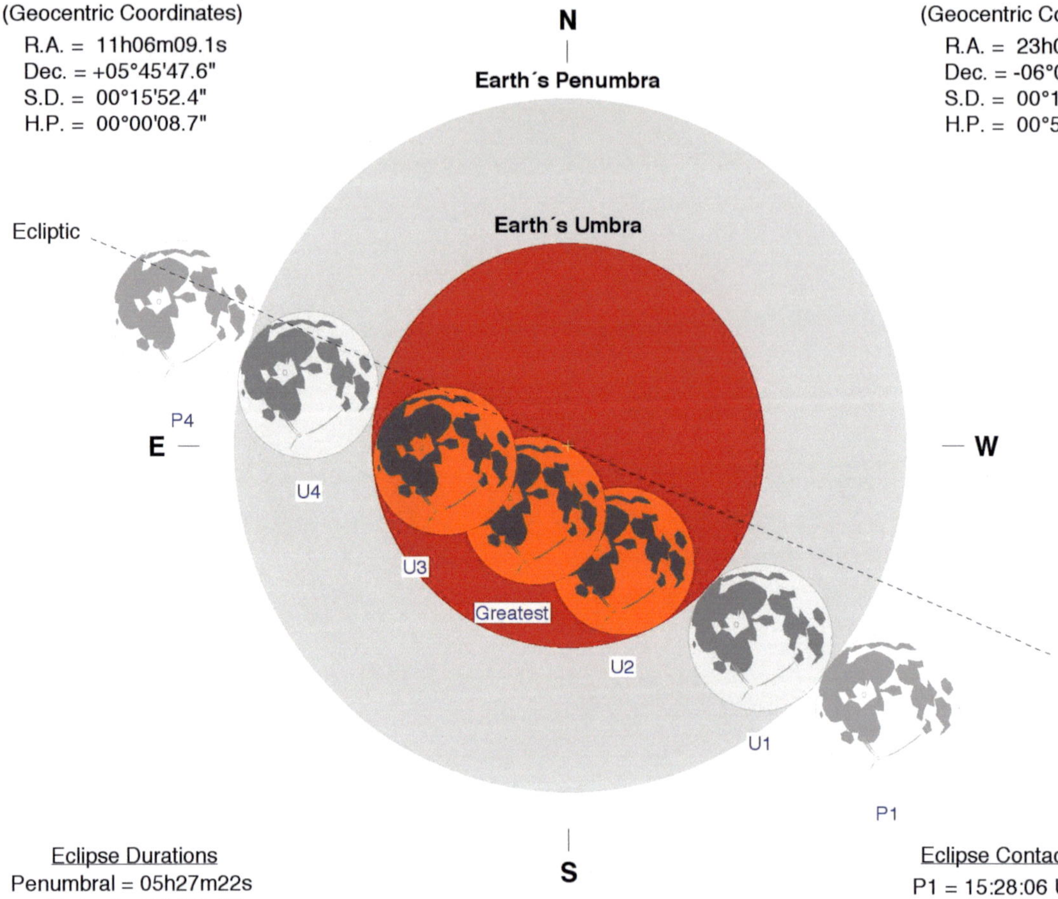

Eclipse Durations
Penumbral = 05h27m22s
Umbral = 03h30m02s
Total = 01h22m41s

Eph. = JPL DE430
Rule = Herald-Sinnott
ΔT = 72 s

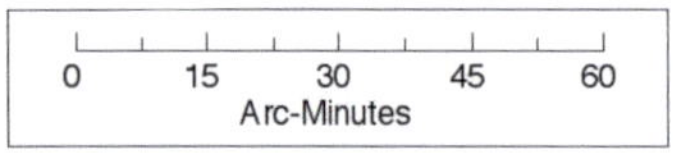

©2020 F. Espenak, www.EclipseWise.com

Eclipse Contacts
P1 = 15:28:06 UT1
U1 = 16:26:51 UT1
U2 = 17:30:37 UT1
U3 = 18:53:18 UT1
U4 = 19:56:53 UT1
P4 = 20:55:28 UT1

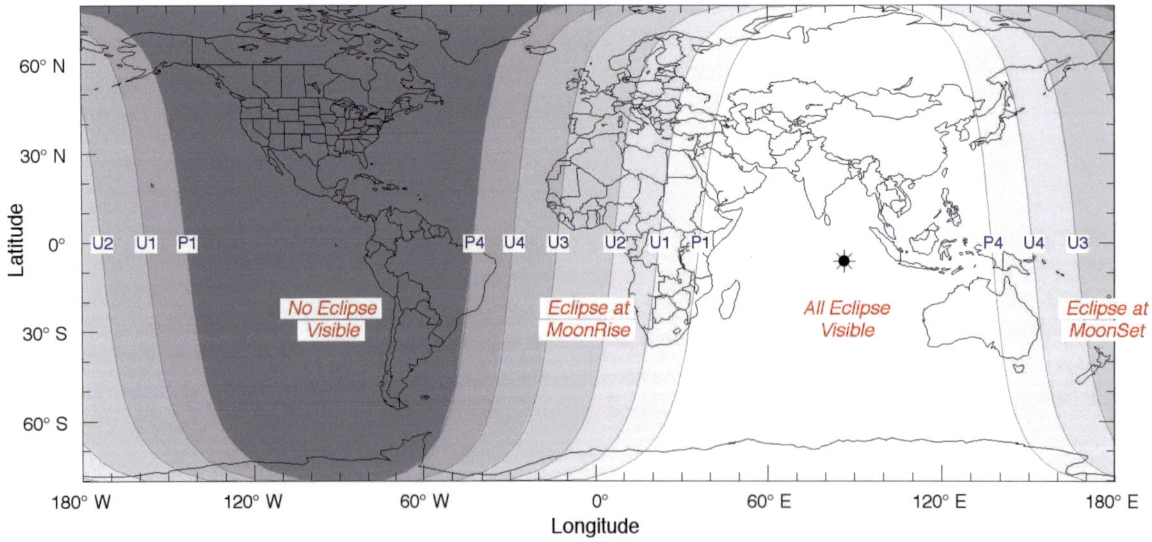

Total Lunar Eclipse of 2026 Mar 03

Greatest Eclipse = 11:34:52.1 TD (= 11:33:40.0 UT1)

Penumbral Magnitude = 2.1858
Umbral Magnitude = 1.1526

Gamma = -0.3765
Axis = 0.3596°

Saros Series = 133
Saros Member = 27 of 71

Sun at Greatest Eclipse
(Geocentric Coordinates)
R.A. = 22h56m56.0s
Dec. = -06°43'06.4"
S.D. = 00°16'08.0"
H.P. = 00°00'08.9"

Moon at Greatest Eclipse
(Geocentric Coordinates)
R.A. = 10h56m15.0s
Dec. = +06°24'05.3"
S.D. = 00°15'37.0"
H.P. = 00°57'18.7"

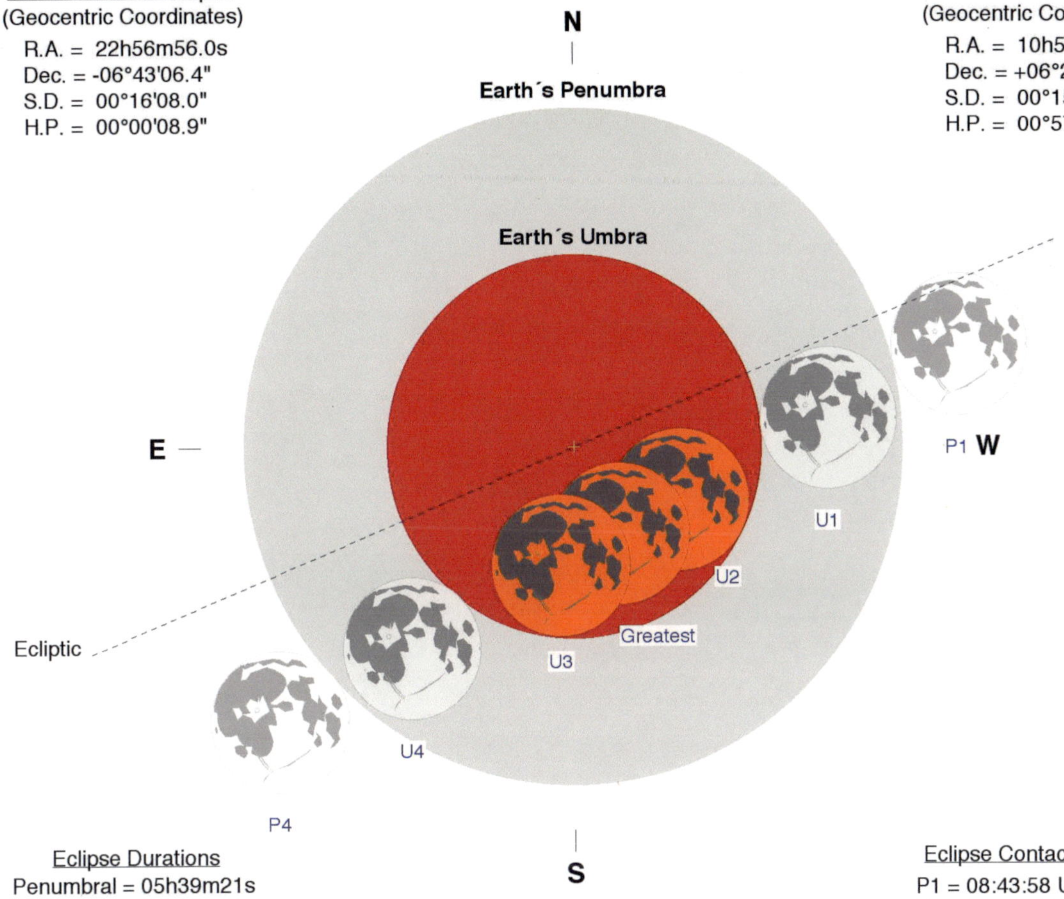

Eclipse Durations
Penumbral = 05h39m21s
Umbral = 03h27m49s
Total = 00h58m58s

Eph. = JPL DE430
Rule = Herald-Sinnott
ΔT = 72 s

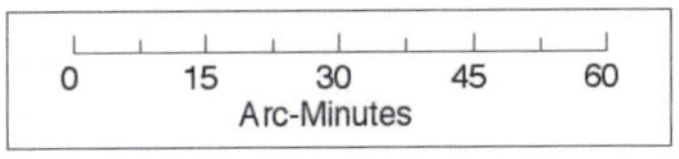

©2020 F. Espenak, www.EclipseWise.com

Eclipse Contacts
P1 = 08:43:58 UT1
U1 = 09:49:37 UT1
U2 = 11:03:54 UT1
U3 = 12:02:53 UT1
U4 = 13:17:26 UT1
P4 = 14:23:19 UT1

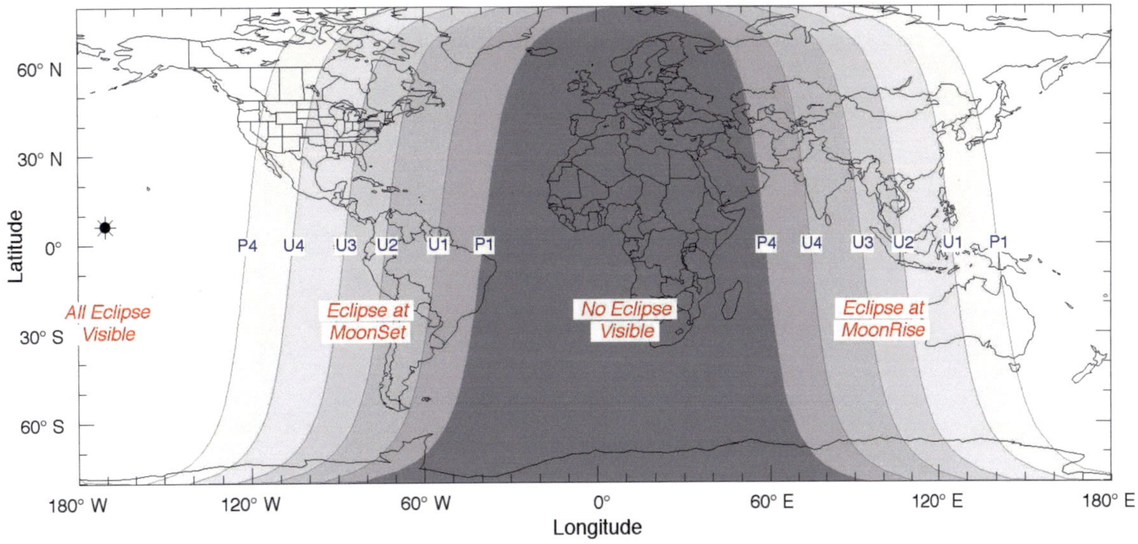

Partial Lunar Eclipse of 2026 Aug 28

Greatest Eclipse = 04:14:04.3 TD (= 04:12:52.0 UT1)

Penumbral Magnitude = 1.9664 Gamma = 0.4964 Saros Series = 138
Umbral Magnitude = 0.9319 Axis = 0.4647° Saros Member = 29 of 82

Sun at Greatest Eclipse
(Geocentric Coordinates)
R.A. = 10h26m57.9s
Dec. = +09°42'52.7"
S.D. = 00°15'50.0"
H.P. = 00°00'08.7"

Moon at Greatest Eclipse
(Geocentric Coordinates)
R.A. = 22h26m06.3s
Dec. = -09°18'03.6"
S.D. = 00°15'18.2"
H.P. = 00°56'09.9"

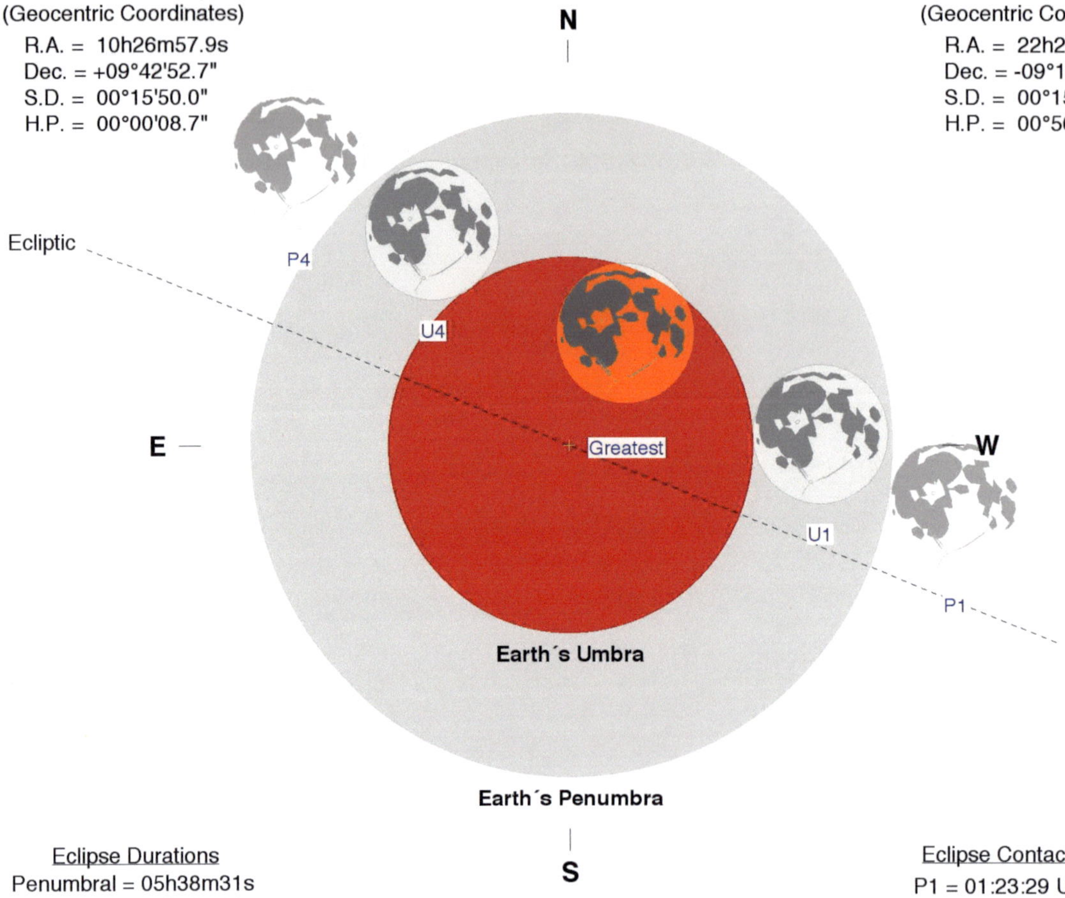

Eclipse Durations
Penumbral = 05h38m31s
Umbral = 03h18m48s

Eclipse Contacts
P1 = 01:23:29 UT1
U1 = 02:33:21 UT1
U4 = 05:52:09 UT1
P4 = 07:02:00 UT1

Eph. = JPL DE430
Rule = Herald-Sinnott
ΔT = 72 s

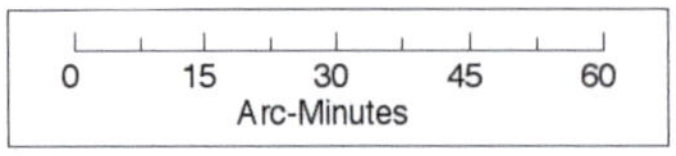

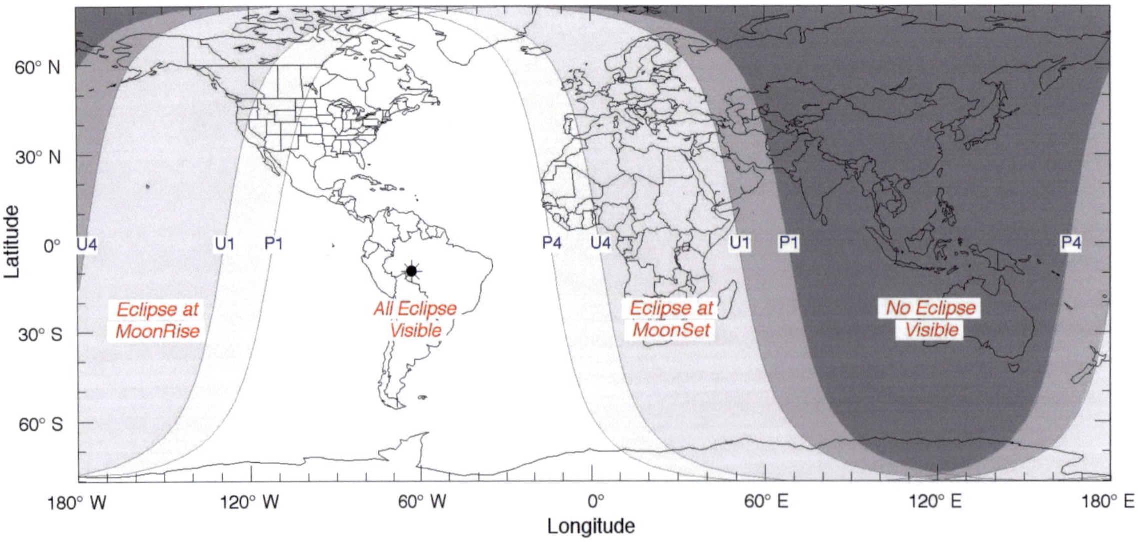

Penumbral Lunar Eclipse of 2027 Feb 20

Greatest Eclipse = 23:14:05.3 TD (= 23:12:52.7 UT1)

Penumbral Magnitude = 0.9286 | Gamma = -1.0480 | Saros Series = 143
Umbral Magnitude = -0.0549 | Axis = 1.0542° | Saros Member = 18 of 72

Sun at Greatest Eclipse
(Geocentric Coordinates)
R.A. = 22h16m18.3s
Dec. = -10°43'53.9"
S.D. = 00°16'10.5"
H.P. = 00°00'08.9"

Moon at Greatest Eclipse
(Geocentric Coordinates)
R.A. = 10h14m23.7s
Dec. = +09°47'16.6"
S.D. = 00°16'26.8"
H.P. = 01°00'21.6"

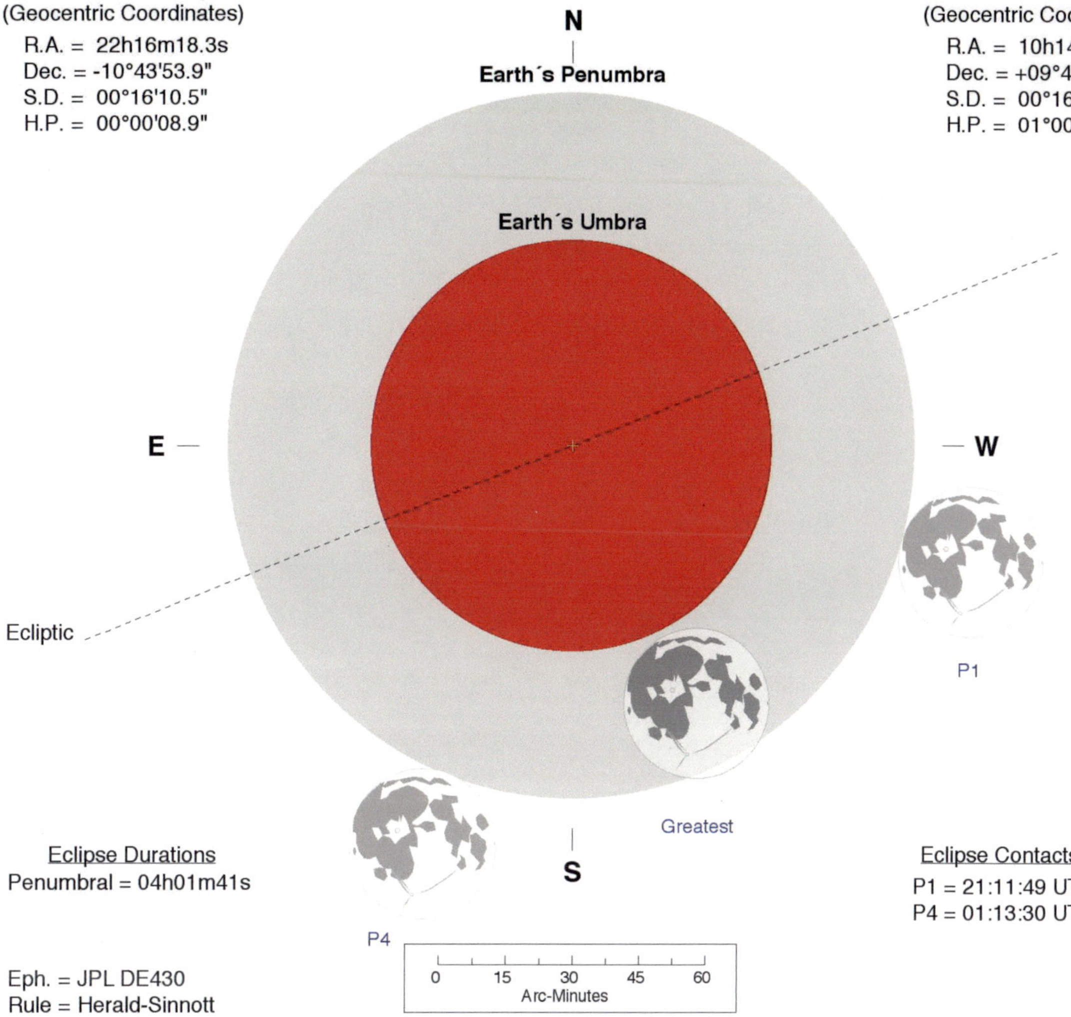

Eclipse Durations
Penumbral = 04h01m41s

Eclipse Contacts
P1 = 21:11:49 UT1
P4 = 01:13:30 UT1

Eph. = JPL DE430
Rule = Herald-Sinnott
ΔT = 73 s

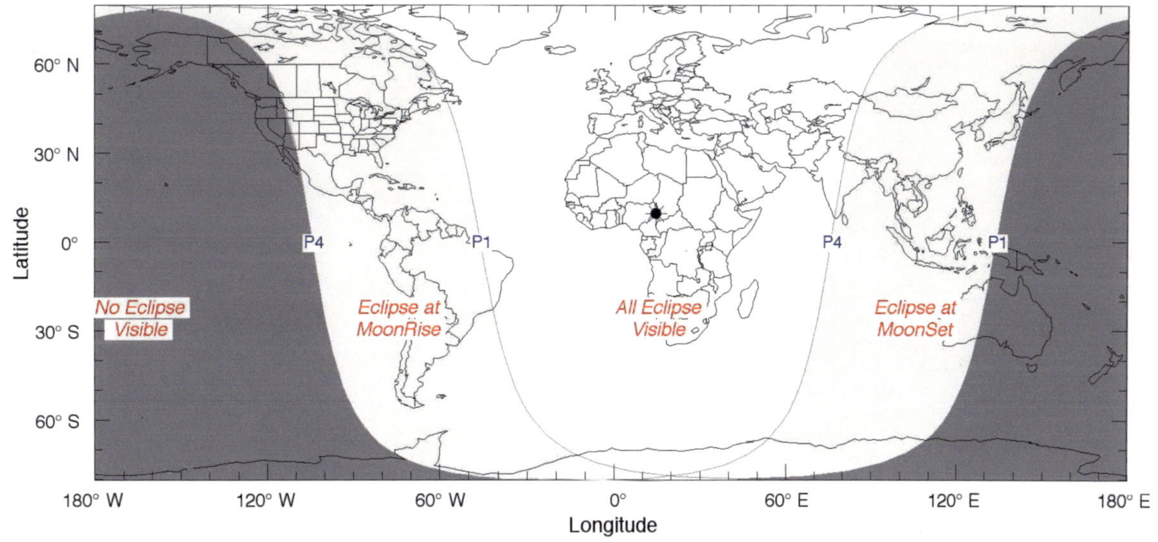

Penumbral Lunar Eclipse of 2027 Jul 18

Greatest Eclipse = 16:04:10.5 TD (= 16:02:57.8 UT1)

Penumbral Magnitude = 0.0032 Gamma = -1.5759 Saros Series = 110
Umbral Magnitude = -1.0662 Axis = 1.4184° Saros Member = 72 of 72

<u>Sun at Greatest Eclipse</u>
(Geocentric Coordinates)
R.A. = 07h51m14.4s
Dec. = +20°58'43.6"
S.D. = 00°15'44.3"
H.P. = 00°00'08.7"

<u>Moon at Greatest Eclipse</u>
(Geocentric Coordinates)
R.A. = 19h52m57.2s
Dec. = -22°20'25.3"
S.D. = 00°14'43.0"
H.P. = 00°54'00.6"

N

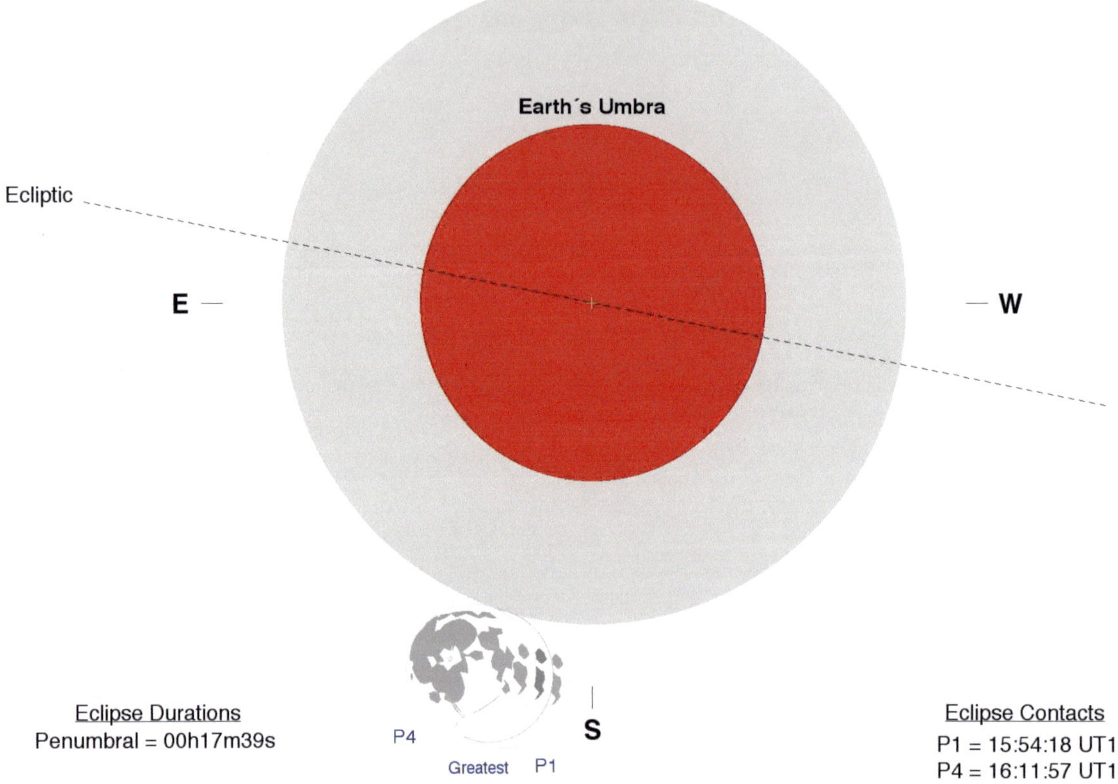

<u>Eclipse Durations</u>
Penumbral = 00h17m39s

<u>Eclipse Contacts</u>
P1 = 15:54:18 UT1
P4 = 16:11:57 UT1

0 15 30 45 60
Arc-Minutes

Eph. = JPL DE430
Rule = Herald-Sinnott
ΔT = 73 s

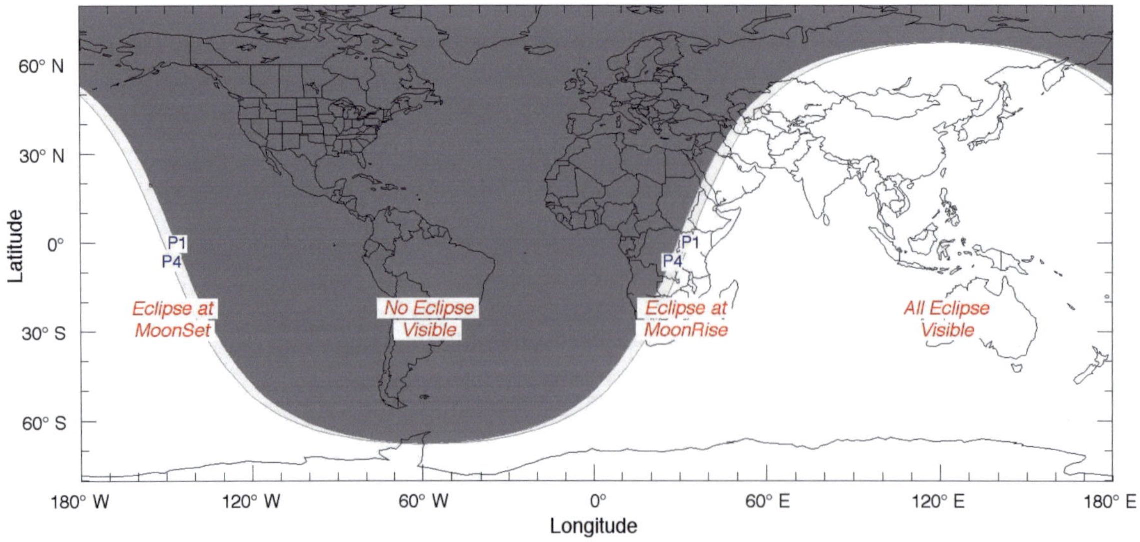

Penumbral Lunar Eclipse of 2027 Aug 17

Greatest Eclipse = 07:14:57.9 TD (= 07:13:45.1 UT1)

Penumbral Magnitude = 0.5476 Gamma = 1.2797 Saros Series = 148
Umbral Magnitude = -0.5234 Axis = 1.1544° Saros Member = 4 of 70

Sun at Greatest Eclipse
(Geocentric Coordinates)
R.A. = 09h45m58.6s
Dec. = +13°27'30.2"
S.D. = 00°15'47.8"
H.P. = 00°00'08.7"

Moon at Greatest Eclipse
(Geocentric Coordinates)
R.A. = 21h43m58.8s
Dec. = -12°24'40.9"
S.D. = 00°14'44.9"
H.P. = 00°54'07.8"

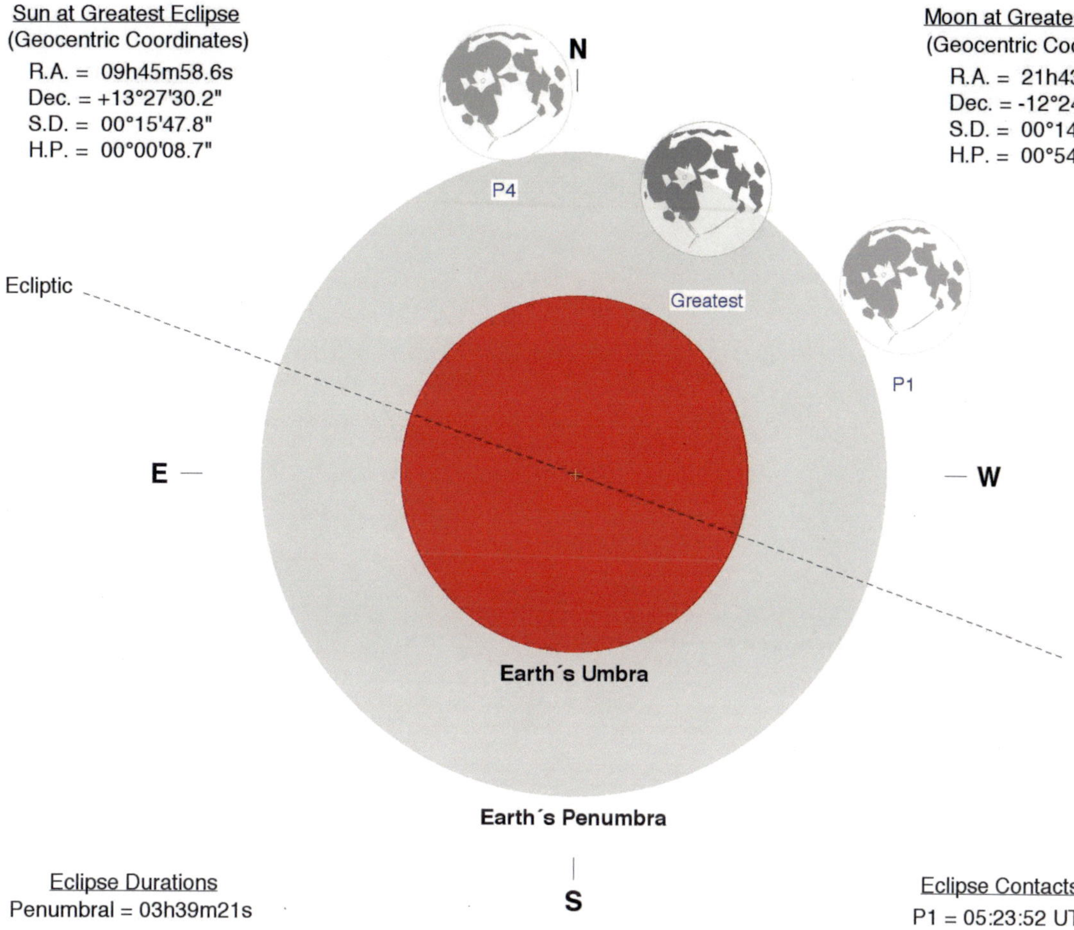

Eclipse Durations
Penumbral = 03h39m21s

Eclipse Contacts
P1 = 05:23:52 UT1
P4 = 09:03:14 UT1

Eph. = JPL DE430
Rule = Herald-Sinnott
ΔT = 73 s

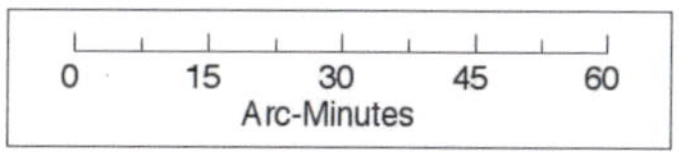

©2020 F. Espenak, www.EclipseWise.com

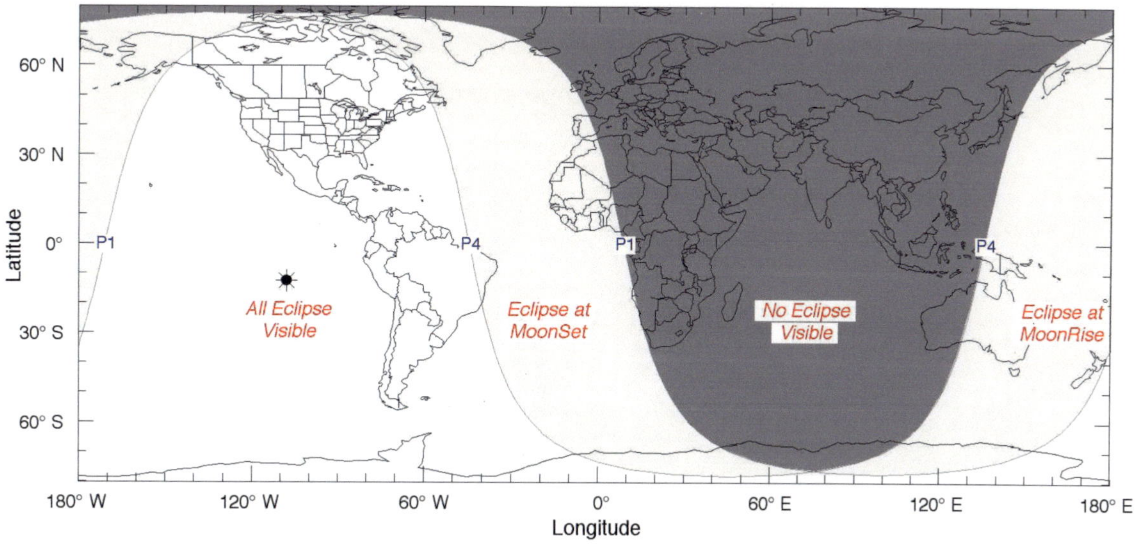

Partial Lunar Eclipse of 2028 Jan 12

Greatest Eclipse = 04:14:13.4 TD (= 04:13:00.5 UT1)

Penumbral Magnitude = 1.0485
Umbral Magnitude = 0.0679

Gamma = 0.9818
Axis = 0.9958°

Saros Series = 115
Saros Member = 58 of 72

Sun at Greatest Eclipse
(Geocentric Coordinates)
R.A. = 19h32m47.8s
Dec. = -21°43'29.4"
S.D. = 00°16'15.8"
H.P. = 00°00'08.9"

Moon at Greatest Eclipse
(Geocentric Coordinates)
R.A. = 07h33m53.0s
Dec. = +22°41'18.2"
S.D. = 00°16'35.1"
H.P. = 01°00'52.0"

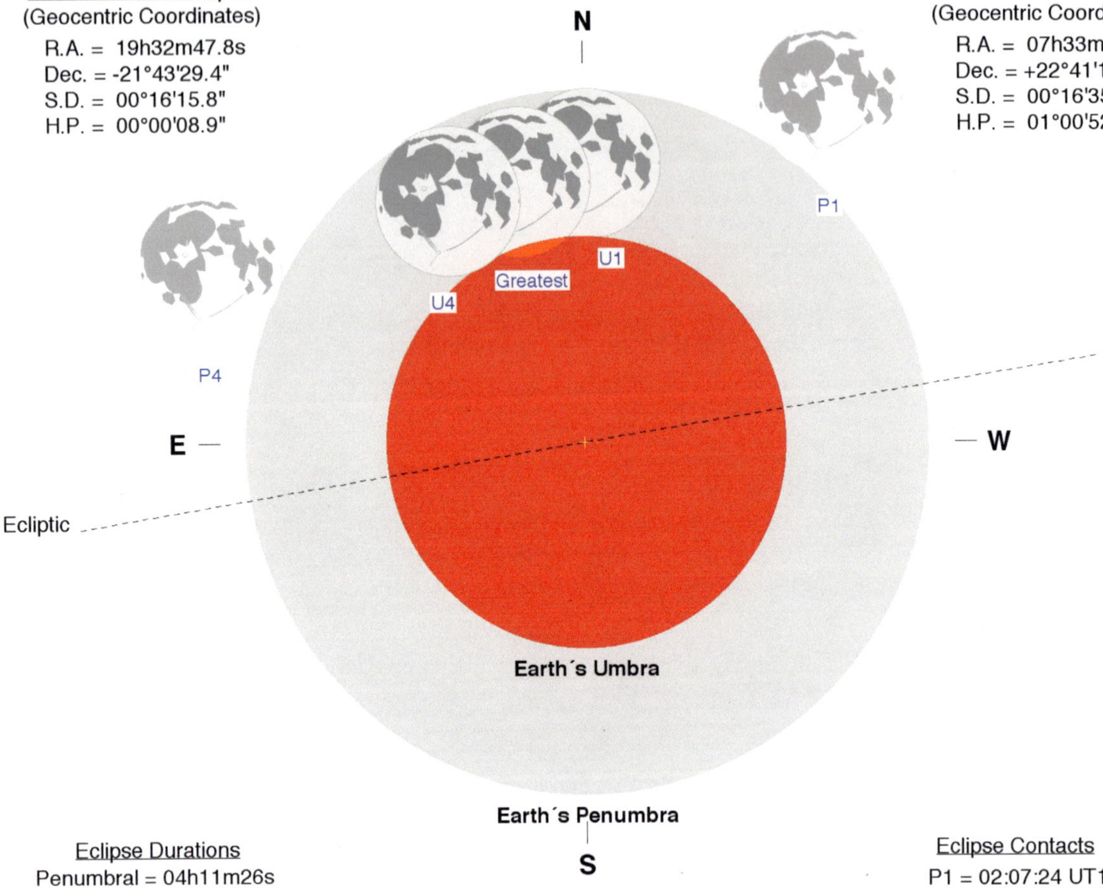

Eclipse Durations
Penumbral = 04h11m26s
Umbral = 00h56m53s

Eclipse Contacts
P1 = 02:07:24 UT1
U1 = 03:44:46 UT1
U4 = 04:41:39 UT1
P4 = 06:18:50 UT1

Eph. = JPL DE430
Rule = Herald-Sinnott
ΔT = 73 s

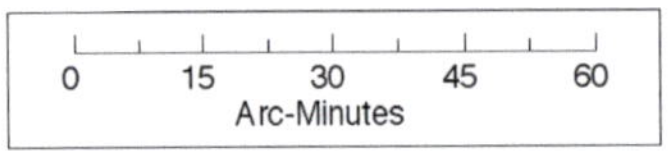

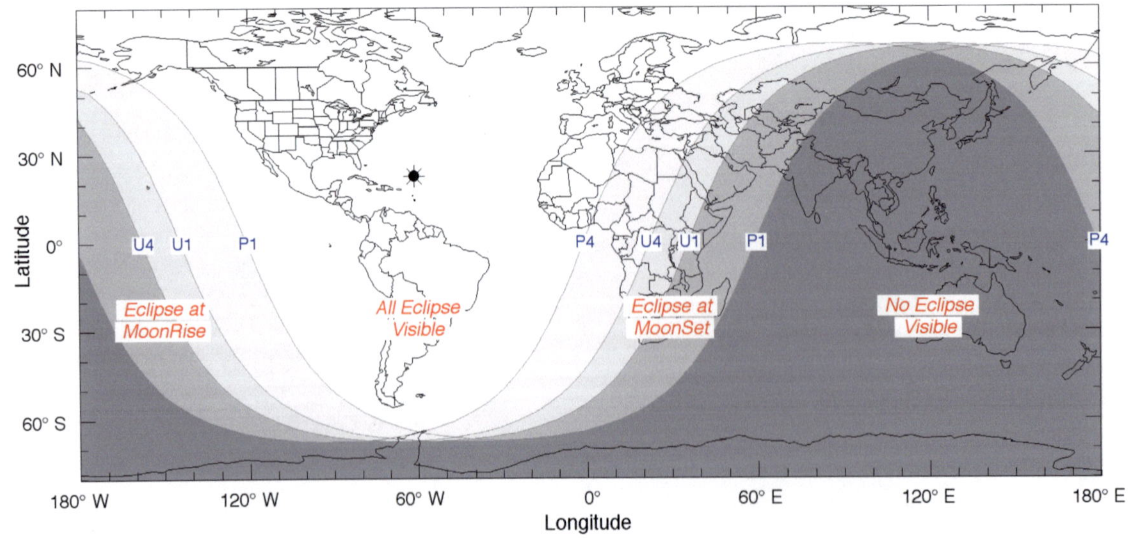

Partial Lunar Eclipse of 2028 Jul 06

Greatest Eclipse = 18:20:57.2 TD (= 18:19:44.0 UT1)

Penumbral Magnitude = 1.4282 Gamma = -0.7904 Saros Series = 120
Umbral Magnitude = 0.3908 Axis = 0.7331° Saros Member = 58 of 83

Sun at Greatest Eclipse
(Geocentric Coordinates)
R.A. = 07h05m56.7s
Dec. = +22°34'16.5"
S.D. = 00°15'43.9"
H.P. = 00°00'08.6"

Moon at Greatest Eclipse
(Geocentric Coordinates)
R.A. = 19h06m37.0s
Dec. = -23°17'16.4"
S.D. = 00°15'09.9"
H.P. = 00°55'39.4"

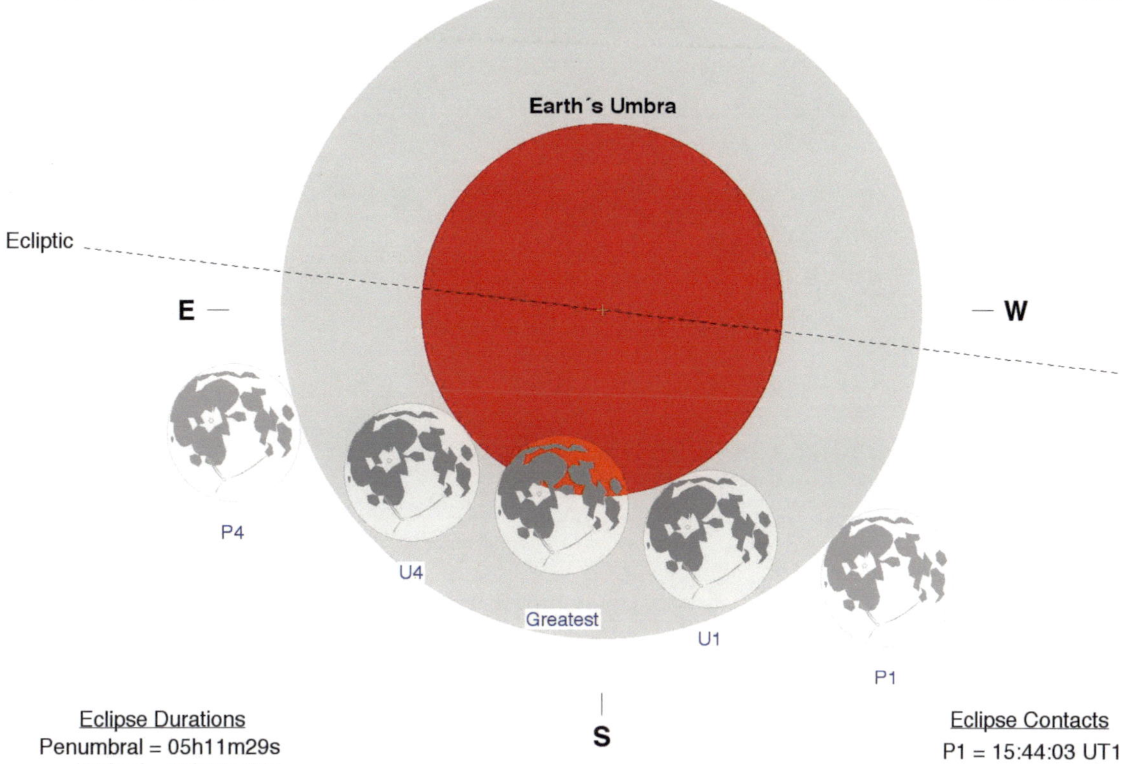

Eclipse Durations
Penumbral = 05h11m29s
Umbral = 02h22m13s

Eclipse Contacts
P1 = 15:44:03 UT1
U1 = 17:08:40 UT1
U4 = 19:30:53 UT1
P4 = 20:55:33 UT1

0 15 30 45 60
Arc-Minutes

Eph. = JPL DE430
Rule = Herald-Sinnott
ΔT = 73 s

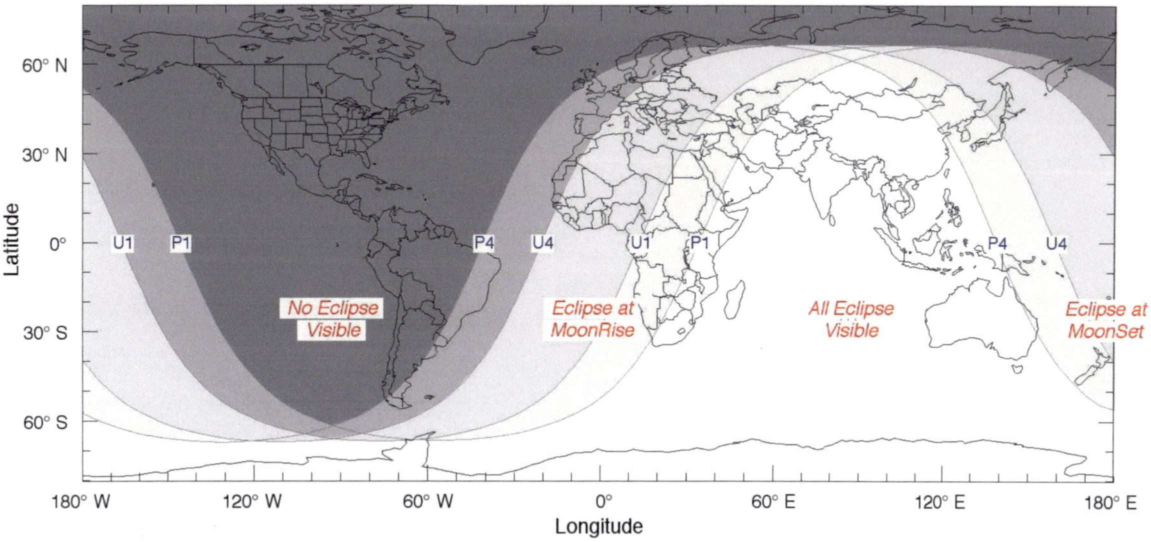

Total Lunar Eclipse of 2028 Dec 31

Greatest Eclipse = 16:53:15.1 TD (= 16:52:01.7 UT1)

Penumbral Magnitude = 2.2758   Gamma = 0.3258   Saros Series = 125
Umbral Magnitude = 1.2478   Axis = 0.3153°   Saros Member = 49 of 72

Sun at Greatest Eclipse
(Geocentric Coordinates)
R.A. = 18h45m53.7s
Dec. = -23°01'00.5"
S.D. = 00°16'15.9"
H.P. = 00°00'08.9"

Moon at Greatest Eclipse
(Geocentric Coordinates)
R.A. = 06h46m08.4s
Dec. = +23°19'37.5"
S.D. = 00°15'49.4"
H.P. = 00°58'04.3"

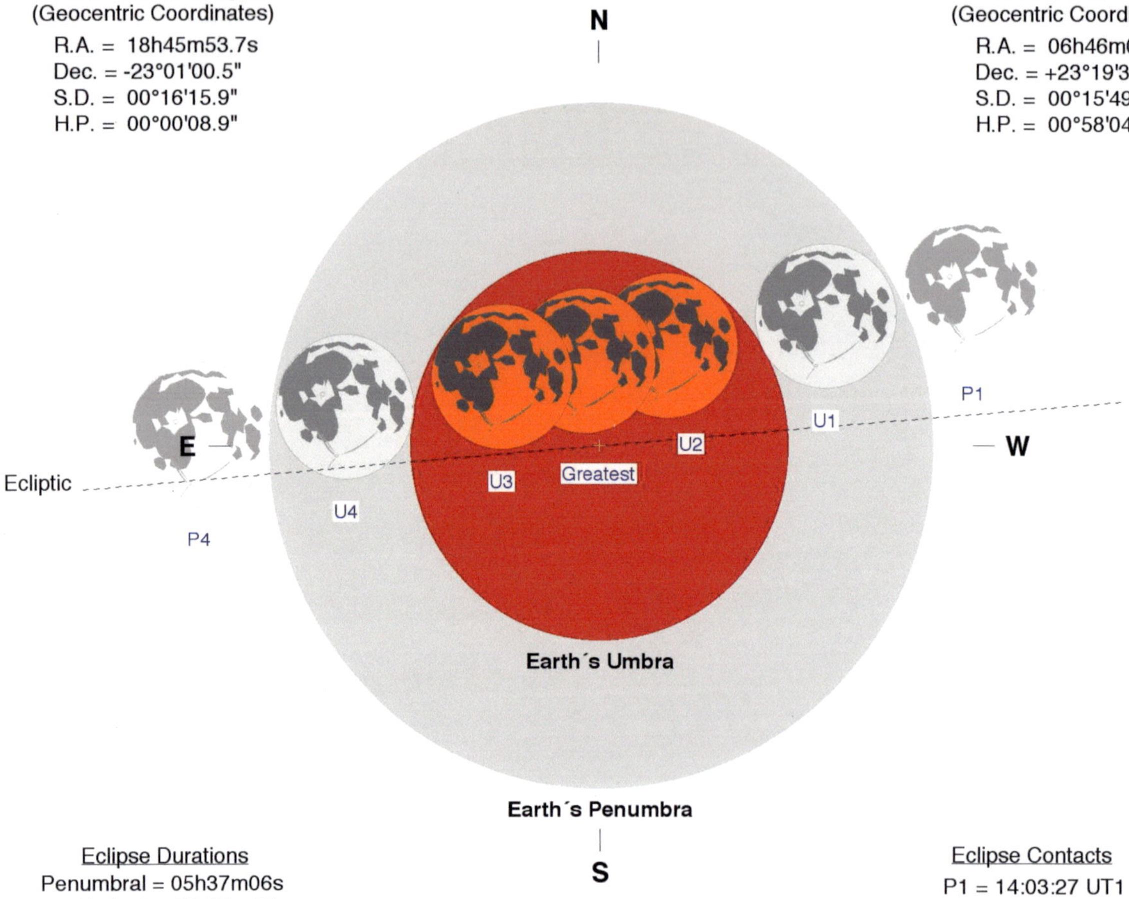

Eclipse Durations
Penumbral = 05h37m06s
Umbral = 03h29m37s
Total = 01h12m02s

Eph. = JPL DE430
Rule = Herald-Sinnott
ΔT = 73 s

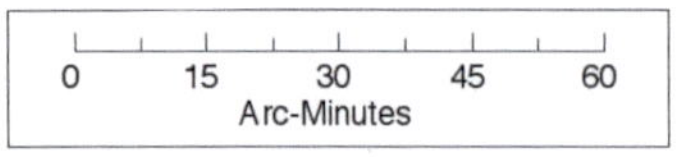

©2020 F. Espenak, www.EclipseWise.com

Eclipse Contacts
P1 = 14:03:27 UT1
U1 = 15:07:17 UT1
U2 = 16:16:07 UT1
U3 = 17:28:08 UT1
U4 = 18:36:54 UT1
P4 = 19:40:34 UT1

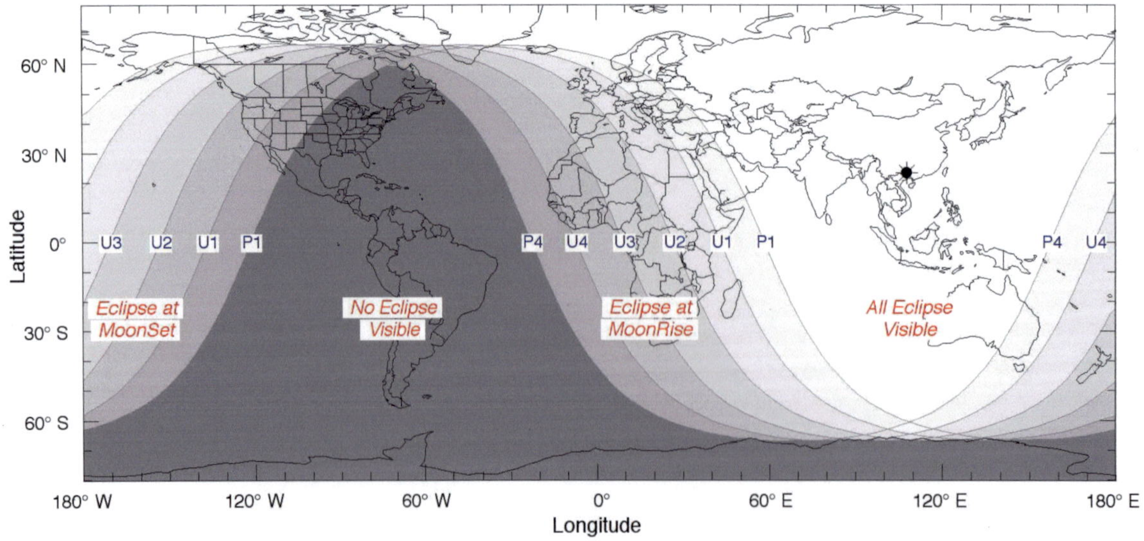

Total Lunar Eclipse of 2029 Jun 26

Greatest Eclipse = 03:23:22.5 TD (= 03:22:08.8 UT1)

Penumbral Magnitude = 2.8282 Gamma = 0.0124 Saros Series = 130
Umbral Magnitude = 1.8452 Axis = 0.0121° Saros Member = 35 of 71

Sun at Greatest Eclipse
(Geocentric Coordinates)
R.A. = 06h21m03.1s
Dec. = +23°20'50.2"
S.D. = 00°15'44.1"
H.P. = 00°00'08.7"

Moon at Greatest Eclipse
(Geocentric Coordinates)
R.A. = 18h21m02.6s
Dec. = -23°20'06.9"
S.D. = 00°16'00.4"
H.P. = 00°58'44.7"

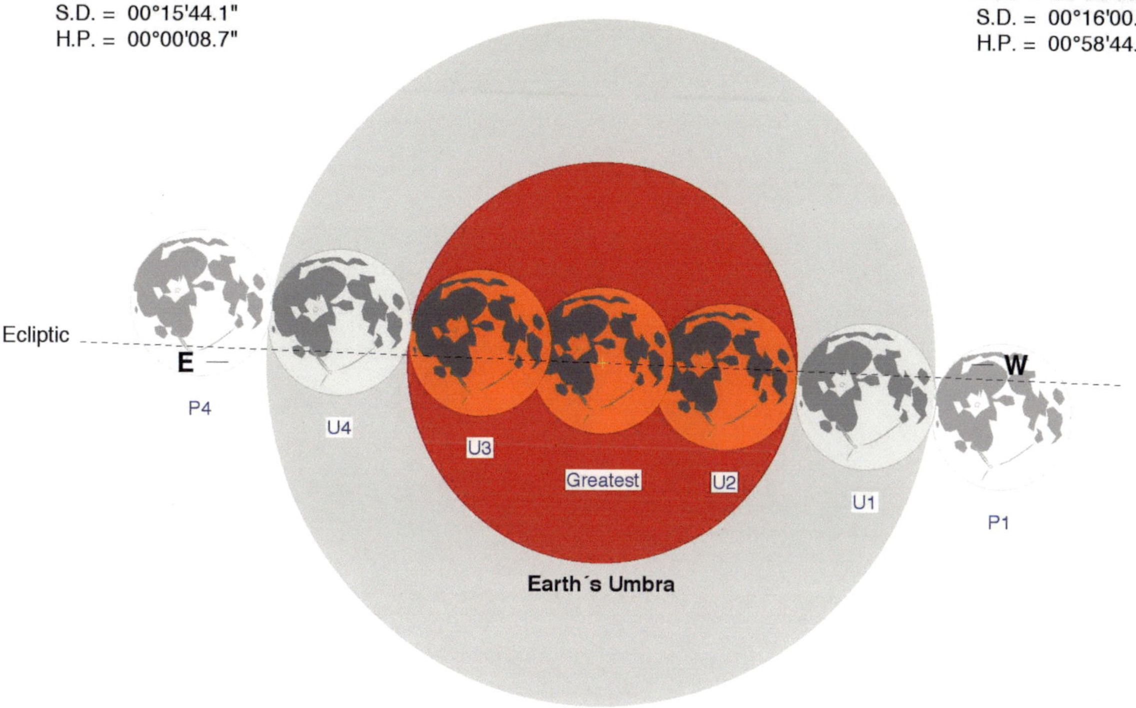

Eclipse Durations
Penumbral = 05h36m01s
Umbral = 03h40m20s
Total = 01h42m40s

Eph. = JPL DE430
Rule = Herald-Sinnott
ΔT = 74 s

0 15 30 45 60
Arc-Minutes

©2020 F. Espenak, www.EclipseWise.com

Eclipse Contacts
P1 = 00:34:11 UT1
U1 = 01:31:58 UT1
U2 = 02:30:48 UT1
U3 = 04:13:28 UT1
U4 = 05:12:18 UT1
P4 = 06:10:12 UT1

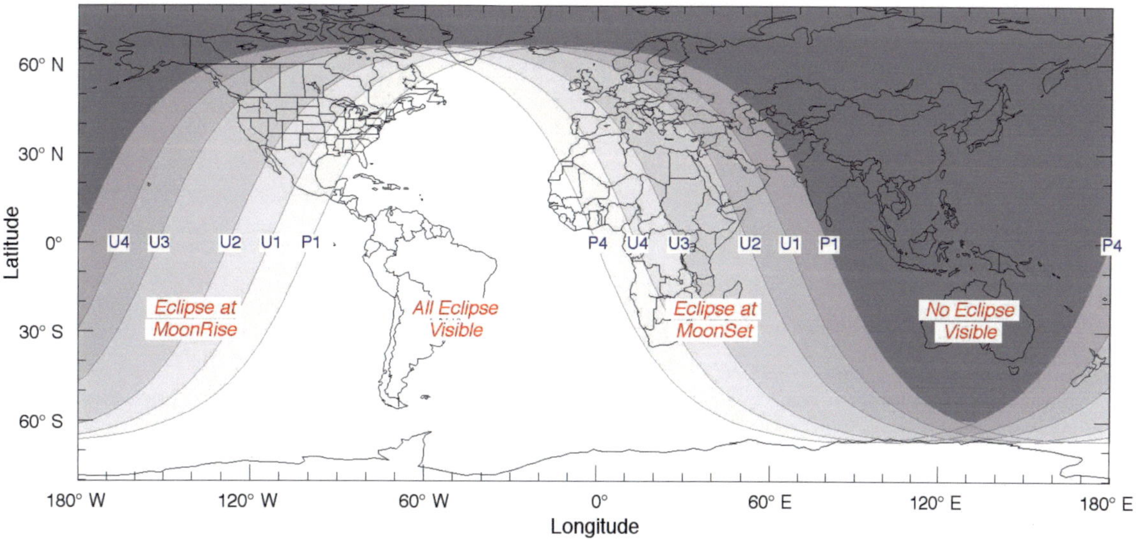

Total Lunar Eclipse of 2029 Dec 20

Greatest Eclipse = 22:43:11.8 TD (= 22:41:57.9 UT1)

Penumbral Magnitude = 2.2023 Gamma = -0.3811 Saros Series = 135
Umbral Magnitude = 1.1190 Axis = 0.3498° Saros Member = 24 of 71

Sun at Greatest Eclipse
(Geocentric Coordinates)
R.A. = 17h57m07.6s
Dec. = -23°26'00.2"
S.D. = 00°16'15.5"
H.P. = 00°00'08.9"

Moon at Greatest Eclipse
(Geocentric Coordinates)
R.A. = 05h56m59.0s
Dec. = +23°05'06.7"
S.D. = 00°15'00.4"
H.P. = 00°55'04.6"

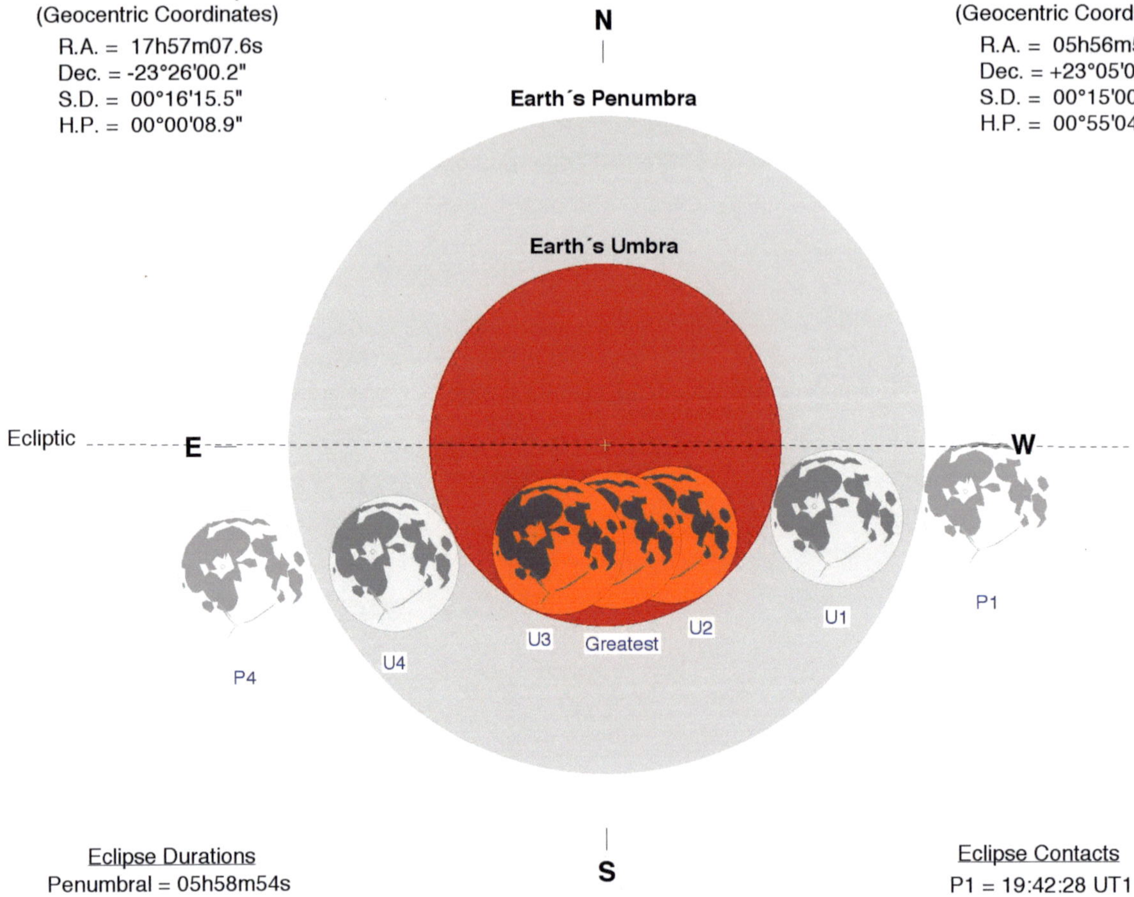

Eclipse Durations
Penumbral = 05h58m54s
Umbral = 03h34m07s
Total = 00h54m23s

Eph. = JPL DE430
Rule = Herald-Sinnott
ΔT = 74 s

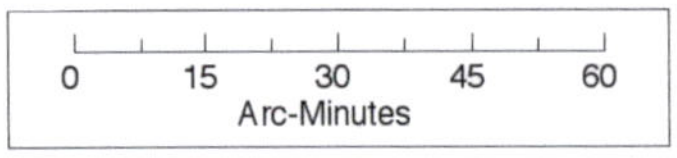

©2020 F. Espenak, www.EclipseWise.com

Eclipse Contacts
P1 = 19:42:28 UT1
U1 = 20:54:54 UT1
U2 = 22:14:44 UT1
U3 = 23:09:07 UT1
U4 = 00:29:01 UT1
P4 = 01:41:22 UT1

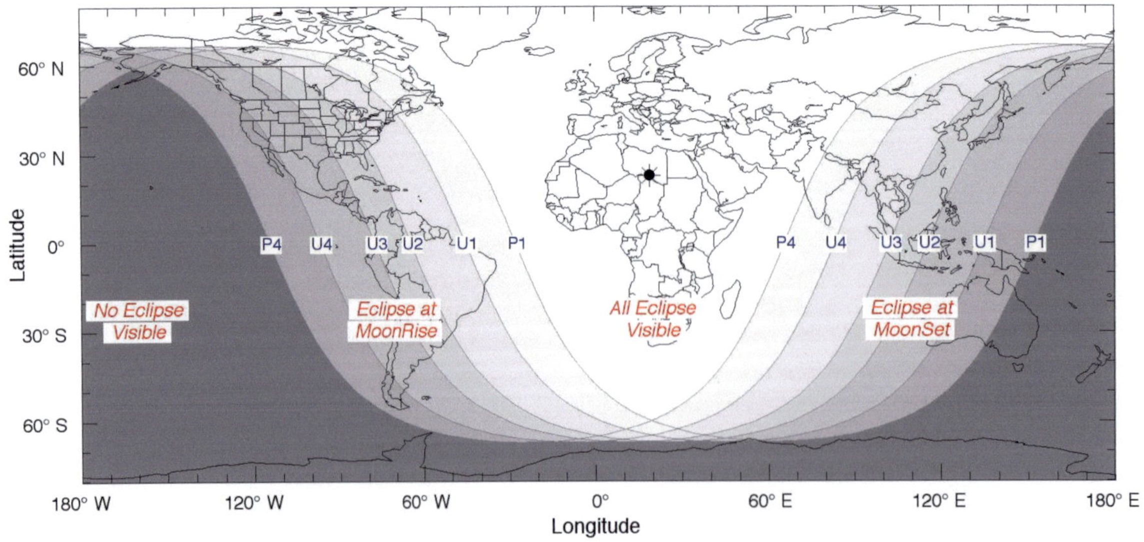

Partial Lunar Eclipse of 2030 Jun 15

Greatest Eclipse = 18:34:33.5 TD (= 18:33:19.4 UT1)

Penumbral Magnitude = 1.4495
Umbral Magnitude = 0.5040

Gamma = 0.7535
Axis = 0.7674°

Saros Series = 140
Saros Member = 25 of 77

Sun at Greatest Eclipse
(Geocentric Coordinates)
R.A. = 05h36m57.6s
Dec. = +23°19'44.0"
S.D. = 00°15'44.7"
H.P. = 00°00'08.7"

Moon at Greatest Eclipse
(Geocentric Coordinates)
R.A. = 17h36m46.1s
Dec. = -22°33'45.8"
S.D. = 00°16'39.2"
H.P. = 01°01'07.1"

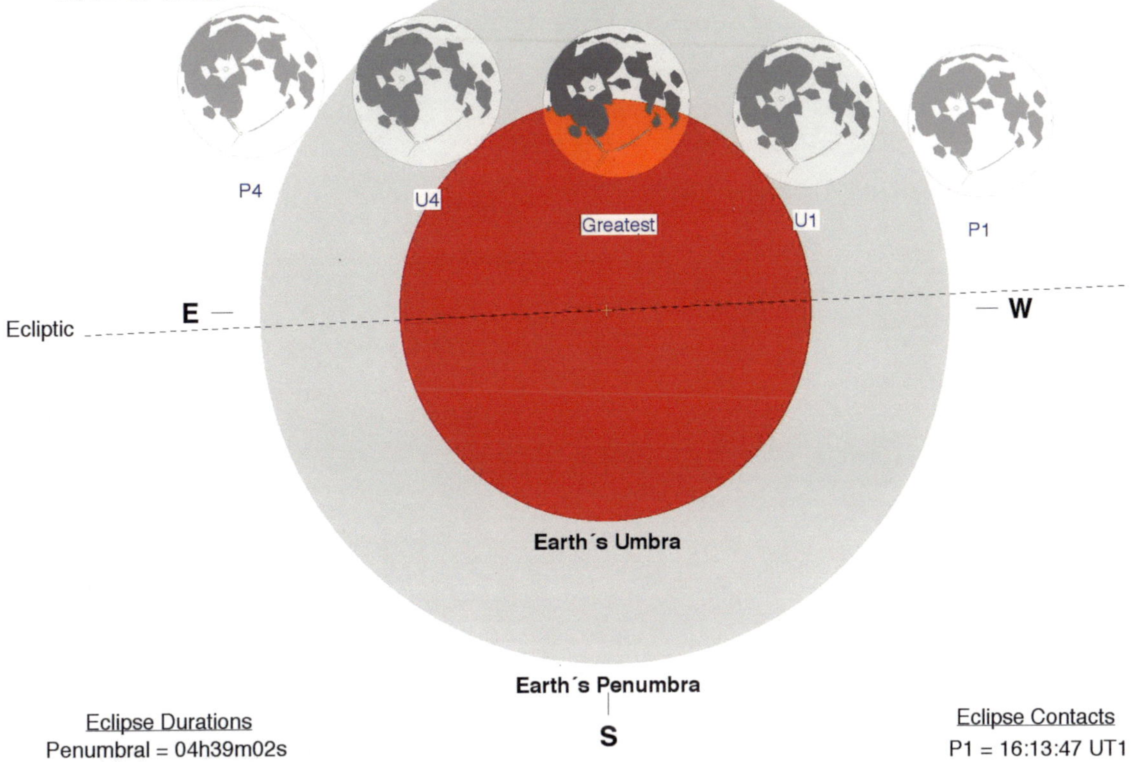

Eclipse Durations
Penumbral = 04h39m02s
Umbral = 02h25m02s

Eclipse Contacts
P1 = 16:13:47 UT1
U1 = 17:20:46 UT1
U4 = 19:45:47 UT1
P4 = 20:52:49 UT1

Eph. = JPL DE430
Rule = Herald-Sinnott
ΔT = 74 s

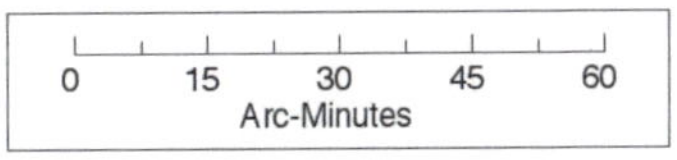

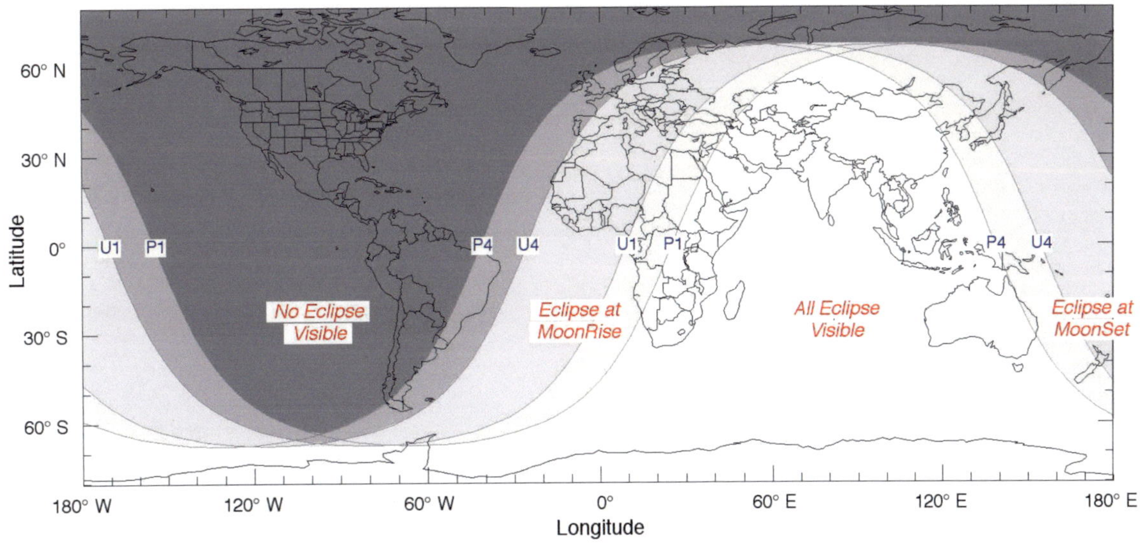

Penumbral Lunar Eclipse of 2030 Dec 09

Greatest Eclipse = 22:28:51.1 TD (= 22:27:36.7 UT1)

Penumbral Magnitude = 0.9430
Umbral Magnitude = -0.1613

Gamma = -1.0732
Axis = 0.9652°

Saros Series = 145
Saros Member = 12 of 71

<u>Sun at Greatest Eclipse</u>
(Geocentric Coordinates)
R.A. = 17h07m21.3s
Dec. = -22°52'57.8"
S.D. = 00°16'14.4"
H.P. = 00°00'08.9"

<u>Moon at Greatest Eclipse</u>
(Geocentric Coordinates)
R.A. = 05h07m19.1s
Dec. = +21°55'03.1"
S.D. = 00°14'42.3"
H.P. = 00°53'58.2"

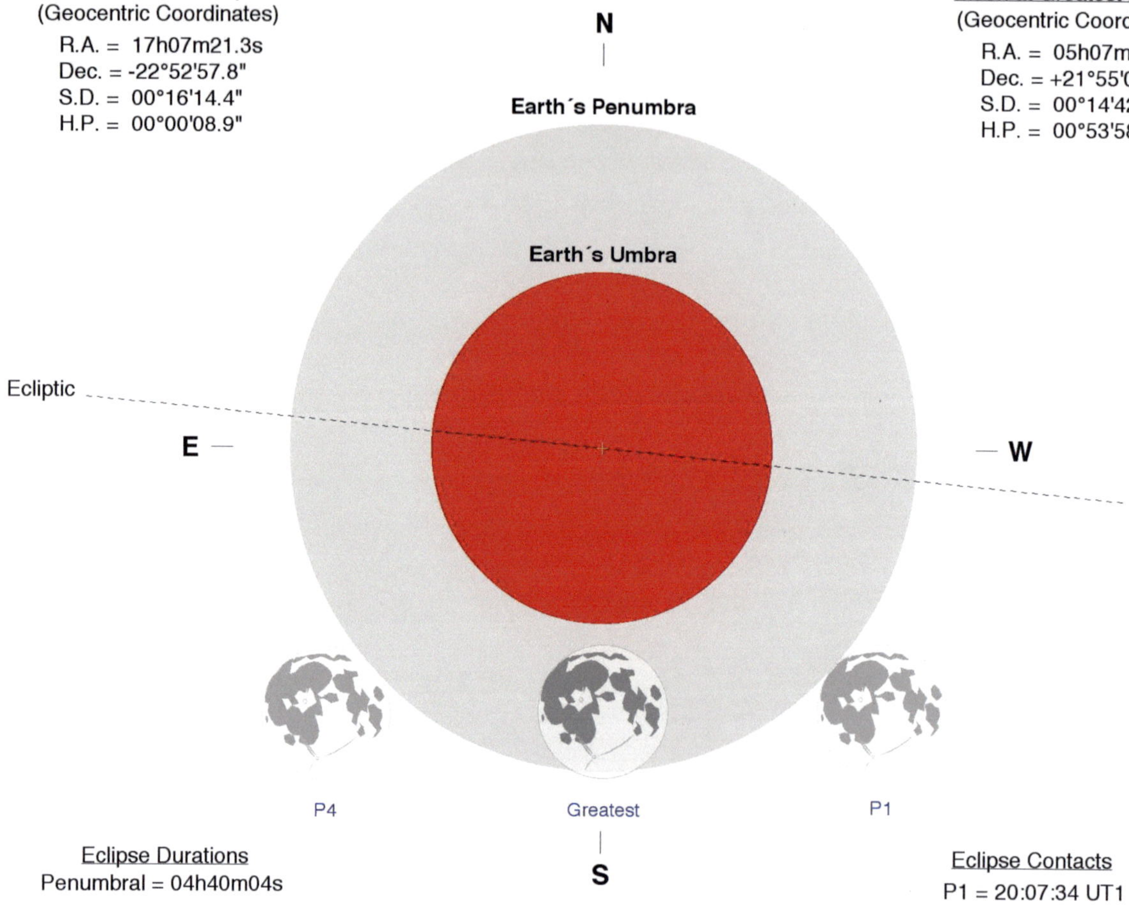

<u>Eclipse Durations</u>
Penumbral = 04h40m04s

<u>Eclipse Contacts</u>
P1 = 20:07:34 UT1
P4 = 00:47:38 UT1

Eph. = JPL DE430
Rule = Herald-Sinnott
ΔT = 74 s

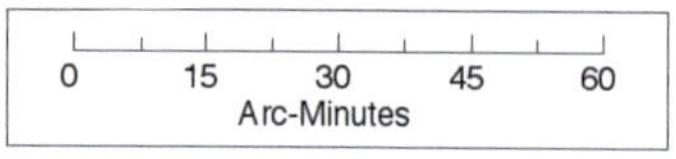

©2020 F. Espenak, www.EclipseWise.com

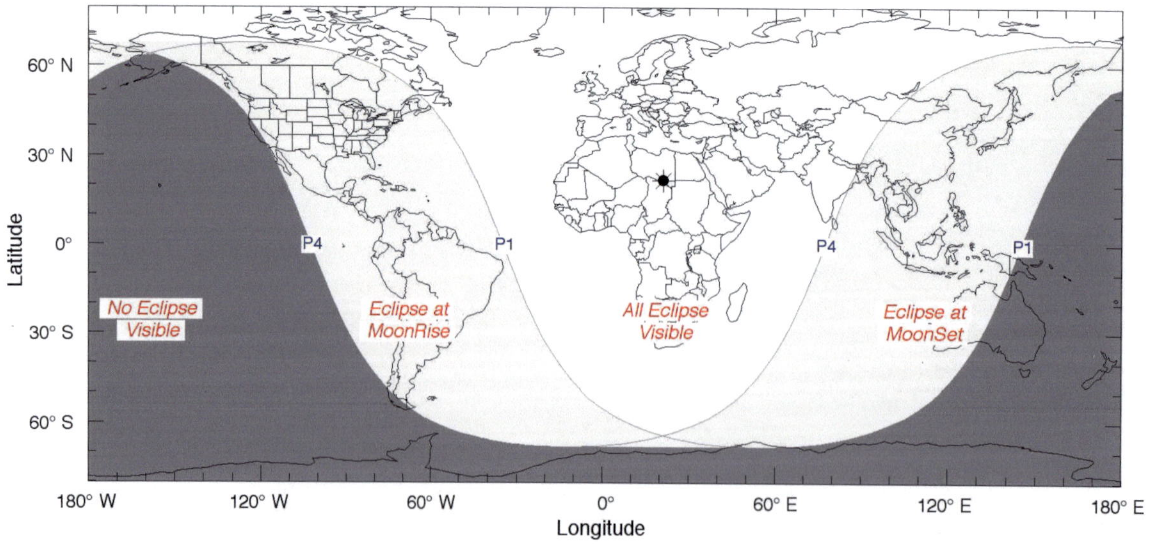

Penumbral Lunar Eclipse of 2031 May 07

Greatest Eclipse = 03:52:01.3 TD (= 03:50:46.7 UT1)

Penumbral Magnitude = 0.8827   Gamma = -1.0695   Saros Series = 112
Umbral Magnitude = -0.0892   Axis = 1.0670°   Saros Member = 66 of 72

Sun at Greatest Eclipse
(Geocentric Coordinates)
R.A. = 02h55m49.7s
Dec. = +16°44'40.2"
S.D. = 00°15'51.2"
H.P. = 00°00'08.7"

Moon at Greatest Eclipse
(Geocentric Coordinates)
R.A. = 14h54m58.0s
Dec. = -17°47'29.4"
S.D. = 00°16'18.7"
H.P. = 00°59'52.0"

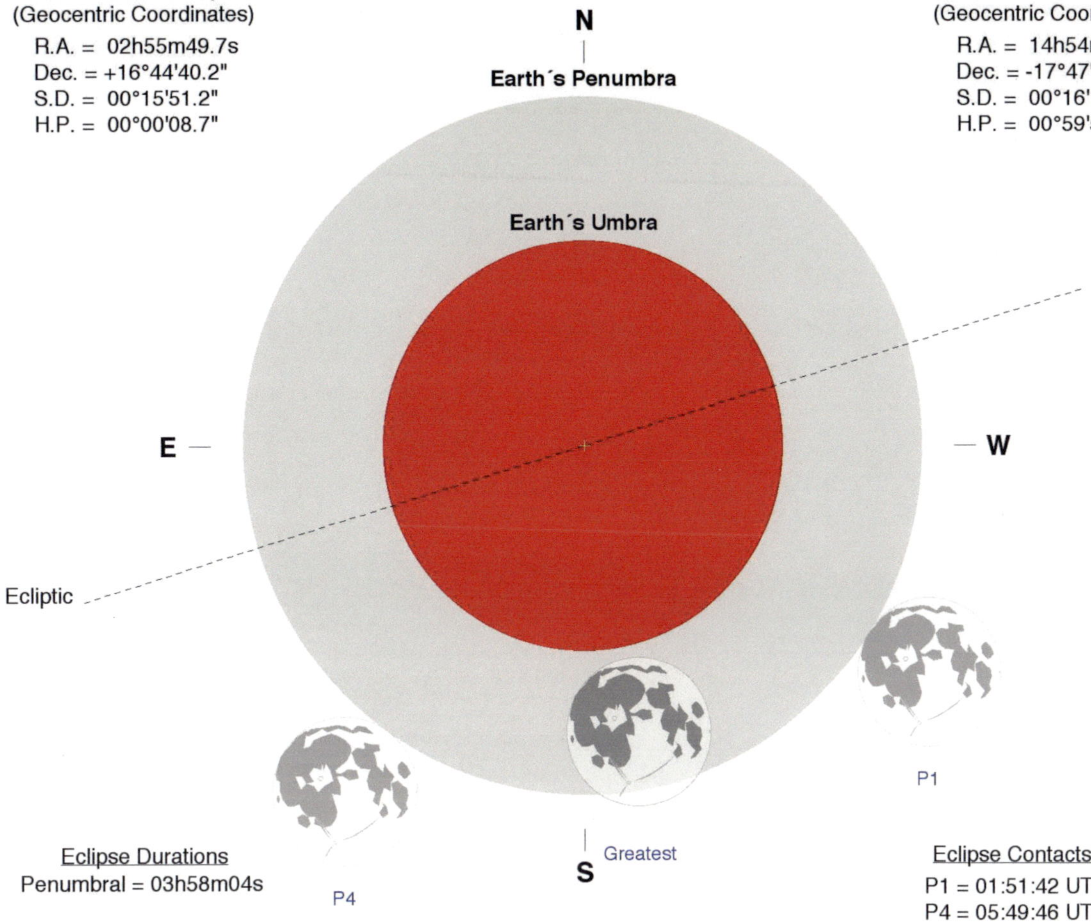

Eclipse Durations
Penumbral = 03h58m04s

Eclipse Contacts
P1 = 01:51:42 UT1
P4 = 05:49:46 UT1

Eph. = JPL DE430
Rule = Herald-Sinnott
ΔT = 75 s

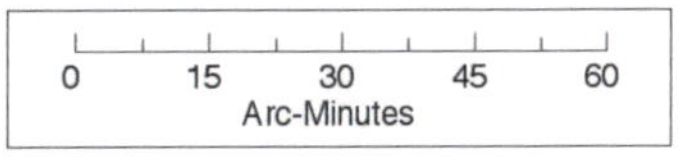

©2020 F. Espenak, www.EclipseWise.com

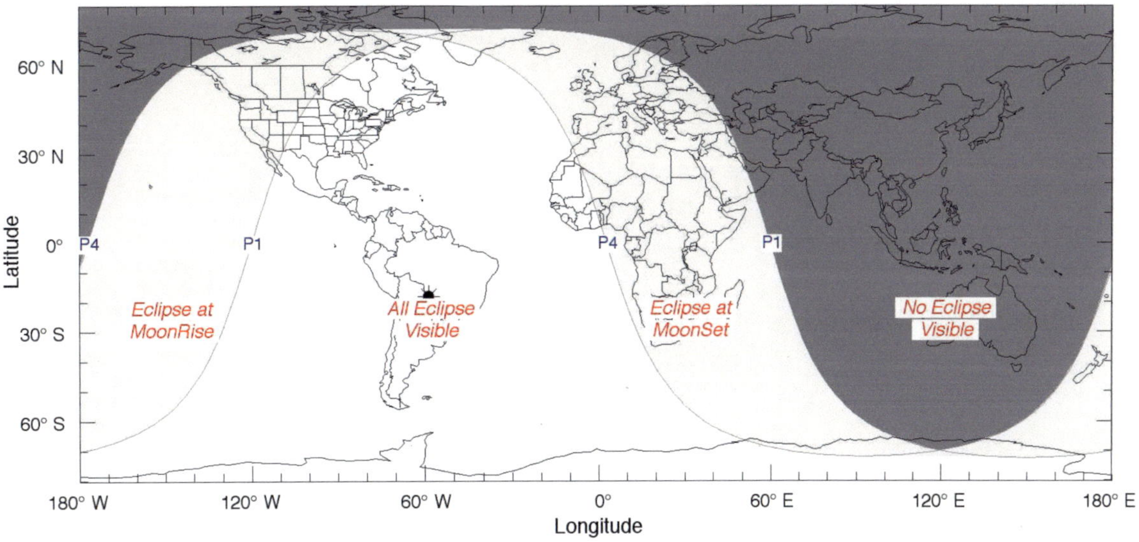

Penumbral Lunar Eclipse of 2031 Jun 05

Greatest Eclipse = 11:45:17.4 TD (= 11:44:02.8 UT1)

Penumbral Magnitude = 0.1306
Umbral Magnitude = -0.8185

Gamma = 1.4732
Axis = 1.4967°

Saros Series = 150
Saros Member = 2 of 71

Sun at Greatest Eclipse
(Geocentric Coordinates)
R.A. = 04h53m21.6s
Dec. = +22°33'01.5"
S.D. = 00°15'45.9"
H.P. = 00°00'08.7"

Moon at Greatest Eclipse
(Geocentric Coordinates)
R.A. = 16h53m29.4s
Dec. = -21°03'14.0"
S.D. = 00°16'36.6"
H.P. = 01°00'57.7"

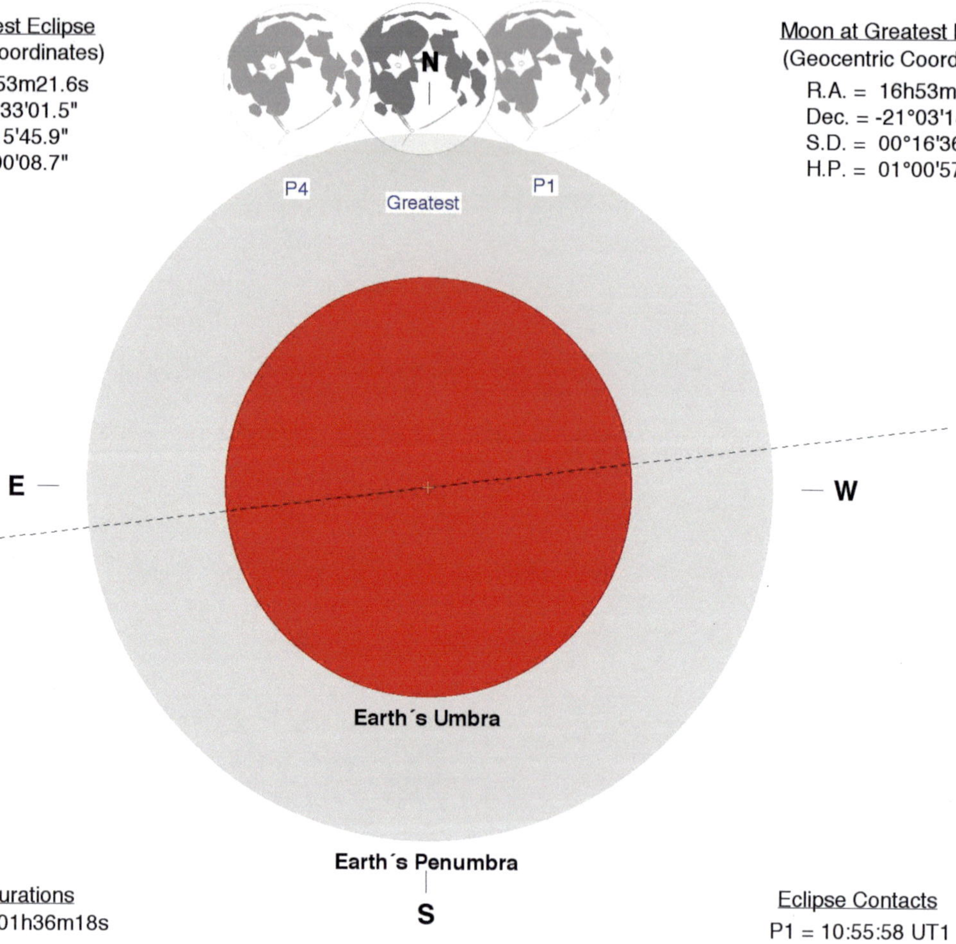

Ecliptic

Eclipse Durations
Penumbral = 01h36m18s

Eclipse Contacts
P1 = 10:55:58 UT1
P4 = 12:32:17 UT1

Eph. = JPL DE430
Rule = Herald-Sinnott
ΔT = 75 s

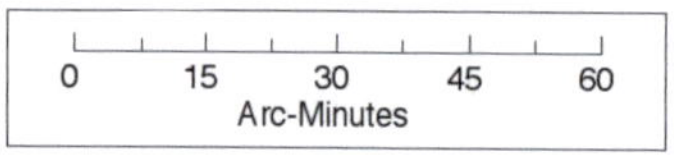

©2020 F. Espenak, www.EclipseWise.com

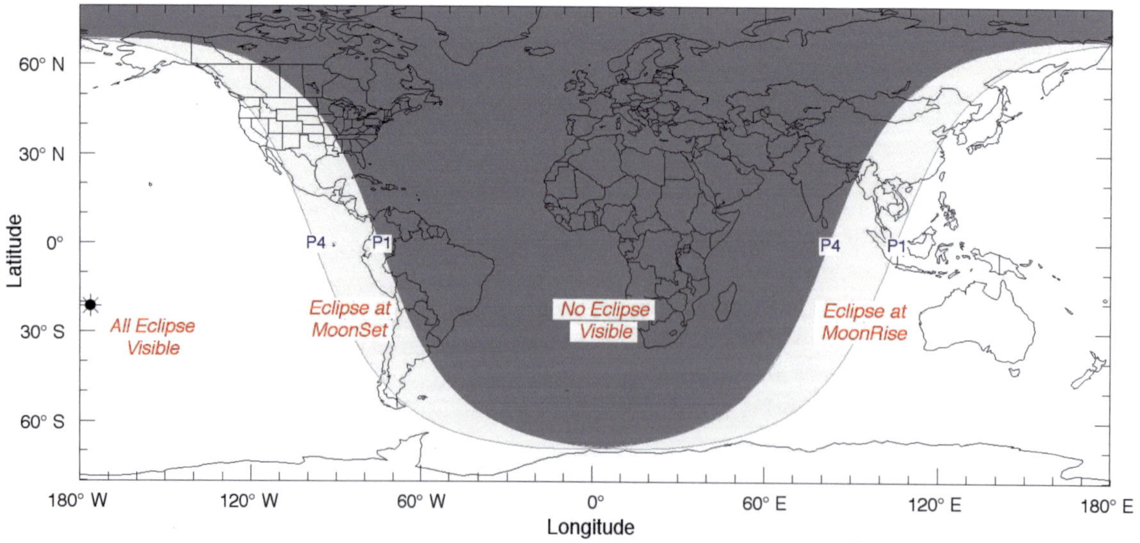

Penumbral Lunar Eclipse of 2031 Oct 30

Greatest Eclipse = 07:46:44.2 TD (= 07:45:29.4 UT1)

Penumbral Magnitude = 0.7173
Umbral Magnitude = -0.3193

Gamma = 1.1774
Axis = 1.1188°

Saros Series = 117
Saros Member = 53 of 71

Sun at Greatest Eclipse
(Geocentric Coordinates)
R.A. = 14h17m25.0s
Dec. = -13°44'38.7"
S.D. = 00°16'06.3"
H.P. = 00°00'08.9"

Moon at Greatest Eclipse
(Geocentric Coordinates)
R.A. = 02h16m19.7s
Dec. = +14°49'53.3"
S.D. = 00°15'32.2"
H.P. = 00°57'01.3"

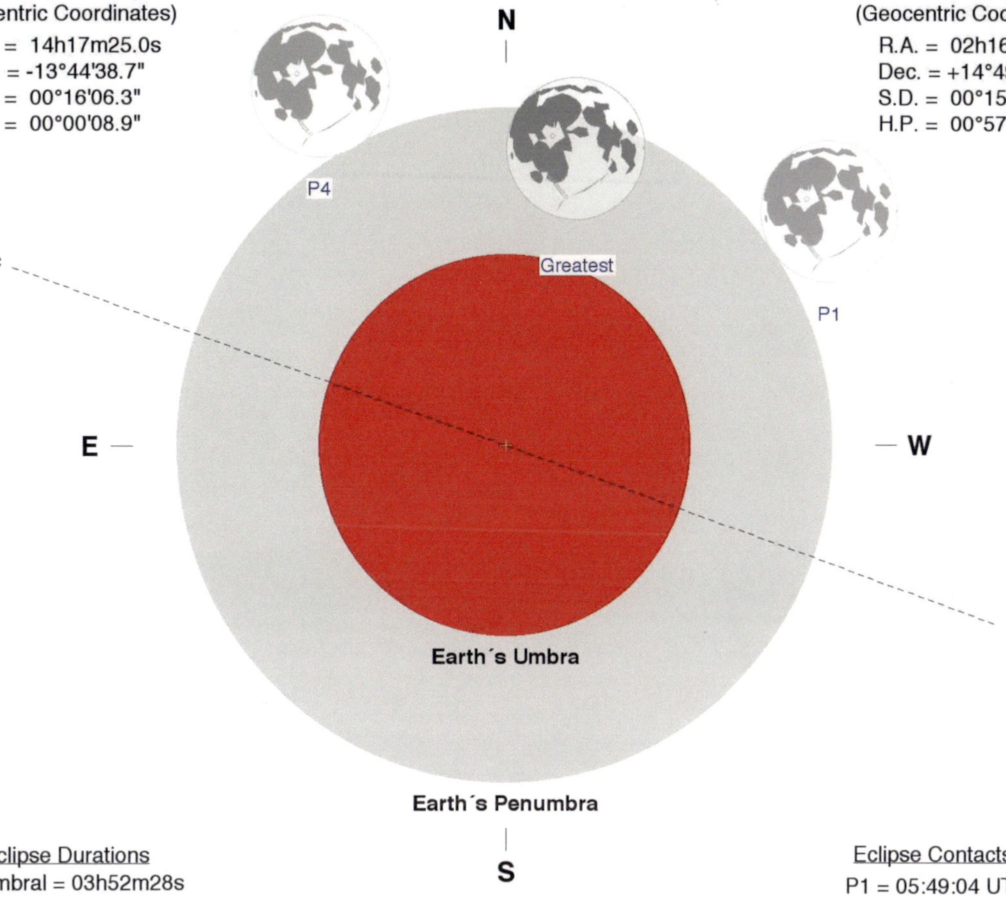

Eclipse Durations
Penumbral = 03h52m28s

Eclipse Contacts
P1 = 05:49:04 UT1
P4 = 09:41:32 UT1

Eph. = JPL DE430
Rule = Herald-Sinnott
ΔT = 75 s

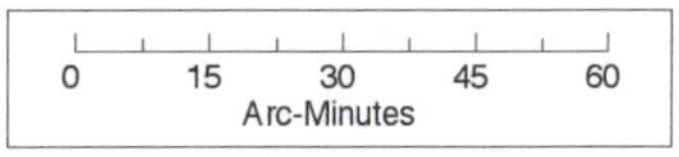

©2020 F. Espenak, www.EclipseWise.com

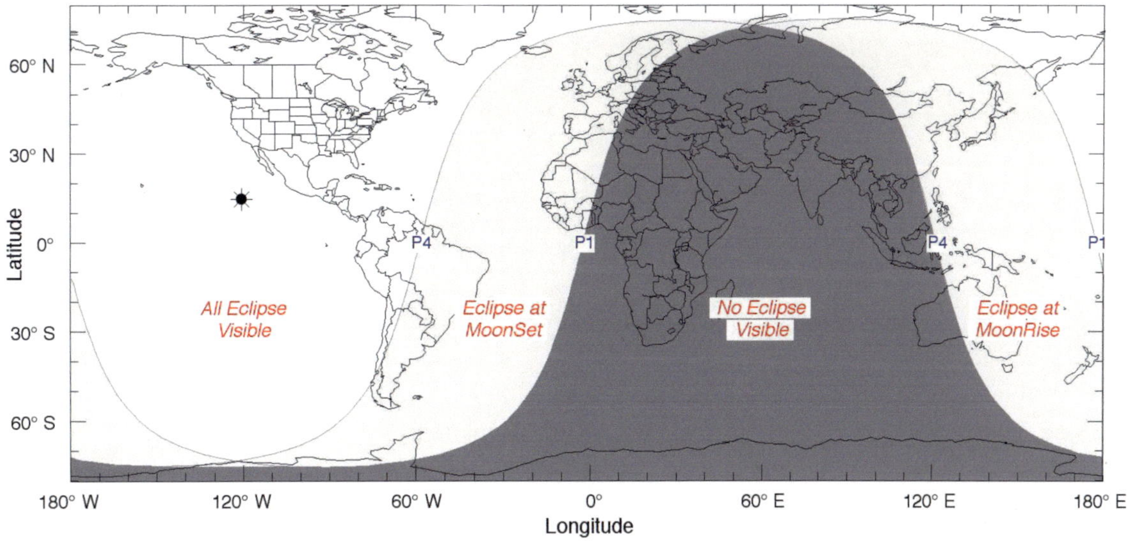

Total Lunar Eclipse of 2032 Apr 25

Greatest Eclipse = 15:14:50.6 TD (= 15:13:35.5 UT1)

Penumbral Magnitude = 2.2204 Gamma = -0.3558 Saros Series = 122
Umbral Magnitude = 1.1925 Axis = 0.3365° Saros Member = 57 of 74

Sun at Greatest Eclipse
(Geocentric Coordinates)
R.A. = 02h14m38.2s
Dec. = +13°30'28.8"
S.D. = 00°15'53.8"
H.P. = 00°00'08.7"

Moon at Greatest Eclipse
(Geocentric Coordinates)
R.A. = 14h14m18.6s
Dec. = -13°50'06.1"
S.D. = 00°15'27.9"
H.P. = 00°56'45.4"

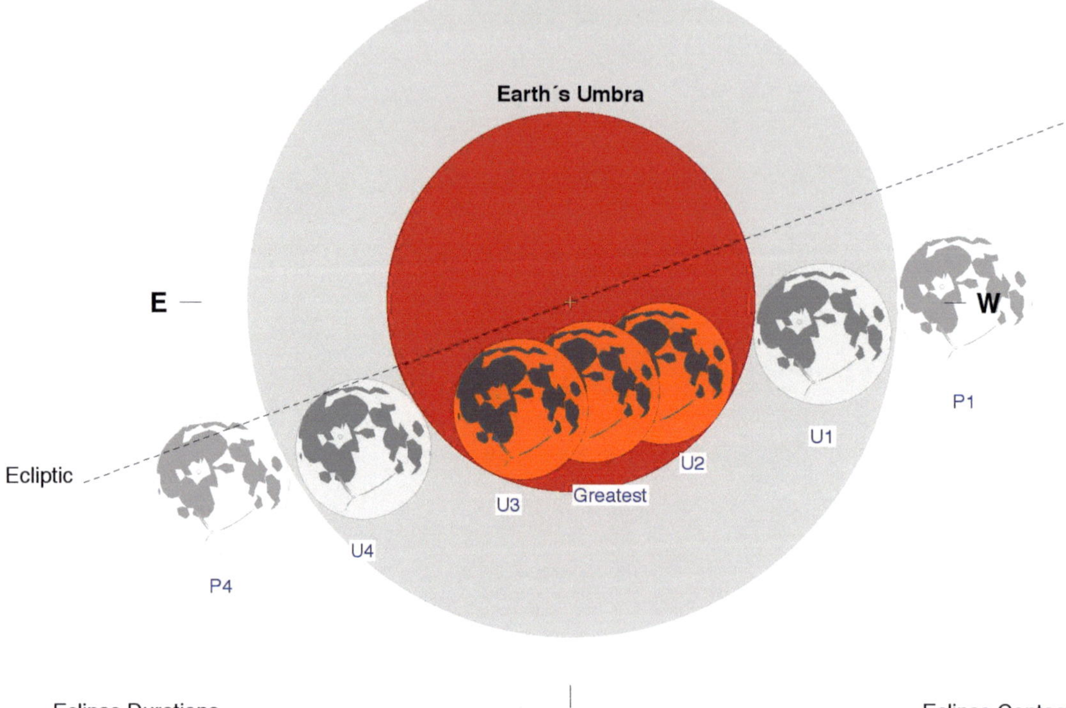

Eclipse Durations
Penumbral = 05h43m18s
Umbral = 03h31m58s
Total = 01h06m10s

Eph. = JPL DE430
Rule = Herald-Sinnott
ΔT = 75 s

0 15 30 45 60
Arc-Minutes

©2020 F. Espenak, www.EclipseWise.com

Eclipse Contacts
P1 = 12:21:52 UT1
U1 = 13:27:35 UT1
U2 = 14:40:24 UT1
U3 = 15:46:34 UT1
U4 = 16:59:33 UT1
P4 = 18:05:10 UT1

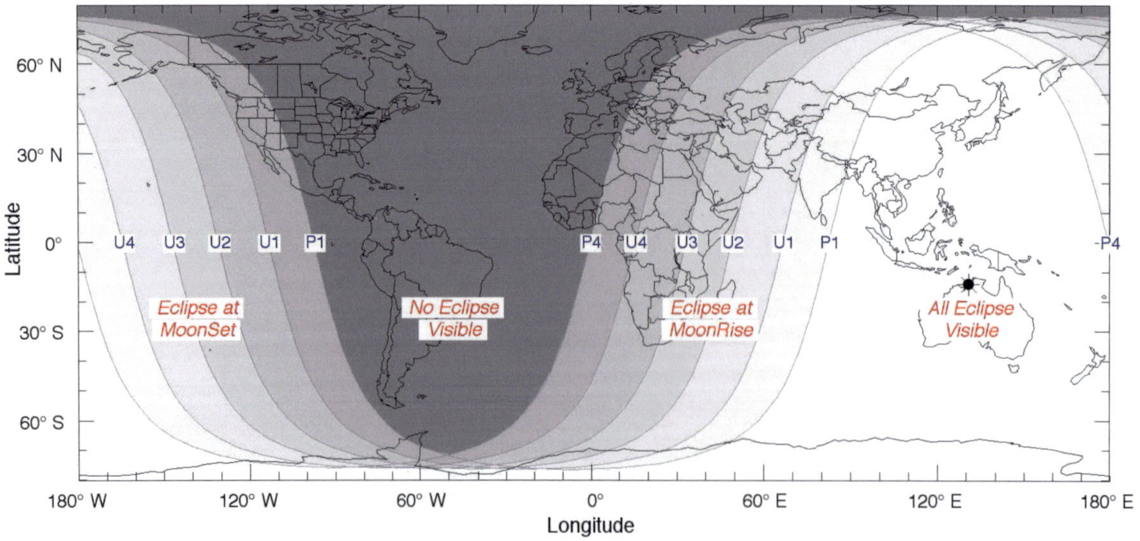

Total Lunar Eclipse of 2032 Oct 18

Greatest Eclipse = 19:03:40.2 TD (= 19:02:24.9 UT1)

Penumbral Magnitude = 2.0841	Gamma = 0.4169	Saros Series = 127
Umbral Magnitude = 1.1039	Axis = 0.4177°	Saros Member = 43 of 72

Sun at Greatest Eclipse
(Geocentric Coordinates)
R.A. = 13h36m15.4s
Dec. = -10°01'20.8"
S.D. = 00°16'03.4"
H.P. = 00°00'08.8"

Moon at Greatest Eclipse
(Geocentric Coordinates)
R.A. = 01h35m47.9s
Dec. = +10°25'28.7"
S.D. = 00°16'22.8"
H.P. = 01°00'07.0"

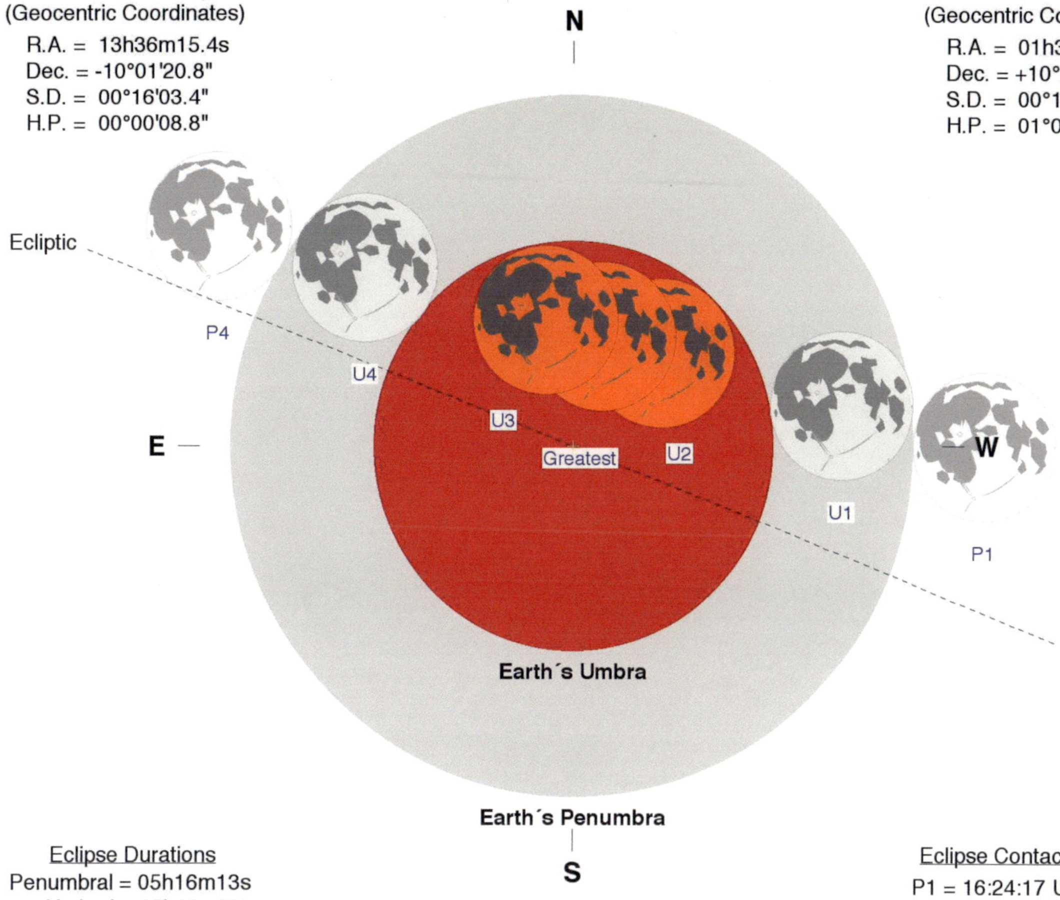

Eclipse Durations
Penumbral = 05h16m13s
Umbral = 03h16m38s
Total = 00h47m40s

Eph. = JPL DE430
Rule = Herald-Sinnott
ΔT = 75 s

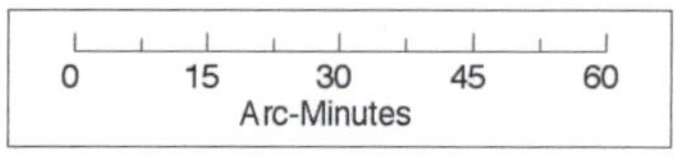

Eclipse Contacts
P1 = 16:24:17 UT1
U1 = 17:24:00 UT1
U2 = 18:38:24 UT1
U3 = 19:26:04 UT1
U4 = 20:40:38 UT1
P4 = 21:40:30 UT1

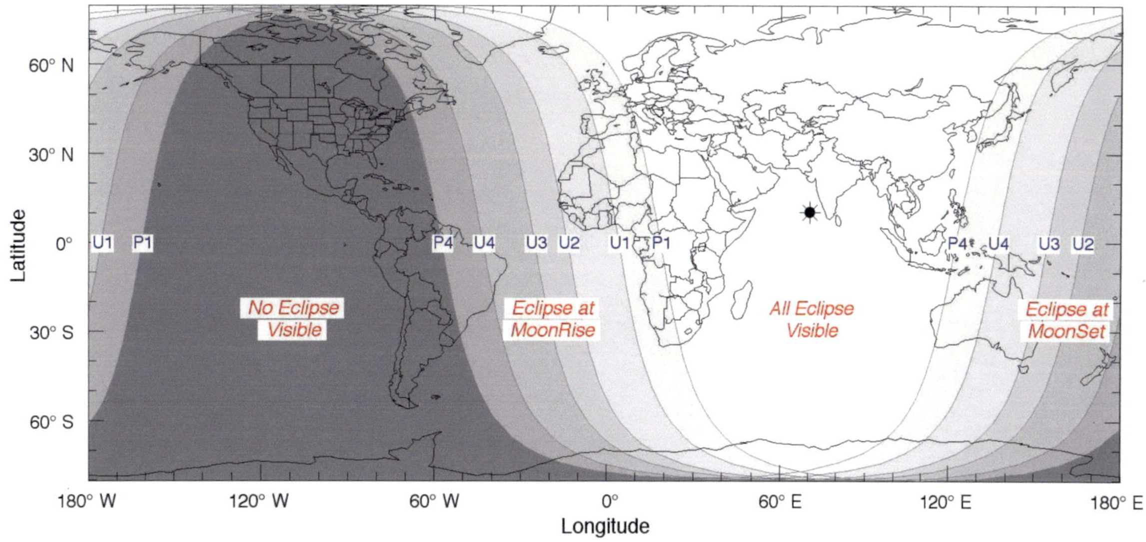

Total Lunar Eclipse of 2033 Apr 14

Greatest Eclipse = 19:13:51.4 TD (= 19:12:35.9 UT1)

Penumbral Magnitude = 2.1722
Umbral Magnitude = 1.0955

Gamma = 0.3954
Axis = 0.3581°

Saros Series = 132
Saros Member = 31 of 71

Sun at Greatest Eclipse
(Geocentric Coordinates)
R.A. = 01h33m13.7s
Dec. = +09°43'50.2"
S.D. = 00°15'56.7"
H.P. = 00°00'08.8"

Moon at Greatest Eclipse
(Geocentric Coordinates)
R.A. = 13h33m37.2s
Dec. = -09°23'08.7"
S.D. = 00°14'48.5"
H.P. = 00°54'21.0"

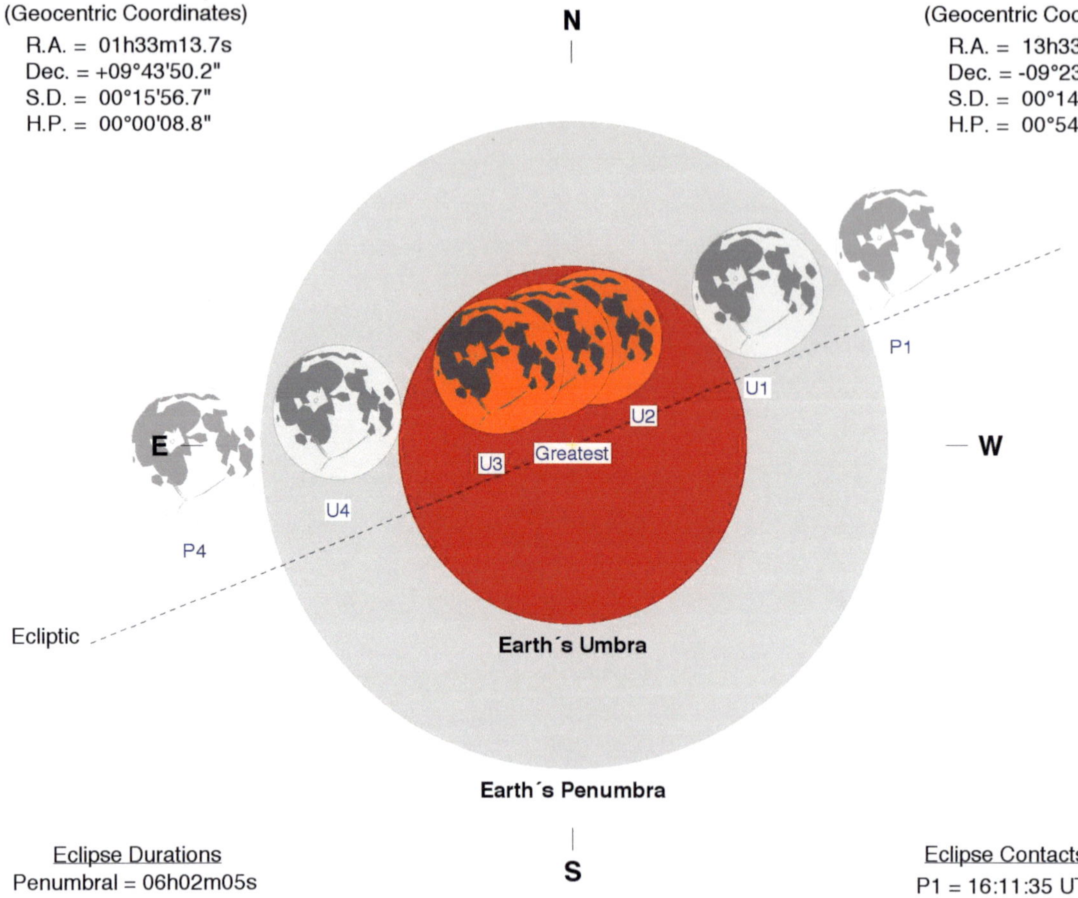

Eclipse Durations
Penumbral = 06h02m05s
Umbral = 03h35m47s
Total = 00h49m50s

Eph. = JPL DE430
Rule = Herald-Sinnott
ΔT = 76 s

0 15 30 45 60
Arc-Minutes

Eclipse Contacts
P1 = 16:11:35 UT1
U1 = 17:24:48 UT1
U2 = 18:47:53 UT1
U3 = 19:37:42 UT1
U4 = 21:00:35 UT1
P4 = 22:13:40 UT1

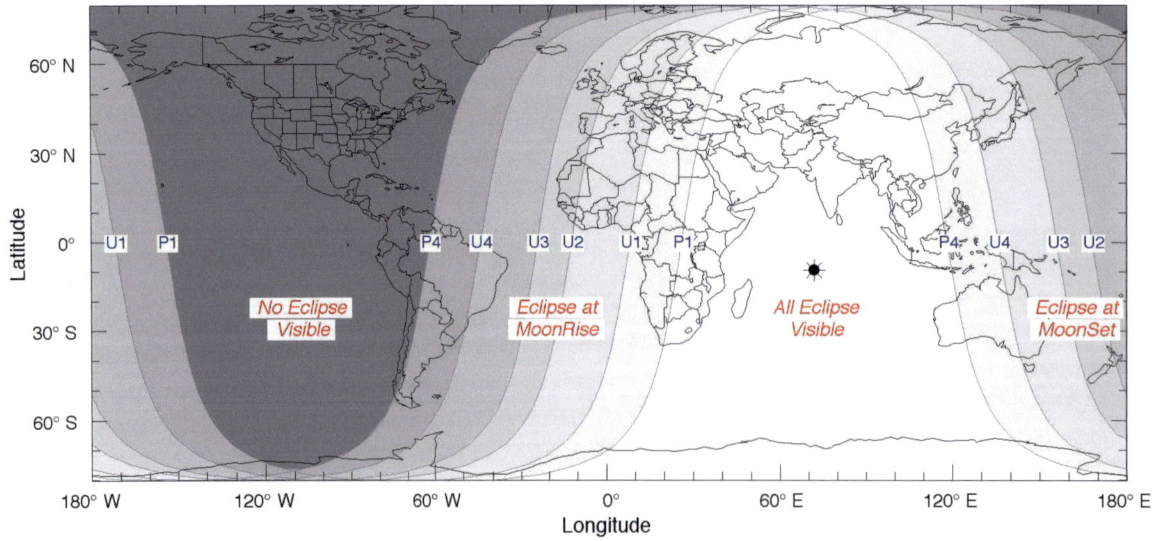

Total Lunar Eclipse of 2033 Oct 08

Greatest Eclipse = 10:56:22.6 TD (= 10:55:06.9 UT1)

Penumbral Magnitude = 2.3068
Umbral Magnitude = 1.3508

Gamma = -0.2889
Axis = 0.2958°

Saros Series = 137
Saros Member = 27 of 78

Sun at Greatest Eclipse
(Geocentric Coordinates)
R.A. = 12h57m01.9s
Dec. = -06°05'34.4"
S.D. = 00°16'00.5"
H.P. = 00°00'08.8"

Moon at Greatest Eclipse
(Geocentric Coordinates)
R.A. = 00h57m22.8s
Dec. = +05°48'36.0"
S.D. = 00°16'44.6"
H.P. = 01°01'27.1"

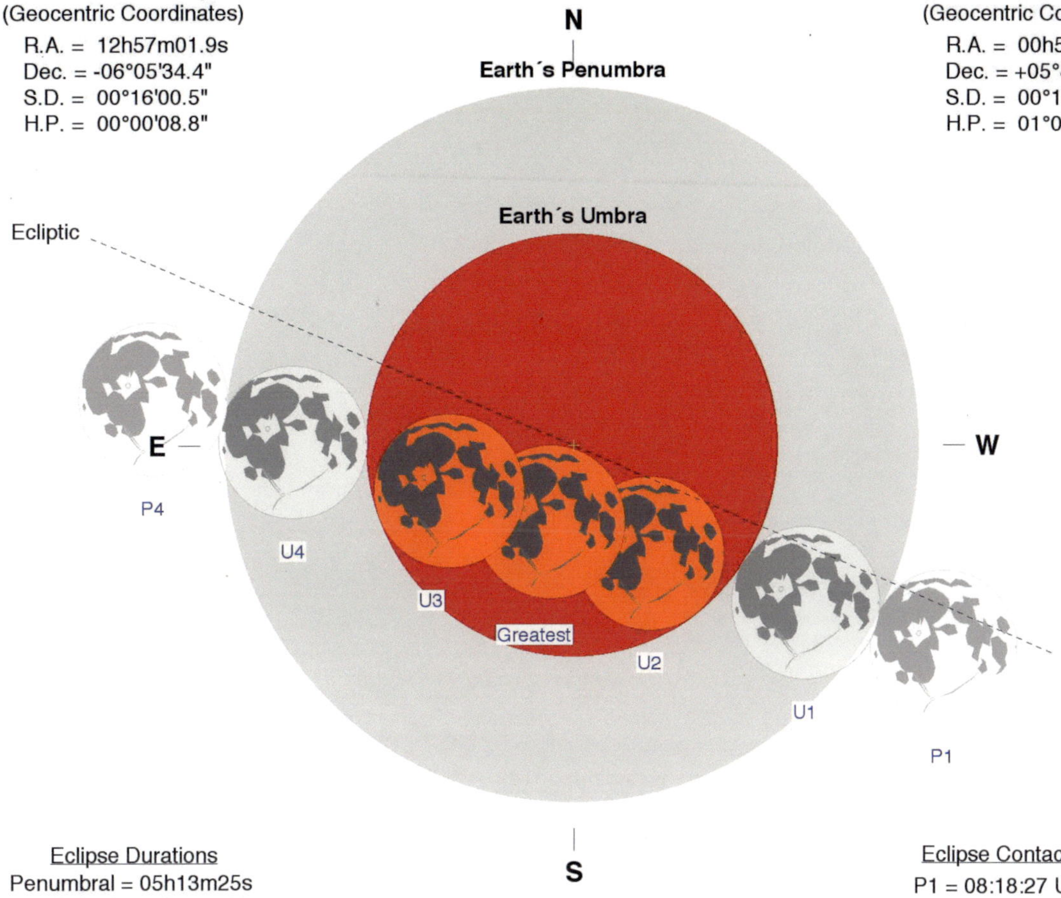

Eclipse Durations
Penumbral = 05h13m25s
Umbral = 03h23m07s
Total = 01h19m25s

Eph. = JPL DE430
Rule = Herald-Sinnott
ΔT = 76 s

0 15 30 45 60
Arc-Minutes

©2020 F. Espenak, www.EclipseWise.com

Eclipse Contacts
P1 = 08:18:27 UT1
U1 = 09:13:37 UT1
U2 = 10:15:31 UT1
U3 = 11:34:56 UT1
U4 = 12:36:43 UT1
P4 = 13:31:51 UT1

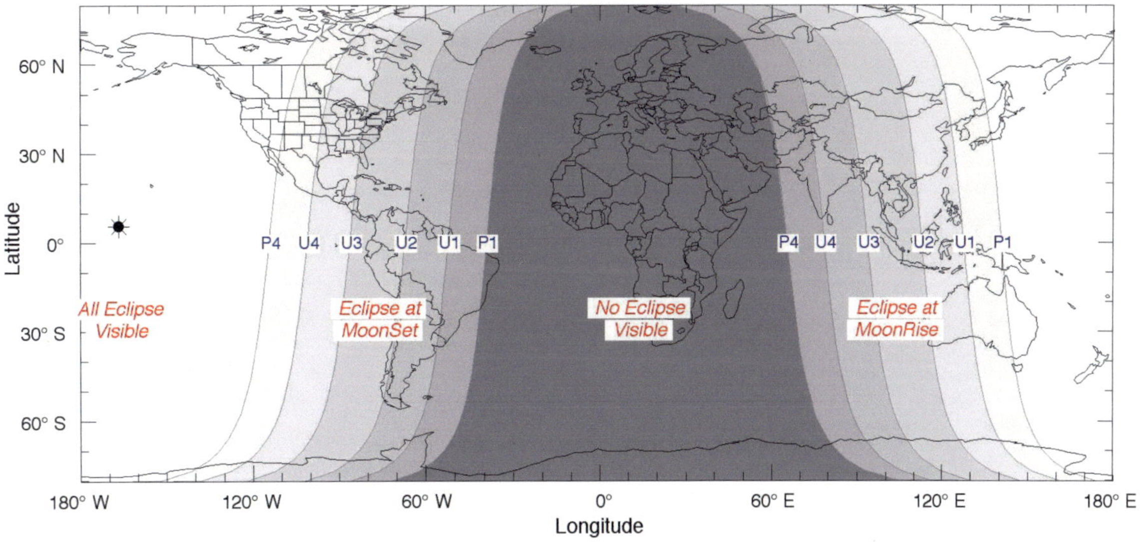

Penumbral Lunar Eclipse of 2034 Apr 03

Greatest Eclipse = 19:06:59.9 TD (= 19:05:43.9 UT1)

Penumbral Magnitude = 0.8557 Gamma = 1.1144 Saros Series = 142
Umbral Magnitude = -0.2263 Axis = 1.0077° Saros Member = 19 of 73

Sun at Greatest Eclipse
(Geocentric Coordinates)
R.A. = 00h51m54.0s
Dec. = +05°33'29.1"
S.D. = 00°15'59.8"
H.P. = 00°00'08.8"

Moon at Greatest Eclipse
(Geocentric Coordinates)
R.A. = 12h53m05.6s
Dec. = -04°35'42.2"
S.D. = 00°14'47.1"
H.P. = 00°54'15.6"

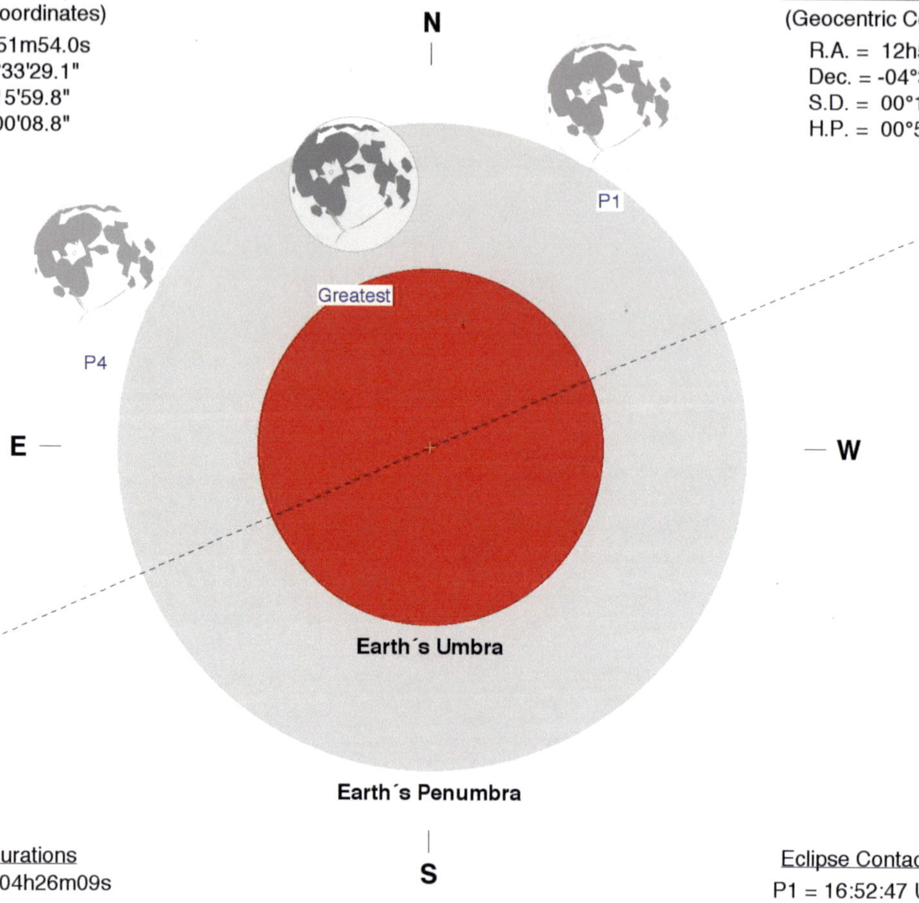

Eclipse Durations
Penumbral = 04h26m09s

Eclipse Contacts
P1 = 16:52:47 UT1
P4 = 21:18:56 UT1

Eph. = JPL DE430
Rule = Herald-Sinnott
ΔT = 76 s

0 15 30 45 60
Arc-Minutes

©2020 F. Espenak, www.EclipseWise.com

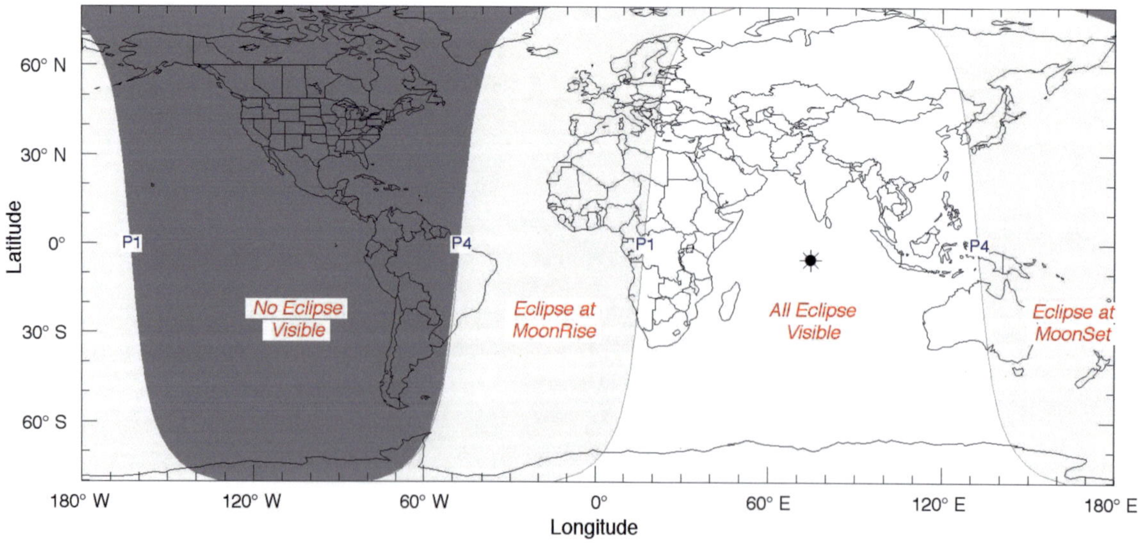

Partial Lunar Eclipse of 2034 Sep 28

Greatest Eclipse = 02:47:37.3 TD (= 02:46:21.1 UT1)

Penumbral Magnitude = 0.9922
Umbral Magnitude = 0.0155

Gamma = -1.0110
Axis = 1.0104°

Saros Series = 147
Saros Member = 9 of 70

Sun at Greatest Eclipse
(Geocentric Coordinates)
R.A. = 12h18m35.8s
Dec. = -02°00'43.0"
S.D. = 00°15'57.6"
H.P. = 00°00'08.8"

Moon at Greatest Eclipse
(Geocentric Coordinates)
R.A. = 00h19m50.0s
Dec. = +01°02'59.4"
S.D. = 00°16'20.4"
H.P. = 00°59'58.2"

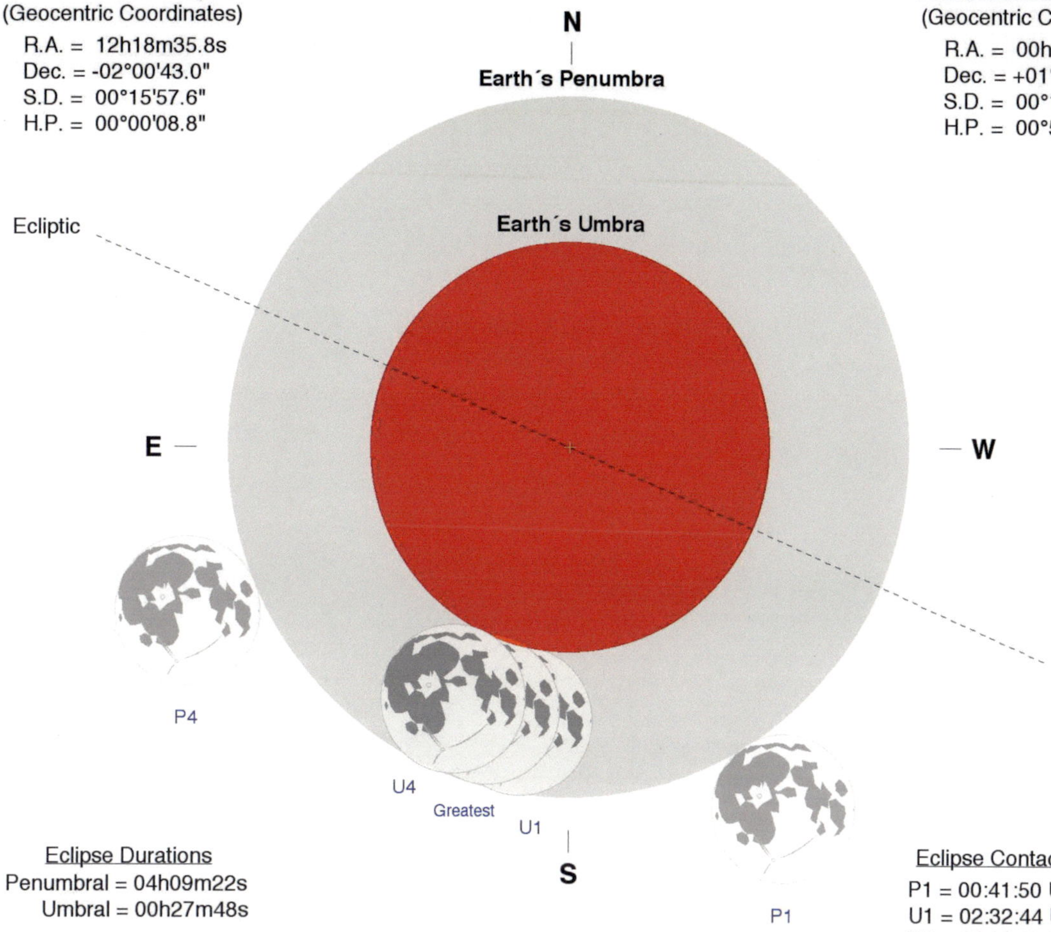

Eclipse Durations
Penumbral = 04h09m22s
Umbral = 00h27m48s

Eclipse Contacts
P1 = 00:41:50 UT1
U1 = 02:32:44 UT1
U4 = 03:00:33 UT1
P4 = 04:51:12 UT1

Eph. = JPL DE430
Rule = Herald-Sinnott
ΔT = 76 s

0 15 30 45 60
Arc-Minutes

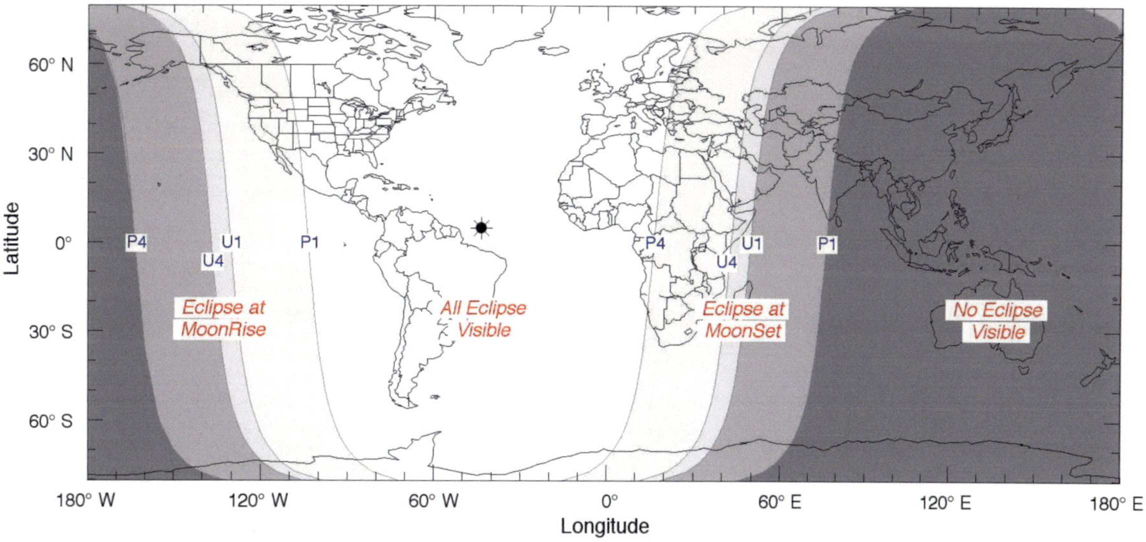

Penumbral Lunar Eclipse of 2035 Feb 22

Greatest Eclipse = 09:06:11.4 TD (= 09:04:55.0 UT1)

Penumbral Magnitude = 0.9663 Gamma = -1.0367 Saros Series = 114
Umbral Magnitude = -0.0523 Axis = 1.0066° Saros Member = 60 of 71

Sun at Greatest Eclipse
(Geocentric Coordinates)
R.A. = 22h21m54.2s
Dec. = -10°11'53.9"
S.D. = 00°16'10.2"
H.P. = 00°00'08.9"

Moon at Greatest Eclipse
(Geocentric Coordinates)
R.A. = 10h20m48.3s
Dec. = +09°13'43.5"
S.D. = 00°15'52.5"
H.P. = 00°58'15.8"

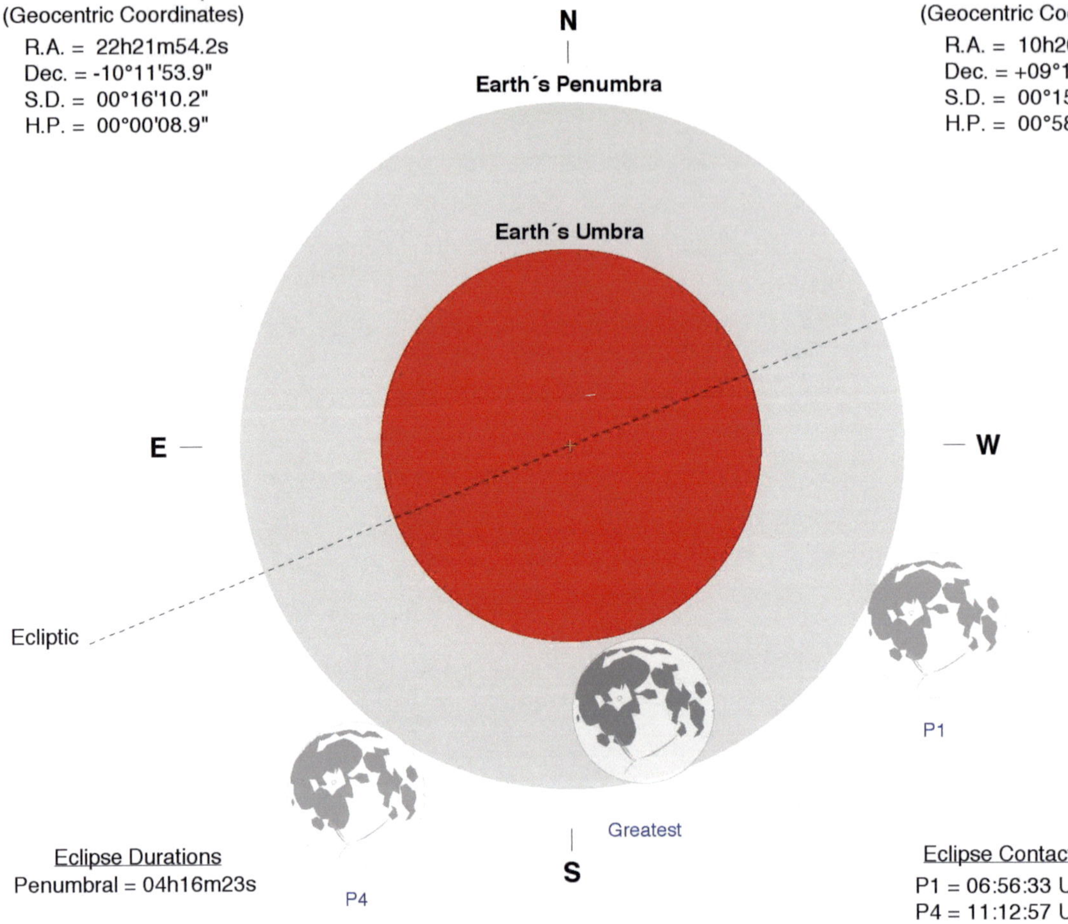

Eclipse Durations
Penumbral = 04h16m23s

Eclipse Contacts
P1 = 06:56:33 UT1
P4 = 11:12:57 UT1

Eph. = JPL DE430
Rule = Herald-Sinnott
ΔT = 76 s

0 15 30 45 60
Arc-Minutes

©2020 F. Espenak, www.EclipseWise.com

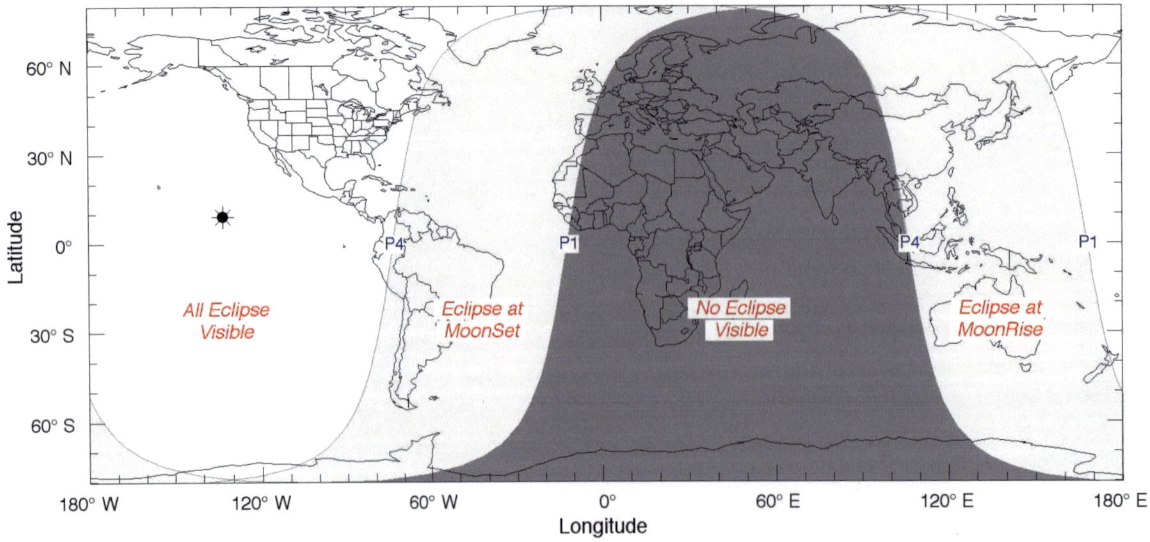

Partial Lunar Eclipse of 2035 Aug 19

Greatest Eclipse = 01:12:15.0 TD (= 01:10:58.2 UT1)

Penumbral Magnitude = 1.1519
Umbral Magnitude = 0.1049

Gamma = 0.9434
Axis = 0.8708°

Saros Series = 119
Saros Member = 62 of 82

Sun at Greatest Eclipse
(Geocentric Coordinates)
R.A. = 09h52m42.8s
Dec. = +12°52'21.3"
S.D. = 00°15'48.1"
H.P. = 00°00'08.7"

Moon at Greatest Eclipse
(Geocentric Coordinates)
R.A. = 21h51m50.7s
Dec. = -12°01'40.6"
S.D. = 00°15'05.5"
H.P. = 00°55'23.4"

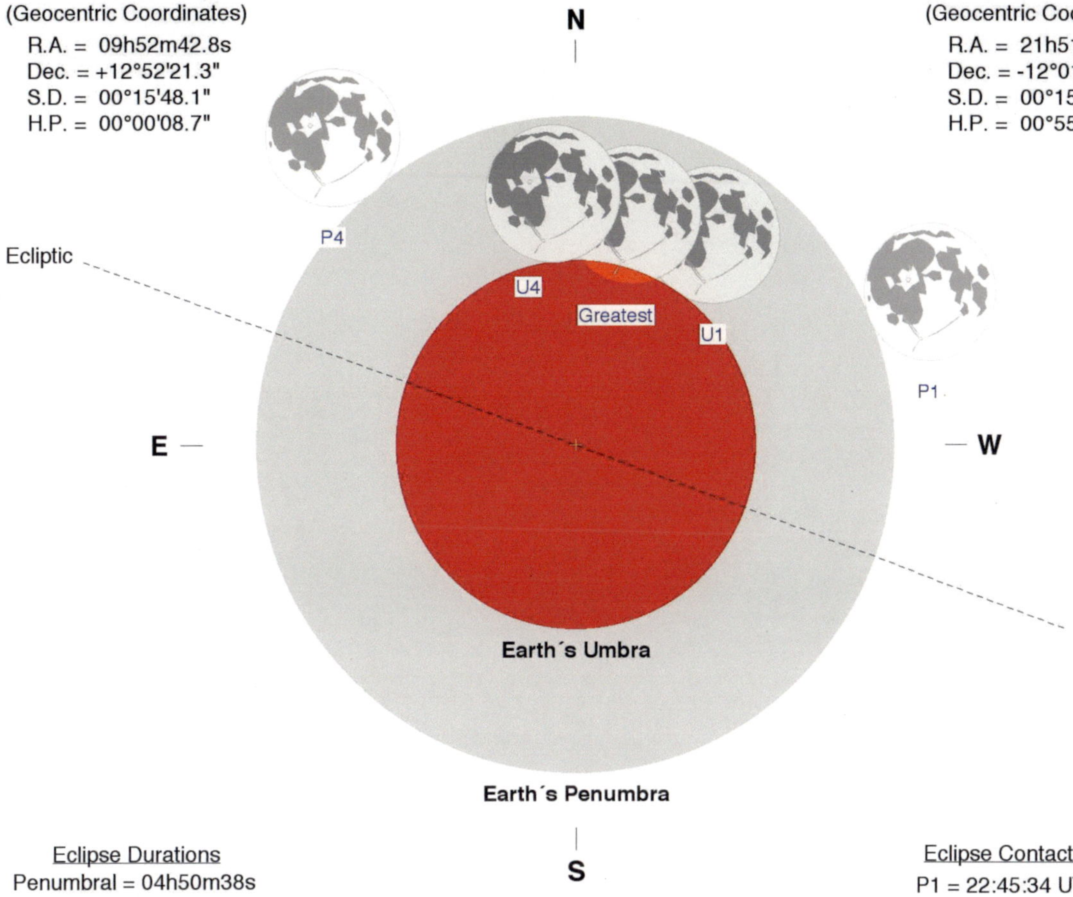

Eclipse Durations
Penumbral = 04h50m38s
Umbral = 01h17m11s

Eclipse Contacts
P1 = 22:45:34 UT1
U1 = 00:32:18 UT1
U4 = 01:49:29 UT1
P4 = 03:36:13 UT1

Eph. = JPL DE430
Rule = Herald-Sinnott
ΔT = 77 s

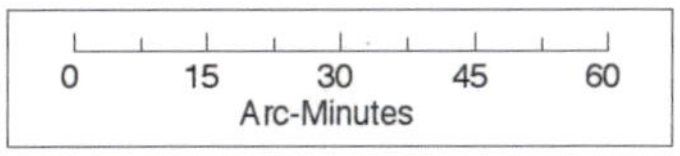

©2020 F. Espenak, www.EclipseWise.com

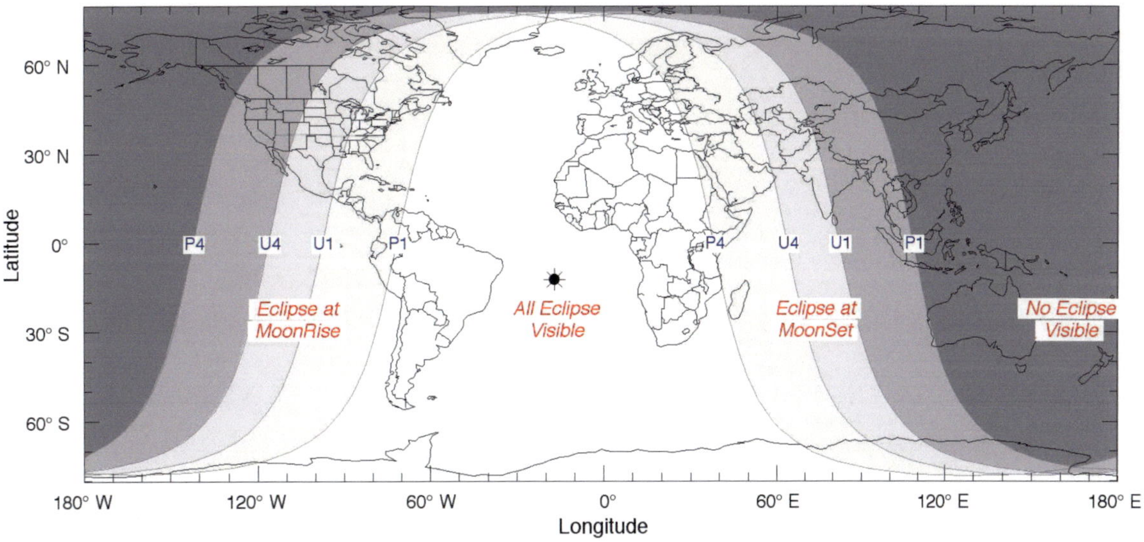

Total Lunar Eclipse of 2036 Feb 11

Greatest Eclipse = 22:13:06.2 TD (= 22:11:49.2 UT1)

Penumbral Magnitude = 2.2762
Umbral Magnitude = 1.3006

Gamma = -0.3110
Axis = 0.3159°

Saros Series = 124
Saros Member = 50 of 73

Sun at Greatest Eclipse
(Geocentric Coordinates)
R.A. = 21h40m25.4s
Dec. = -13°55'30.0"
S.D. = 00°16'12.3"
H.P. = 00°00'08.9"

Moon at Greatest Eclipse
(Geocentric Coordinates)
R.A. = 09h40m07.3s
Dec. = +13°37'03.4"
S.D. = 00°16'36.7"
H.P. = 01°00'57.8"

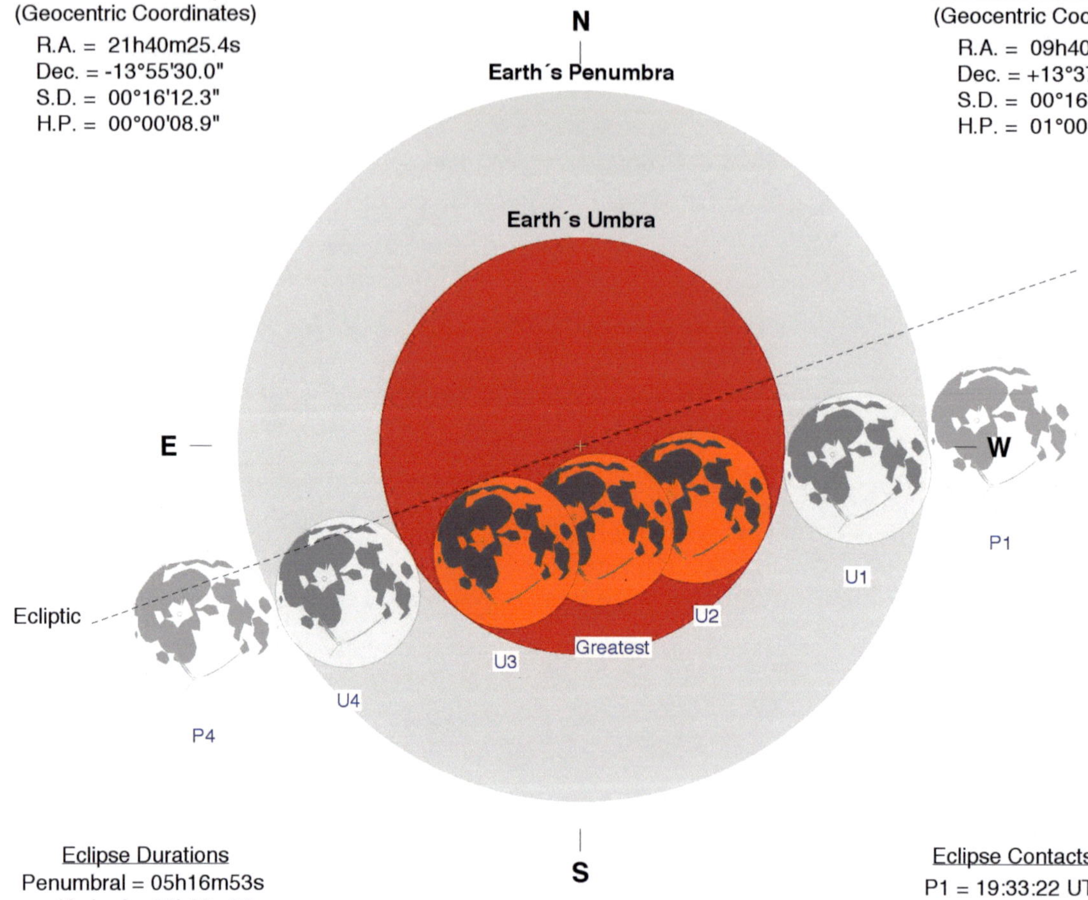

Eclipse Durations
Penumbral = 05h16m53s
Umbral = 03h22m39s
Total = 01h15m06s

Eph. = JPL DE430
Rule = Herald-Sinnott
ΔT = 77 s

0 15 30 45 60
Arc-Minutes

©2020 F. Espenak, www.EclipseWise.com

Eclipse Contacts
P1 = 19:33:22 UT1
U1 = 20:30:26 UT1
U2 = 21:34:10 UT1
U3 = 22:49:16 UT1
U4 = 23:53:05 UT1
P4 = 00:50:15 UT1

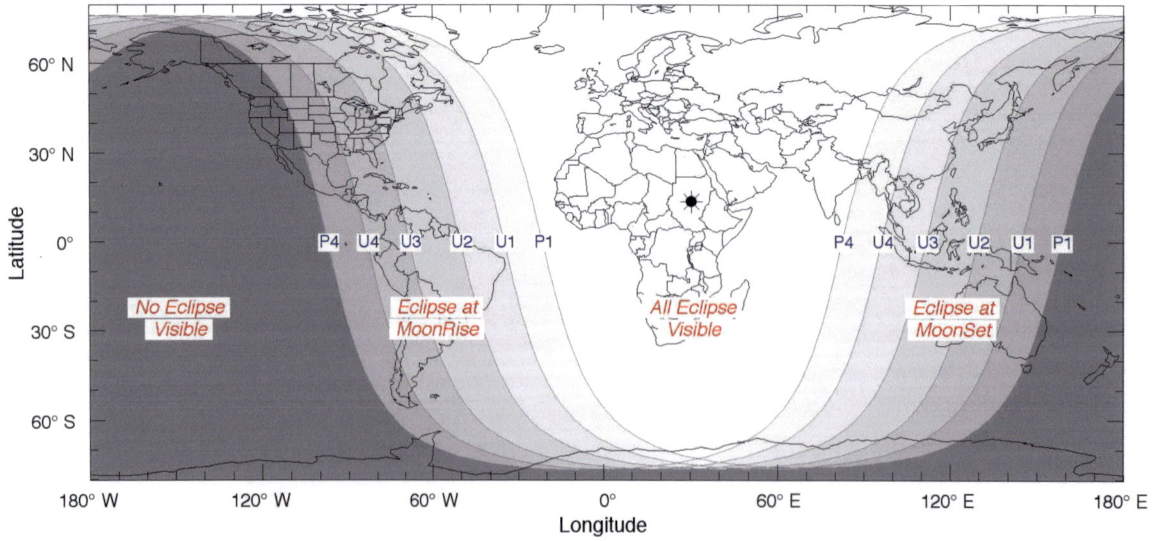

Total Lunar Eclipse of 2036 Aug 07

Greatest Eclipse = 02:52:32.4 TD (= 02:51:15.2 UT1)

Penumbral Magnitude = 2.5279 Gamma = 0.2004 Saros Series = 129
Umbral Magnitude = 1.4556 Axis = 0.1803° Saros Member = 39 of 71

Sun at Greatest Eclipse
(Geocentric Coordinates)
R.A. = 09h10m39.1s
Dec. = +16°16'20.8"
S.D. = 00°15'46.3"
H.P. = 00°00'08.7"

Moon at Greatest Eclipse
(Geocentric Coordinates)
R.A. = 21h10m30.3s
Dec. = -16°05'44.3"
S.D. = 00°14'42.5"
H.P. = 00°53'58.8"

N

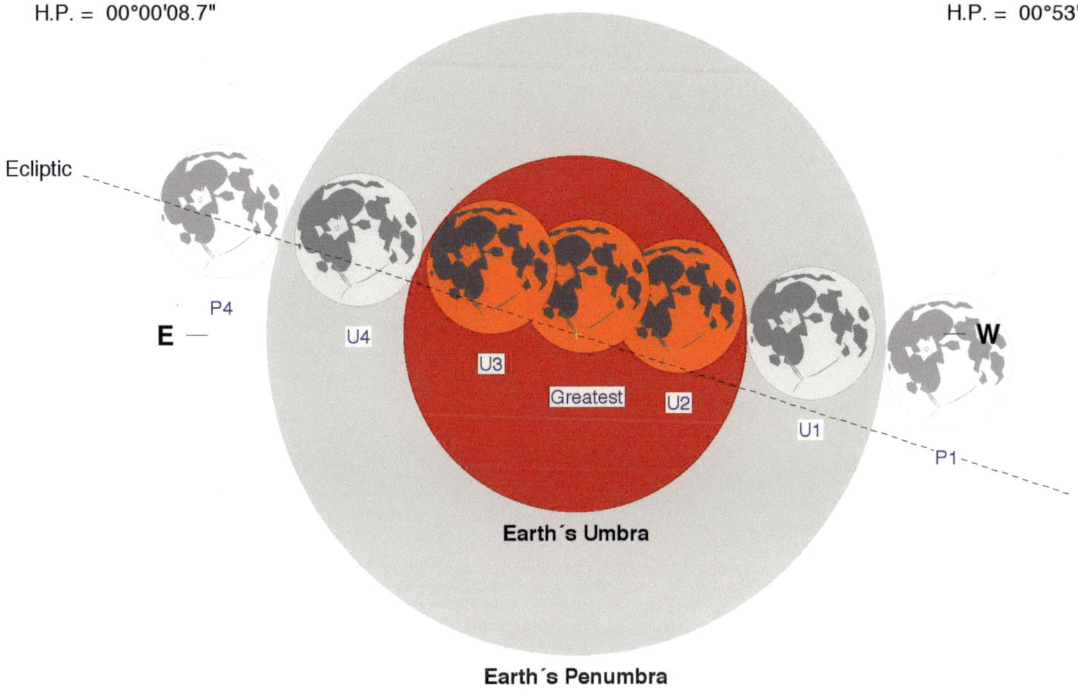

S

Eclipse Durations
Penumbral = 06h13m03s
Umbral = 03h52m11s
Total = 01h36m06s

Eph. = JPL DE430
Rule = Herald-Sinnott
ΔT = 77 s

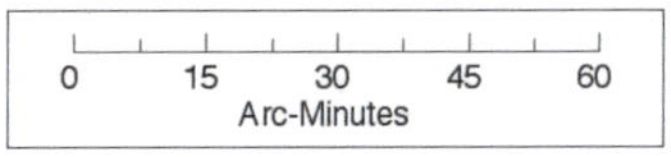

Eclipse Contacts
P1 = 23:44:42 UT1
U1 = 00:55:08 UT1
U2 = 02:03:08 UT1
U3 = 03:39:15 UT1
U4 = 04:47:19 UT1
P4 = 05:57:45 UT1

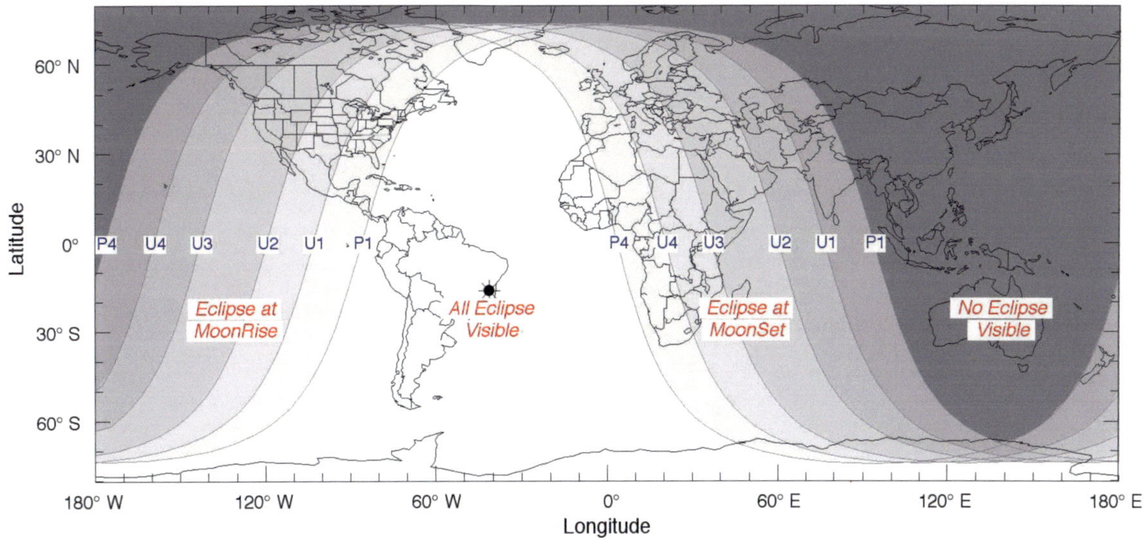

Total Lunar Eclipse of 2037 Jan 31

Greatest Eclipse = 14:01:38.4 TD (= 14:00:20.9 UT1)

Penumbral Magnitude = 2.1815 | Gamma = 0.3619 | Saros Series = 134
Umbral Magnitude = 1.2086 | Axis = 0.3693° | Saros Member = 28 of 72

Sun at Greatest Eclipse
(Geocentric Coordinates)
R.A. = 20h57m58.6s
Dec. = -17°10'47.4"
S.D. = 00°16'14.0"
H.P. = 00°00'08.9"

Moon at Greatest Eclipse
(Geocentric Coordinates)
R.A. = 08h58m15.6s
Dec. = +17°32'34.5"
S.D. = 00°16'41.1"
H.P. = 01°01'14.2"

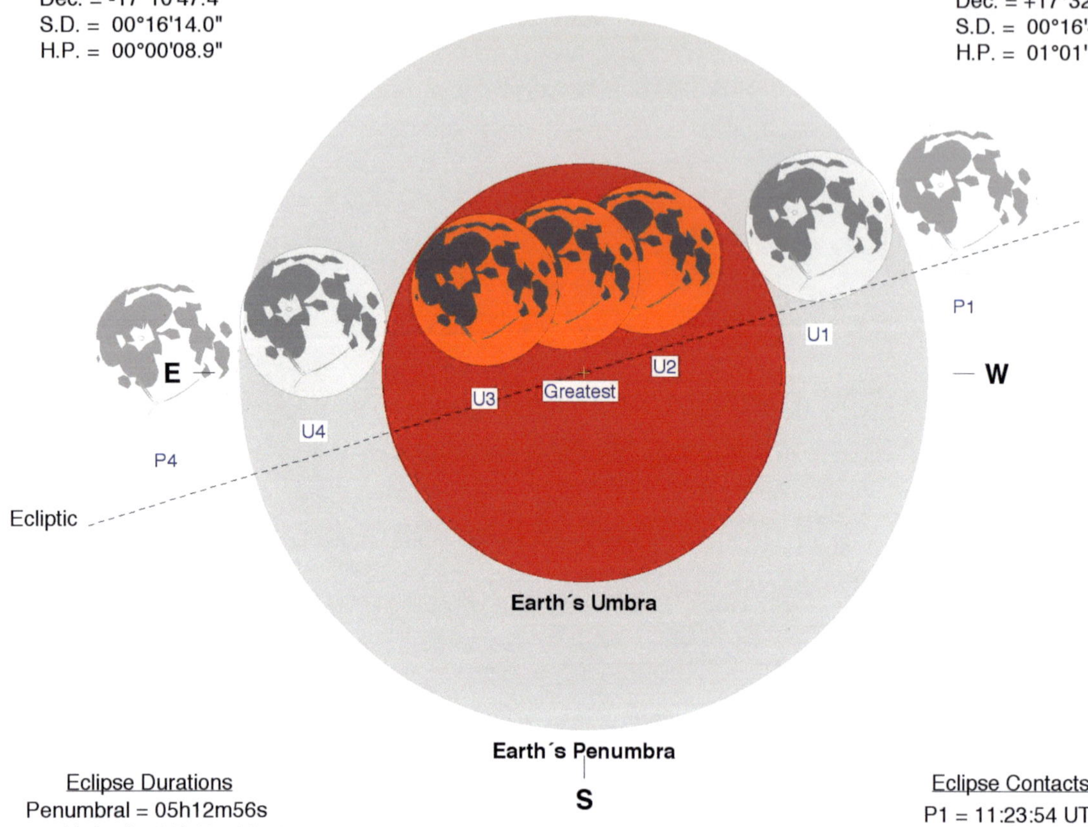

Eclipse Durations
Penumbral = 05h12m56s
Umbral = 03h18m12s
Total = 01h04m17s

Eph. = JPL DE430
Rule = Herald-Sinnott
ΔT = 77 s

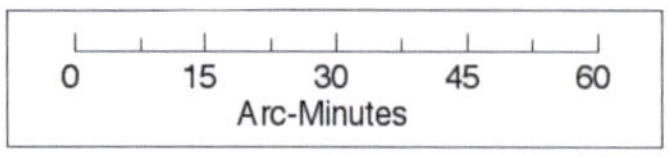

Eclipse Contacts
P1 = 11:23:54 UT1
U1 = 12:21:18 UT1
U2 = 13:28:18 UT1
U3 = 14:32:35 UT1
U4 = 15:39:30 UT1
P4 = 16:36:50 UT1

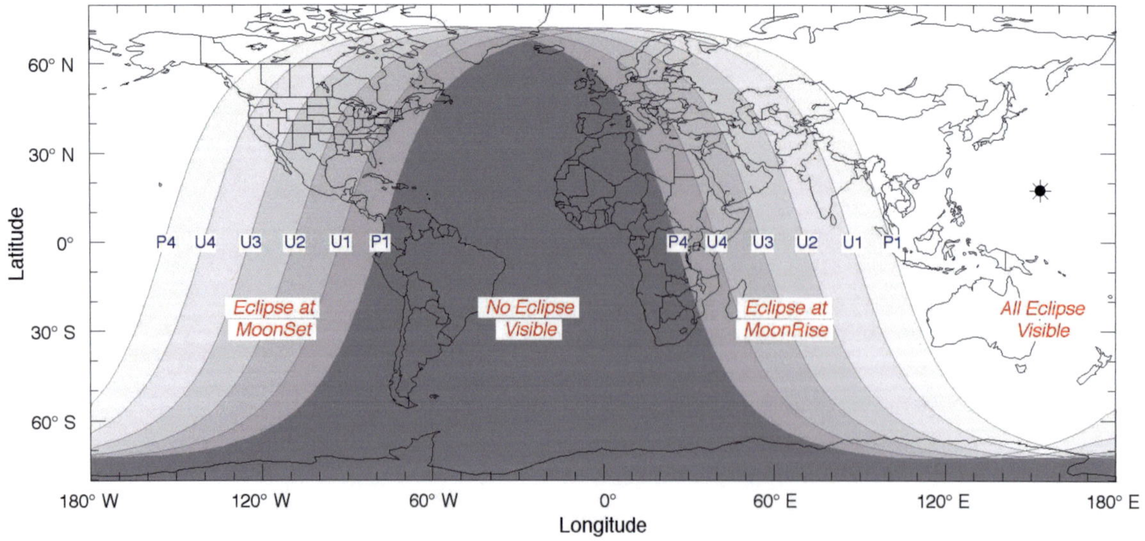

Partial Lunar Eclipse of 2037 Jul 27

Greatest Eclipse = 04:09:53.3 TD (= 04:08:35.6 UT1)

Penumbral Magnitude = 1.8596	Gamma = -0.5582	Saros Series = 139
Umbral Magnitude = 0.8108	Axis = 0.5126°	Saros Member = 22 of 79

Sun at Greatest Eclipse
(Geocentric Coordinates)
R.A. = 08h27m18.9s
Dec. = +19°07'58.8"
S.D. = 00°15'45.0"
H.P. = 00°00'08.7"

Moon at Greatest Eclipse
(Geocentric Coordinates)
R.A. = 20h27m37.3s
Dec. = -19°38'25.9"
S.D. = 00°15'00.9"
H.P. = 00°55'06.5"

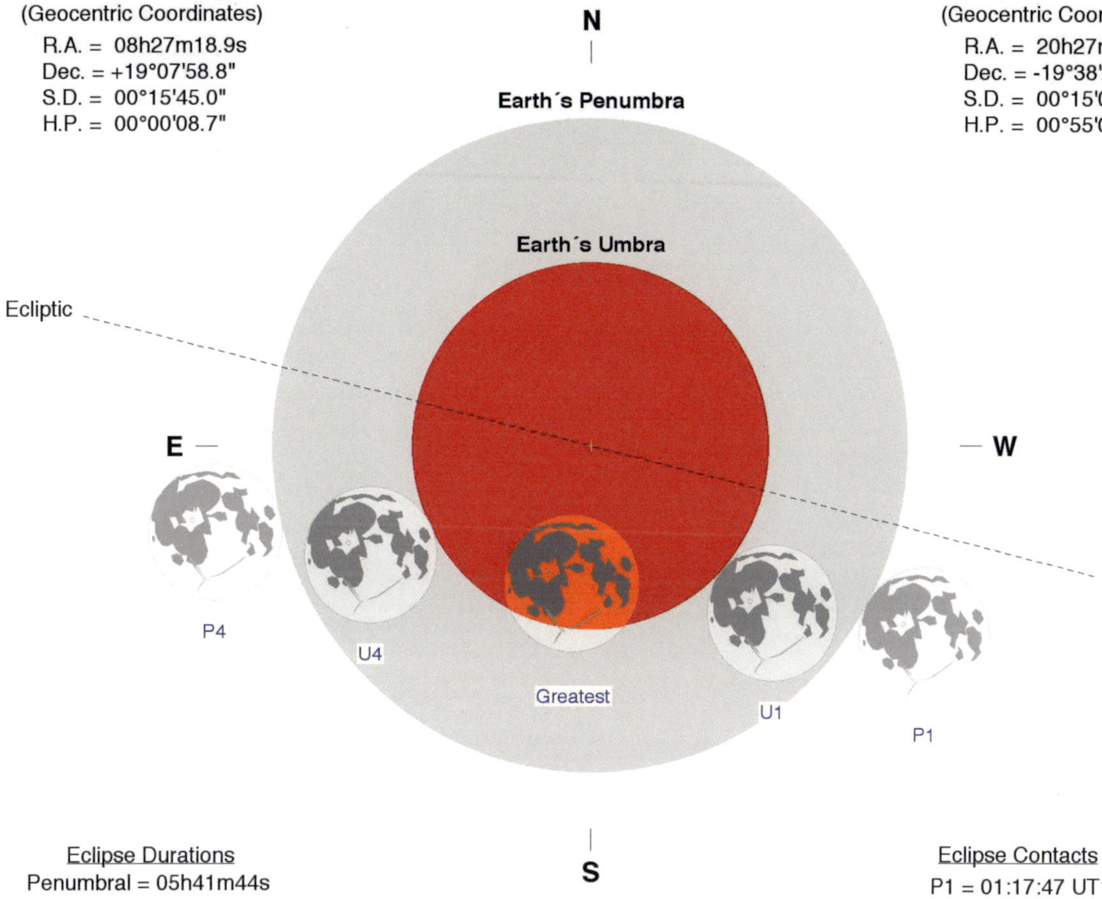

Eclipse Durations
Penumbral = 05h41m44s
Umbral = 03h13m11s

Eclipse Contacts
P1 = 01:17:47 UT1
U1 = 02:32:01 UT1
U4 = 05:45:12 UT1
P4 = 06:59:31 UT1

Eph. = JPL DE430
Rule = Herald-Sinnott
ΔT = 78 s

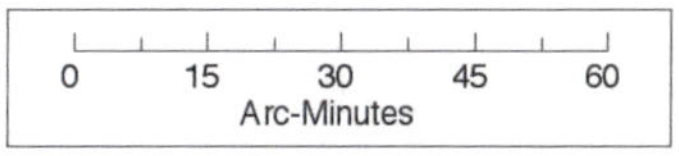

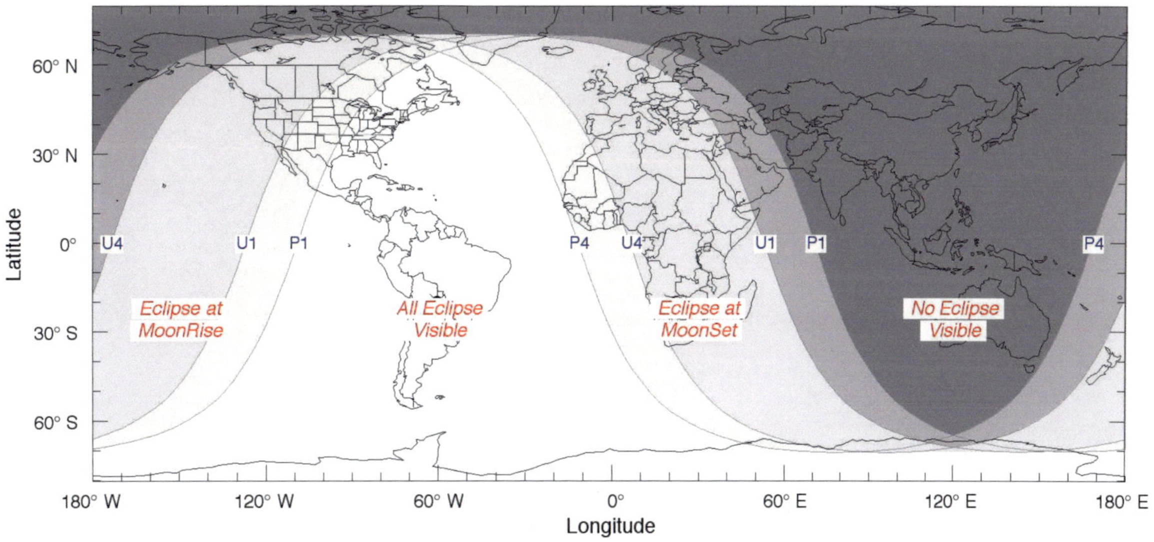

Penumbral Lunar Eclipse of 2038 Jan 21

Greatest Eclipse = 03:49:51.9 TD (= 03:48:33.9 UT1)

Penumbral Magnitude = 0.9009 Gamma = 1.0711 Saros Series = 144
Umbral Magnitude = -0.1127 Axis = 1.0505° Saros Member = 17 of 71

<u>Sun at Greatest Eclipse</u>
(Geocentric Coordinates)
R.A. = 20h13m39.3s
Dec. = -19°53'23.0"
S.D. = 00°16'15.2"
H.P. = 00°00'08.9"

<u>Moon at Greatest Eclipse</u>
(Geocentric Coordinates)
R.A. = 08h14m12.5s
Dec. = +20°55'55.8"
S.D. = 00°16'02.1"
H.P. = 00°58'51.1"

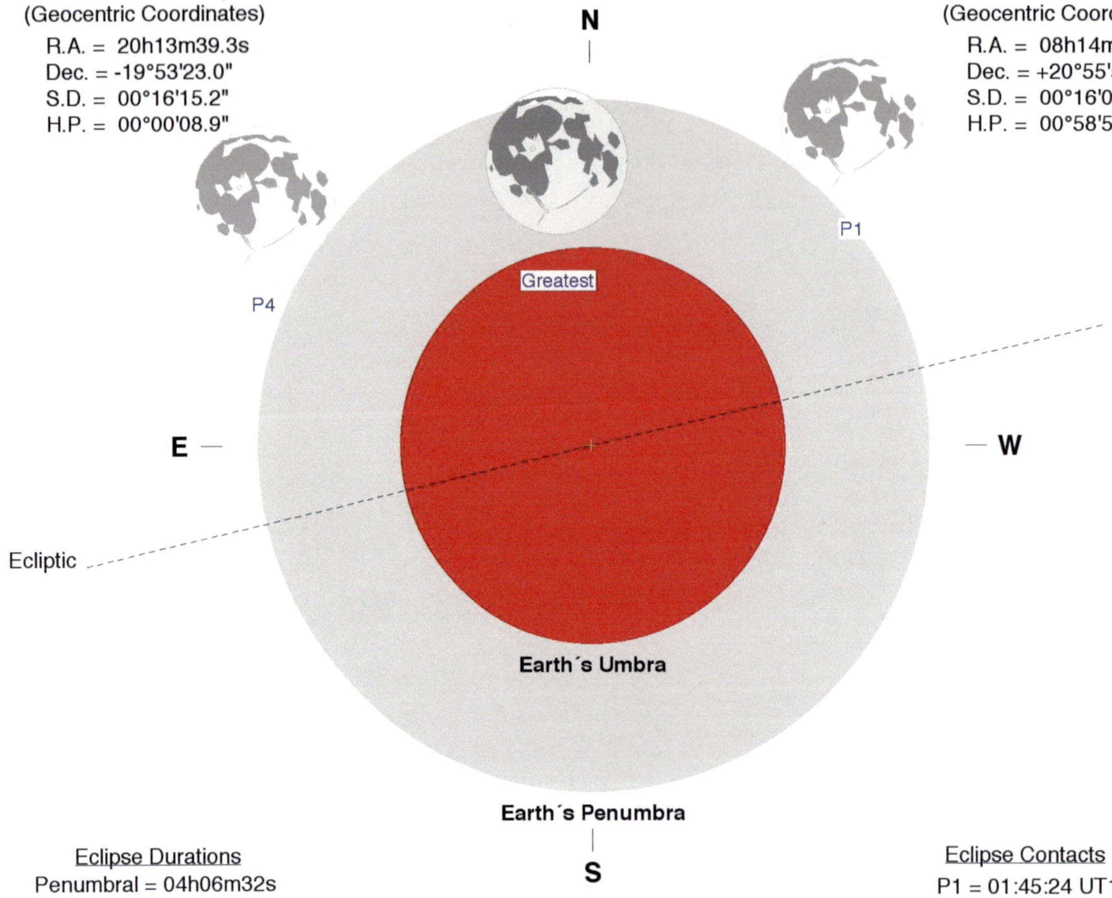

<u>Eclipse Durations</u>
Penumbral = 04h06m32s

<u>Eclipse Contacts</u>
P1 = 01:45:24 UT1
P4 = 05:51:56 UT1

Eph. = JPL DE430
Rule = Herald-Sinnott
ΔT = 78 s

0 15 30 45 60
Arc-Minutes

©2020 F. Espenak, www.EclipseWise.com

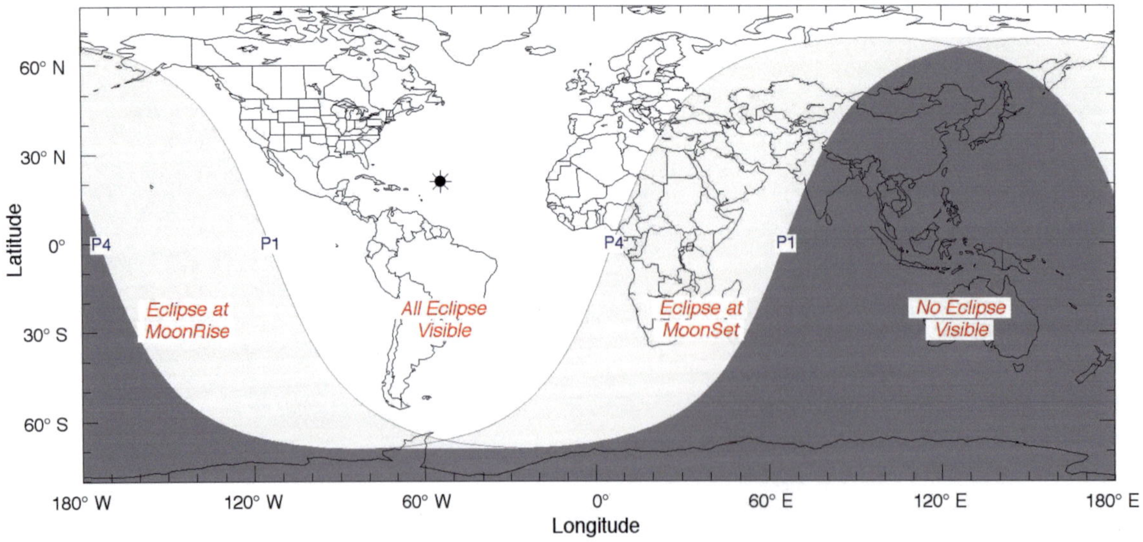

Penumbral Lunar Eclipse of 2038 Jun 17

Greatest Eclipse = 02:45:02.2 TD (= 02:43:44.0 UT1)

Penumbral Magnitude = 0.4438 Gamma = 1.3083 Saros Series = 111
Umbral Magnitude = -0.5259 Axis = 1.2993° Saros Member = 68 of 71

Sun at Greatest Eclipse
(Geocentric Coordinates)
R.A. = 05h42m46.1s
Dec. = +23°22'28.6"
S.D. = 00°15'44.7"
H.P. = 00°00'08.7"

Moon at Greatest Eclipse
(Geocentric Coordinates)
R.A. = 17h43m28.2s
Dec. = -22°05'07.2"
S.D. = 00°16'14.3"
H.P. = 00°59'35.6"

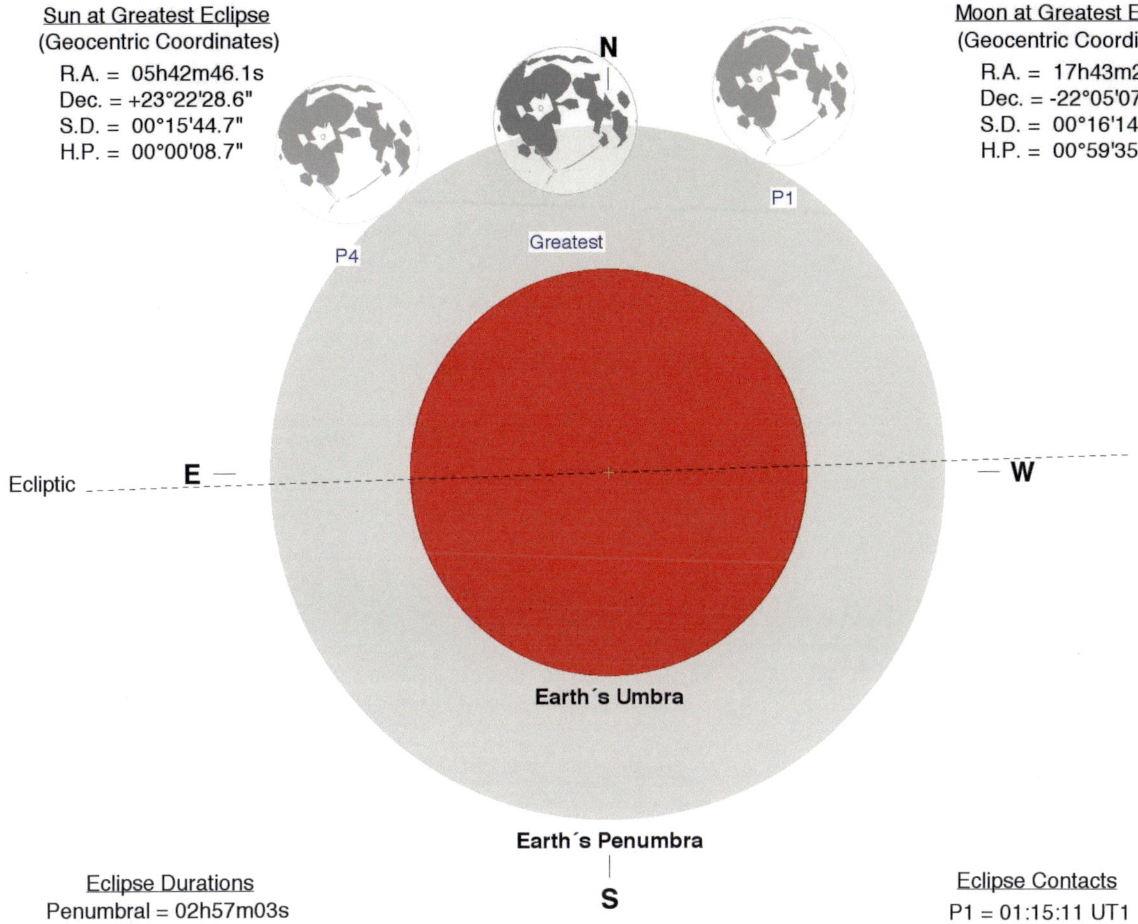

Eclipse Durations
Penumbral = 02h57m03s

Eclipse Contacts
P1 = 01:15:11 UT1
P4 = 04:12:14 UT1

Eph. = JPL DE430
Rule = Herald-Sinnott
ΔT = 78 s

0 15 30 45 60
Arc-Minutes

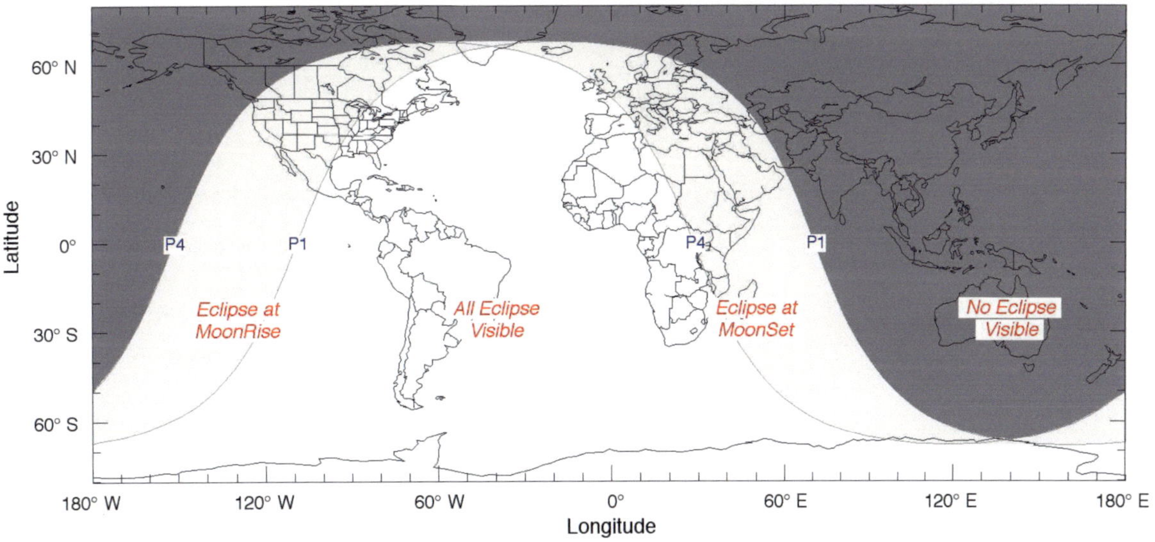

Penumbral Lunar Eclipse of 2038 Jul 16

Greatest Eclipse = 11:35:56.3 TD (= 11:34:38.0 UT1)

Penumbral Magnitude = 0.5012 Gamma = -1.2838 Saros Series = 149
Umbral Magnitude = -0.4938 Axis = 1.2417° Saros Member = 4 of 71

Sun at Greatest Eclipse
(Geocentric Coordinates)
R.A. = 07h43m47.7s
Dec. = +21°17'34.6"
S.D. = 00°15'44.2"
H.P. = 00°00'08.7"

Moon at Greatest Eclipse
(Geocentric Coordinates)
R.A. = 19h44m13.1s
Dec. = -22°31'51.1"
S.D. = 00°15'48.9"
H.P. = 00°58'02.4"

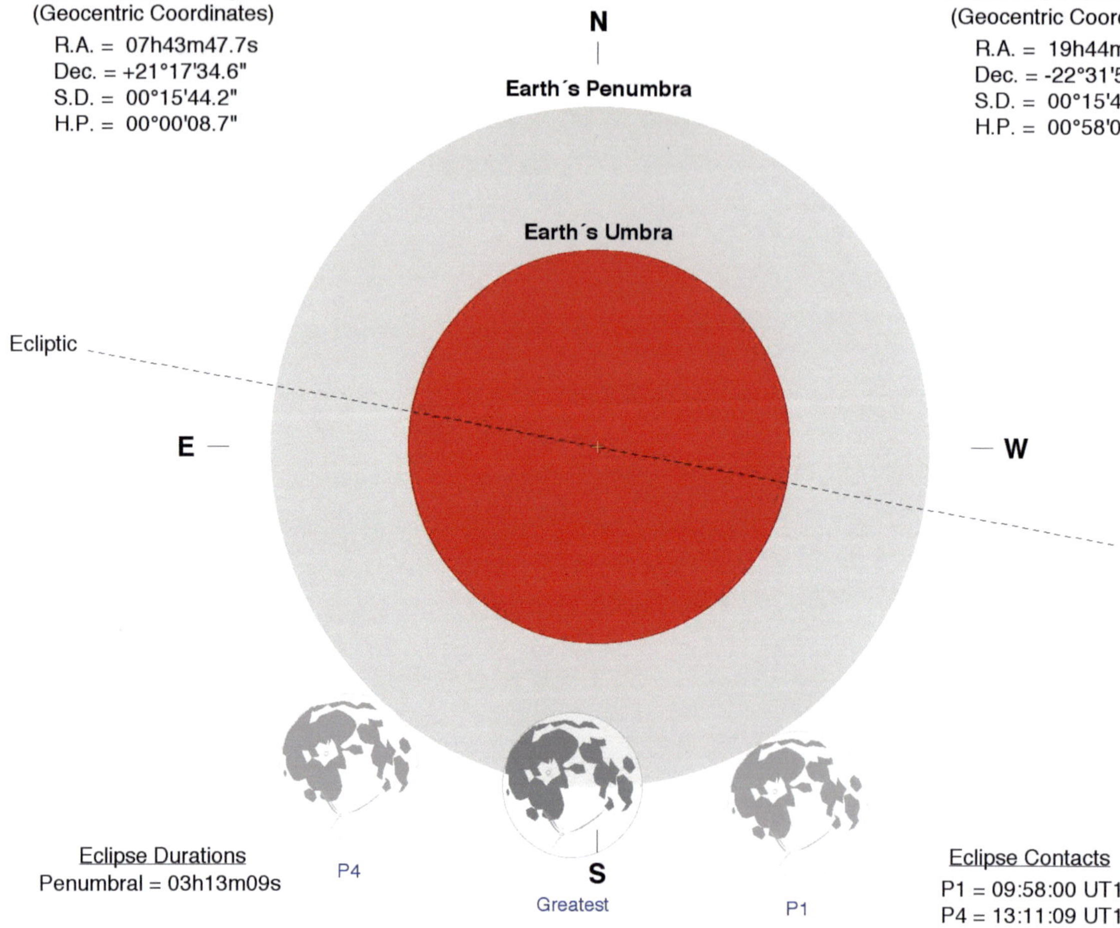

Eclipse Durations
Penumbral = 03h13m09s

Eclipse Contacts
P1 = 09:58:00 UT1
P4 = 13:11:09 UT1

Eph. = JPL DE430
Rule = Herald-Sinnott
ΔT = 78 s

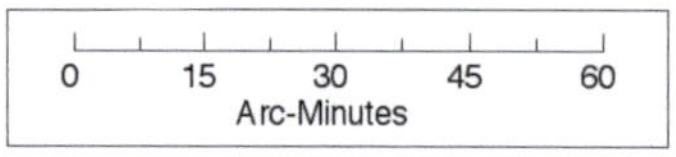

©2020 F. Espenak, www.EclipseWise.com

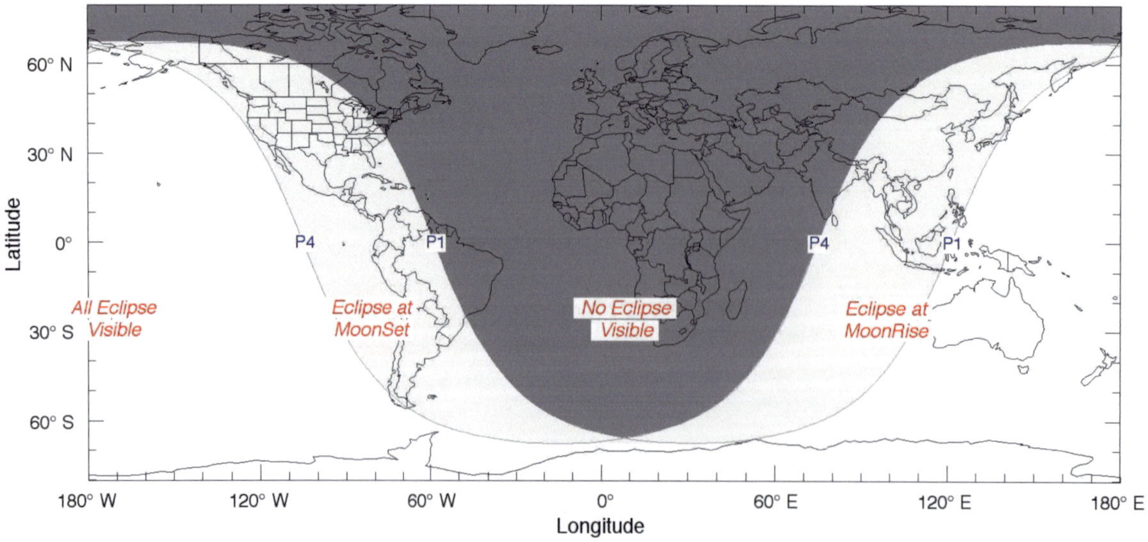

Penumbral Lunar Eclipse of 2038 Dec 11

Greatest Eclipse = 17:45:00.4 TD (= 17:43:41.9 UT1)

Penumbral Magnitude = 0.8062
Umbral Magnitude = -0.2876

Gamma = -1.1449
Axis = 1.0398°

Saros Series = 116
Saros Member = 59 of 73

Sun at Greatest Eclipse
(Geocentric Coordinates)
R.A. = 17h15m29.9s
Dec. = -23°02'24.2"
S.D. = 00°16'14.6"
H.P. = 00°00'08.9"

Moon at Greatest Eclipse
(Geocentric Coordinates)
R.A. = 05h16m16.9s
Dec. = +22°00'57.8"
S.D. = 00°14'51.0"
H.P. = 00°54'29.8"

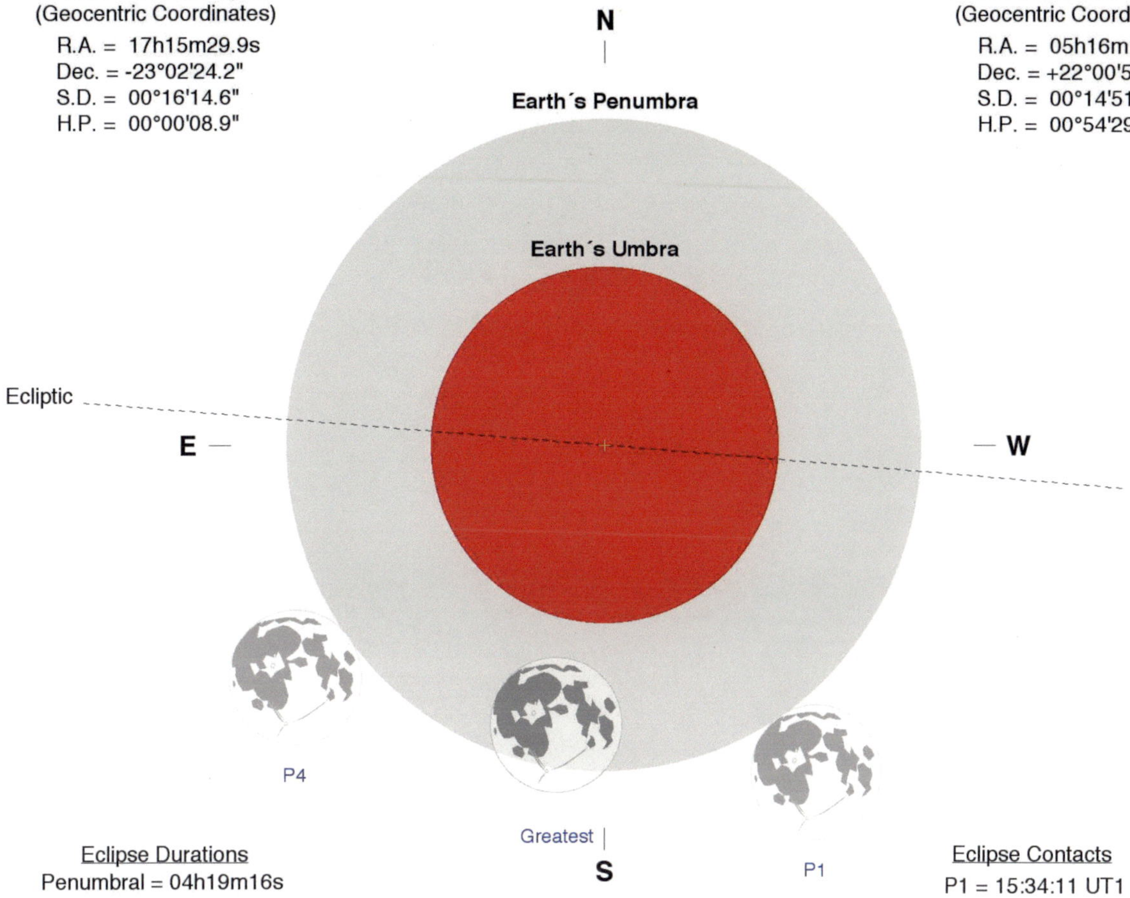

Eclipse Durations
Penumbral = 04h19m16s

Eclipse Contacts
P1 = 15:34:11 UT1
P4 = 19:53:27 UT1

Eph. = JPL DE430
Rule = Herald-Sinnott
ΔT = 79 s

0 15 30 45 60
Arc-Minutes

©2020 F. Espenak, www.EclipseWise.com

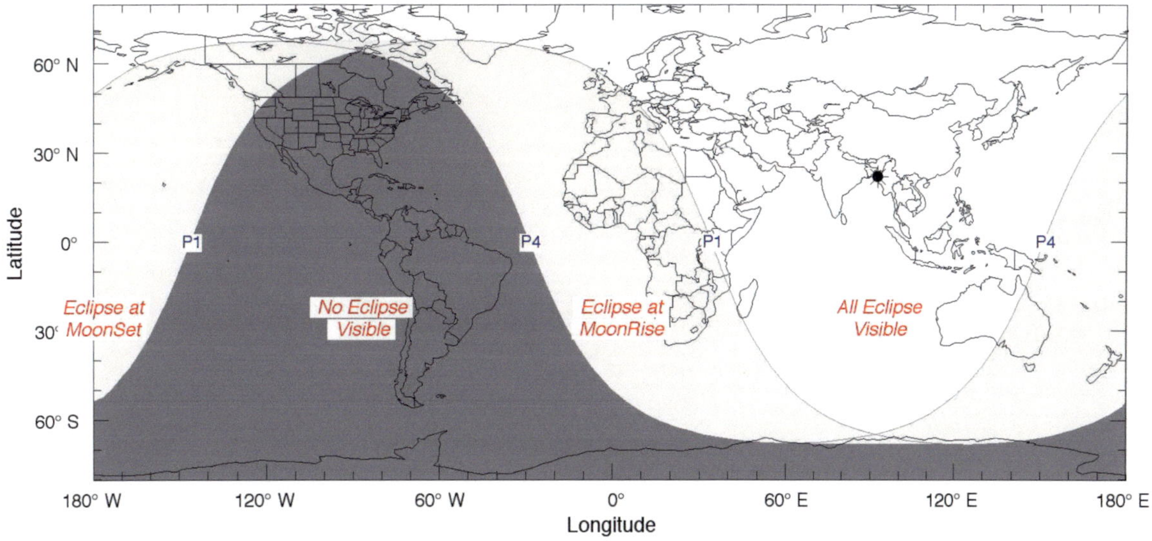

Partial Lunar Eclipse of 2039 Jun 06

Greatest Eclipse = 18:54:25.4 TD (= 18:53:06.7 UT1)

Penumbral Magnitude = 1.8288 Gamma = 0.5460 Saros Series = 121
Umbral Magnitude = 0.8863 Axis = 0.5584° Saros Member = 56 of 82

Sun at Greatest Eclipse
(Geocentric Coordinates)
R.A. = 04h58m56.4s
Dec. = +22°41'33.8"
S.D. = 00°15'45.7"
H.P. = 00°00'08.7"

Moon at Greatest Eclipse
(Geocentric Coordinates)
R.A. = 16h59m25.6s
Dec. = -22°08'44.6"
S.D. = 00°16'43.4"
H.P. = 01°01'22.4"

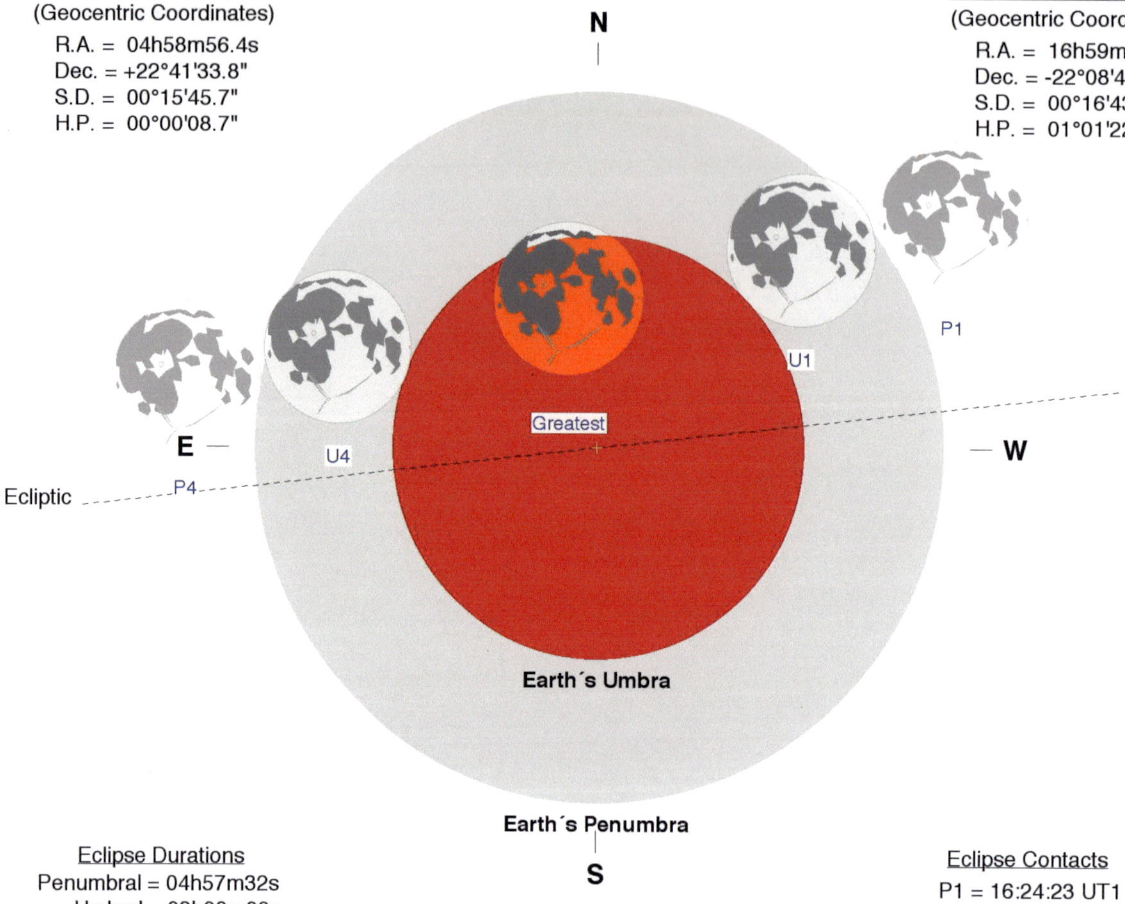

Eclipse Durations
Penumbral = 04h57m32s
Umbral = 03h00m00s

Eclipse Contacts
P1 = 16:24:23 UT1
U1 = 17:23:10 UT1
U4 = 20:23:10 UT1
P4 = 21:21:55 UT1

Eph. = JPL DE430
Rule = Herald-Sinnott
ΔT = 79 s

0 15 30 45 60
Arc-Minutes

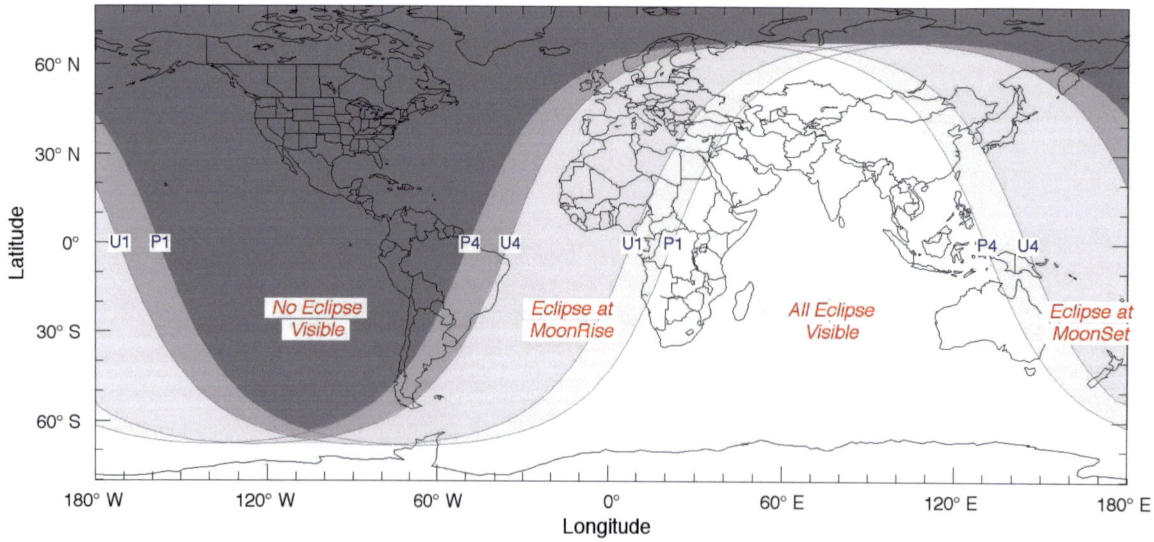

Partial Lunar Eclipse of 2039 Nov 30

Greatest Eclipse = 16:56:27.6 TD (= 16:55:08.5 UT1)

Penumbral Magnitude = 2.0435 Gamma = -0.4721 Saros Series = 126
Umbral Magnitude = 0.9443 Axis = 0.4260° Saros Member = 46 of 70

Sun at Greatest Eclipse
(Geocentric Coordinates)
R.A. = 16h26m20.8s
Dec. = -21°41'27.9"
S.D. = 00°16'13.0"
H.P. = 00°00'08.9"

Moon at Greatest Eclipse
(Geocentric Coordinates)
R.A. = 04h26m48.9s
Dec. = +21°16'45.4"
S.D. = 00°14'45.3"
H.P. = 00°54'08.9"

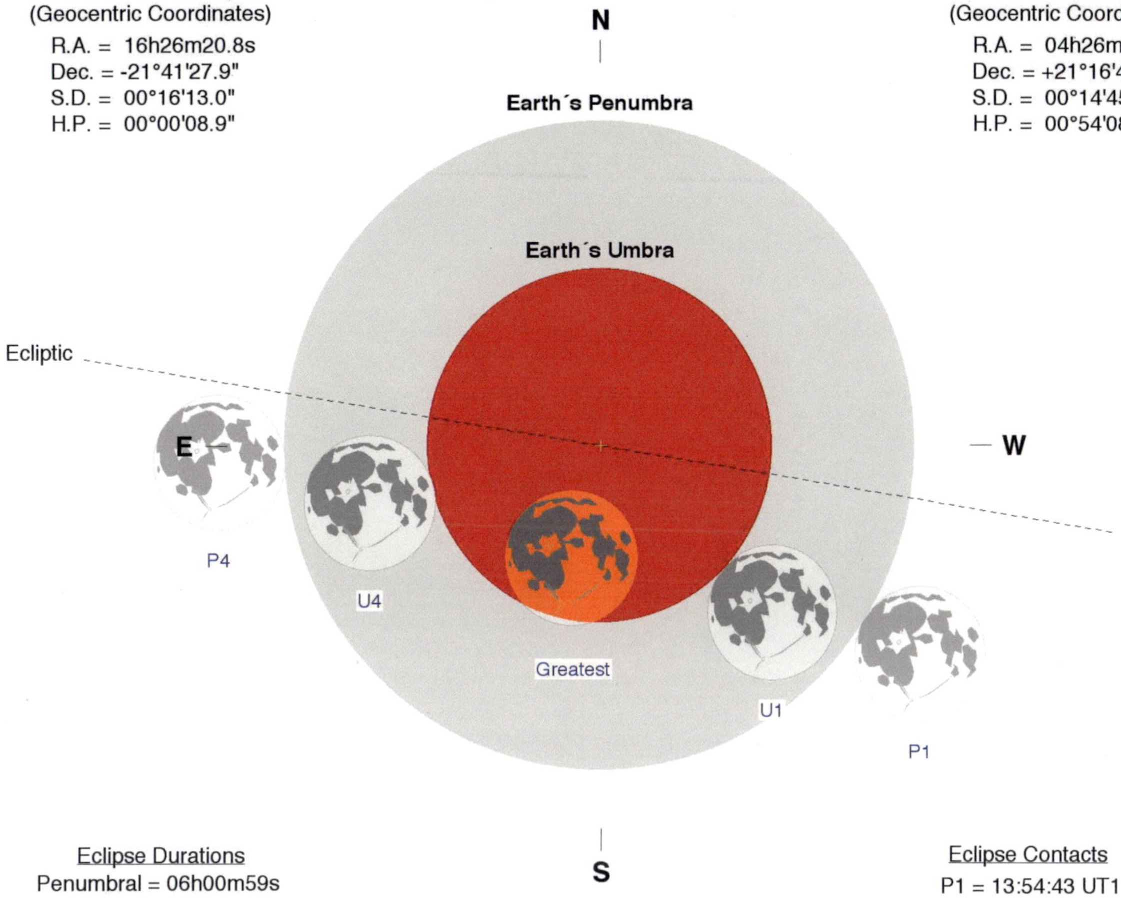

Eclipse Durations
Penumbral = 06h00m59s
Umbral = 03h26m48s

Eclipse Contacts
P1 = 13:54:43 UT1
U1 = 15:11:49 UT1
U4 = 18:38:37 UT1
P4 = 19:55:42 UT1

Eph. = JPL DE430
Rule = Herald-Sinnott
ΔT = 79 s

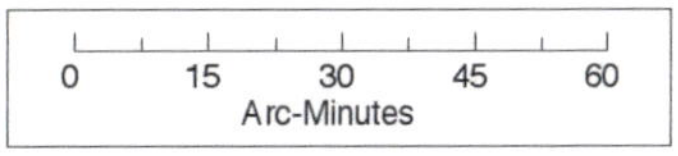

©2020 F. Espenak, www.EclipseWise.com

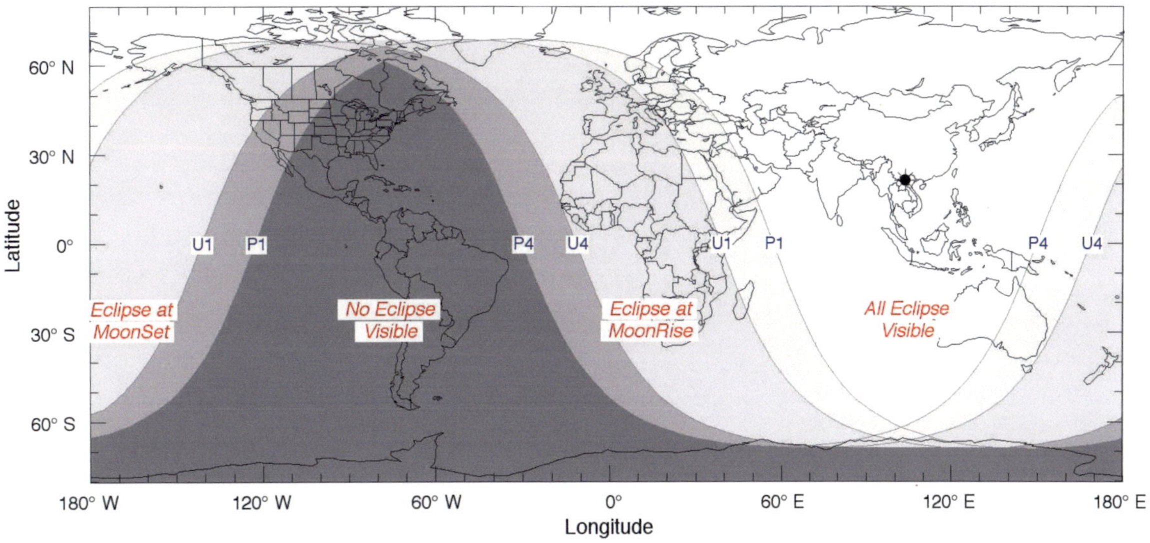

Total Lunar Eclipse of 2040 May 26

Greatest Eclipse = 11:46:21.6 TD (= 11:45:02.2 UT1)

Penumbral Magnitude = 2.4955 Gamma = -0.1872 Saros Series = 131
Umbral Magnitude = 1.5365 Axis = 0.1885° Saros Member = 35 of 72

Sun at Greatest Eclipse
(Geocentric Coordinates)
R.A. = 04h15m46.6s
Dec. = +21°16'35.1"
S.D. = 00°15'47.2"
H.P. = 00°00'08.7"

Moon at Greatest Eclipse
(Geocentric Coordinates)
R.A. = 16h15m33.4s
Dec. = -21°27'28.2"
S.D. = 00°16'27.7"
H.P. = 01°00'24.9"

N
Earth´s Penumbra

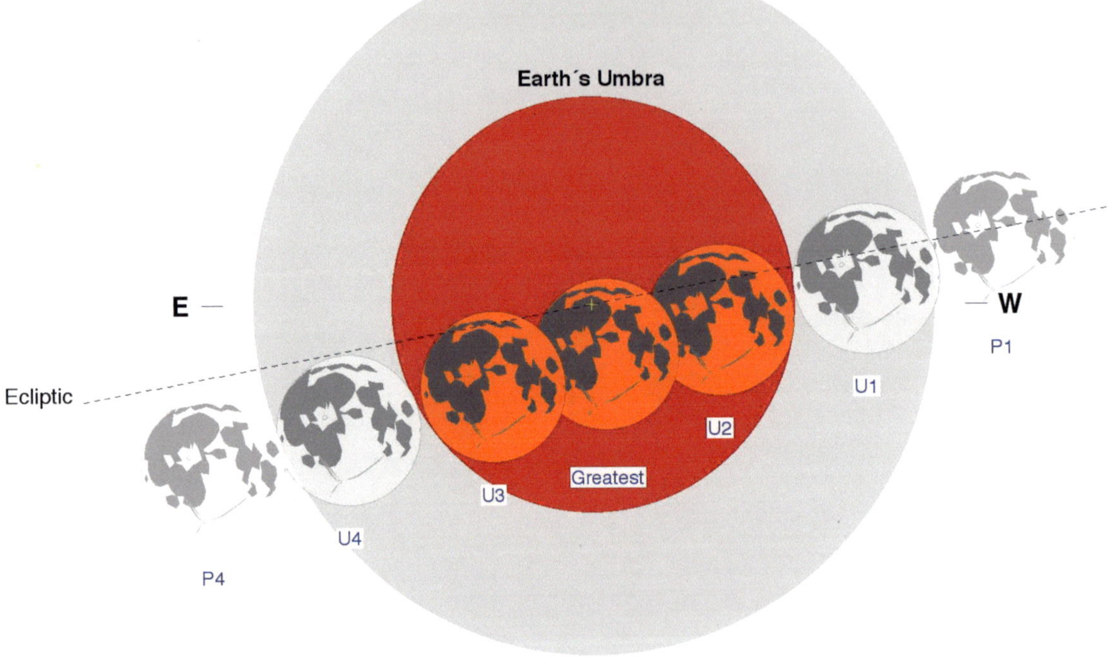

S

Eclipse Durations
Penumbral = 05h22m11s
Umbral = 03h31m28s
Total = 01h32m56s

Eph. = JPL DE430
Rule = Herald-Sinnott
ΔT = 79 s

0 15 30 45 60
Arc-Minutes

Eclipse Contacts
P1 = 09:03:54 UT1
U1 = 09:59:17 UT1
U2 = 10:58:31 UT1
U3 = 12:31:27 UT1
U4 = 13:30:45 UT1
P4 = 14:26:05 UT1

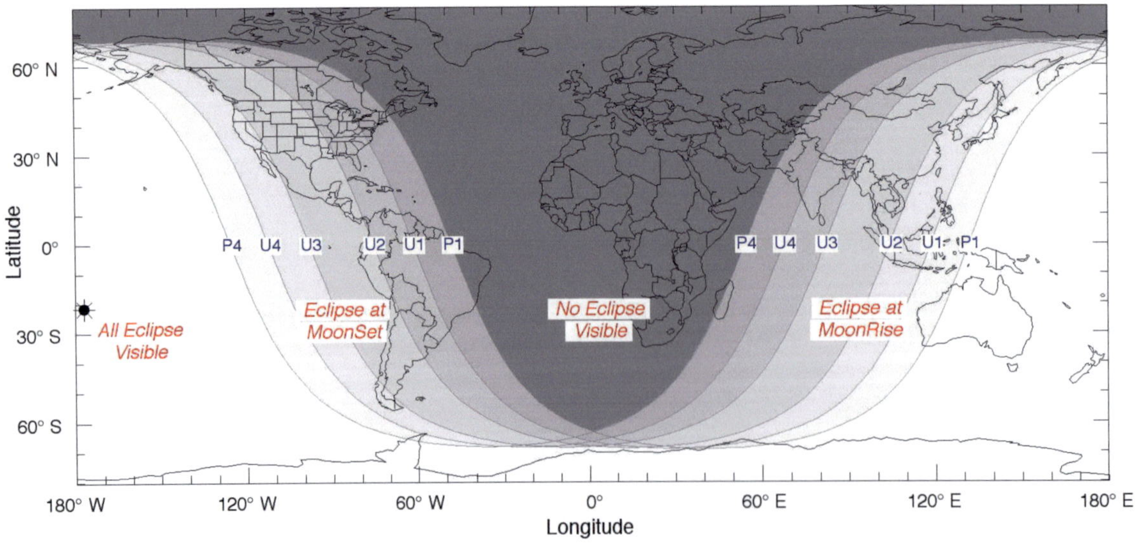

Total Lunar Eclipse of 2040 Nov 18

Greatest Eclipse = 19:04:40.5 TD (= 19:03:20.9 UT1)

Penumbral Magnitude = 2.4543   Gamma = 0.2361   Saros Series = 136
Umbral Magnitude = 1.3991   Axis = 0.2215°   Saros Member = 21 of 72

Sun at Greatest Eclipse
(Geocentric Coordinates)
R.A. = 15h39m03.9s
Dec. = -19°29'49.7"
S.D. = 00°16'11.0"
H.P. = 00°00'08.9"

Moon at Greatest Eclipse
(Geocentric Coordinates)
R.A. = 03h38m45.6s
Dec. = +19°42'23.6"
S.D. = 00°15'20.2"
H.P. = 00°56'17.3"

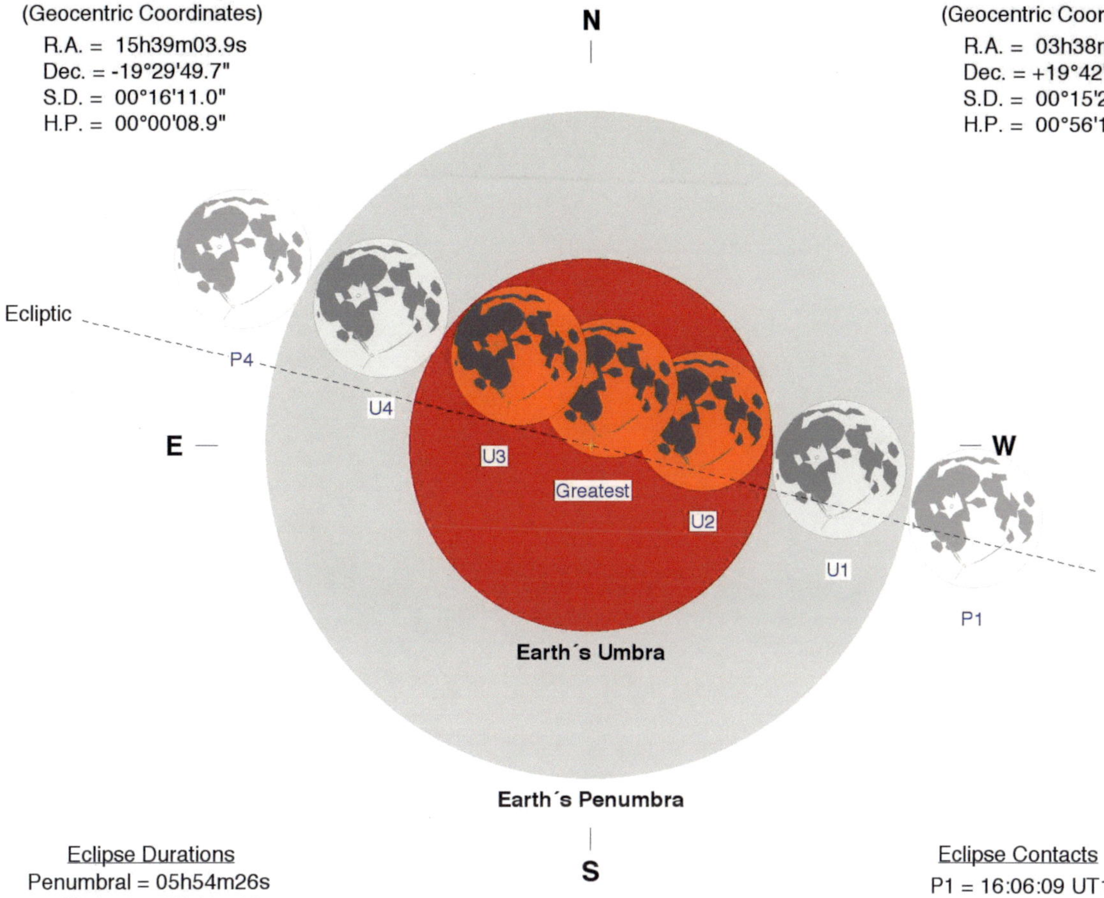

Eclipse Durations
Penumbral = 05h54m26s
Umbral = 03h41m13s
Total = 01h28m33s

Eph. = JPL DE430
Rule = Herald-Sinnott
ΔT = 80 s

0 15 30 45 60
Arc-Minutes

©2020 F. Espenak, www.EclipseWise.com

Eclipse Contacts
P1 = 16:06:09 UT1
U1 = 17:12:40 UT1
U2 = 18:18:57 UT1
U3 = 19:47:30 UT1
U4 = 20:53:53 UT1
P4 = 22:00:35 UT1

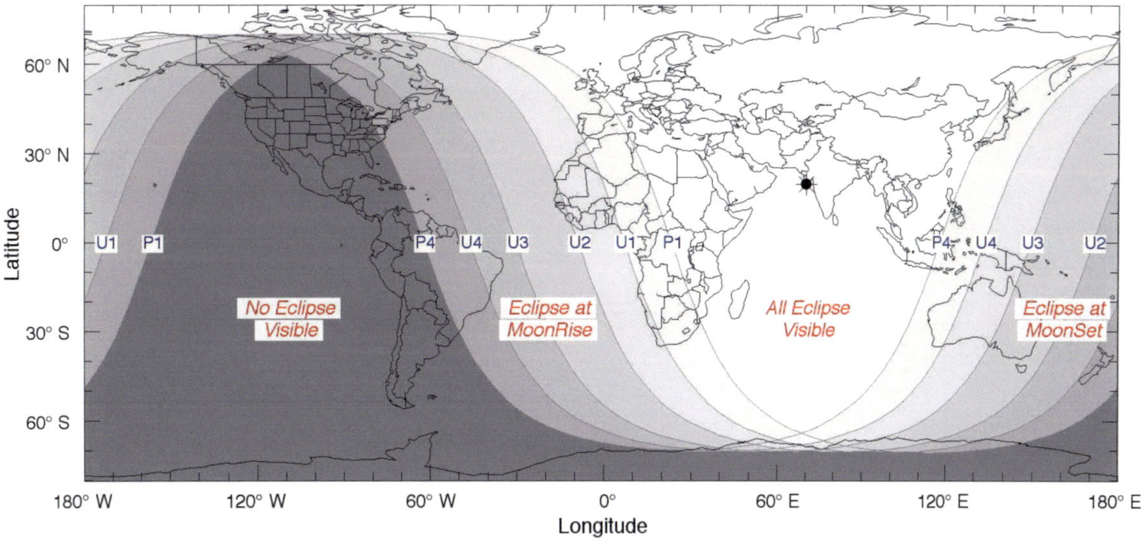

Partial Lunar Eclipse of 2041 May 16

Greatest Eclipse = 00:43:02.1 TD (= 00:41:42.2 UT1)

Penumbral Magnitude = 1.0765
Umbral Magnitude = 0.0663

Gamma = -0.9747
Axis = 0.9335°

Saros Series = 141
Saros Member = 25 of 72

Sun at Greatest Eclipse
(Geocentric Coordinates)
R.A. = 03h32m49.6s
Dec. = +19°08'35.5"
S.D. = 00°15'49.2"
H.P. = 00°00'08.7"

Moon at Greatest Eclipse
(Geocentric Coordinates)
R.A. = 15h31m30.5s
Dec. = -20°01'25.1"
S.D. = 00°15'39.6"
H.P. = 00°57'28.4"

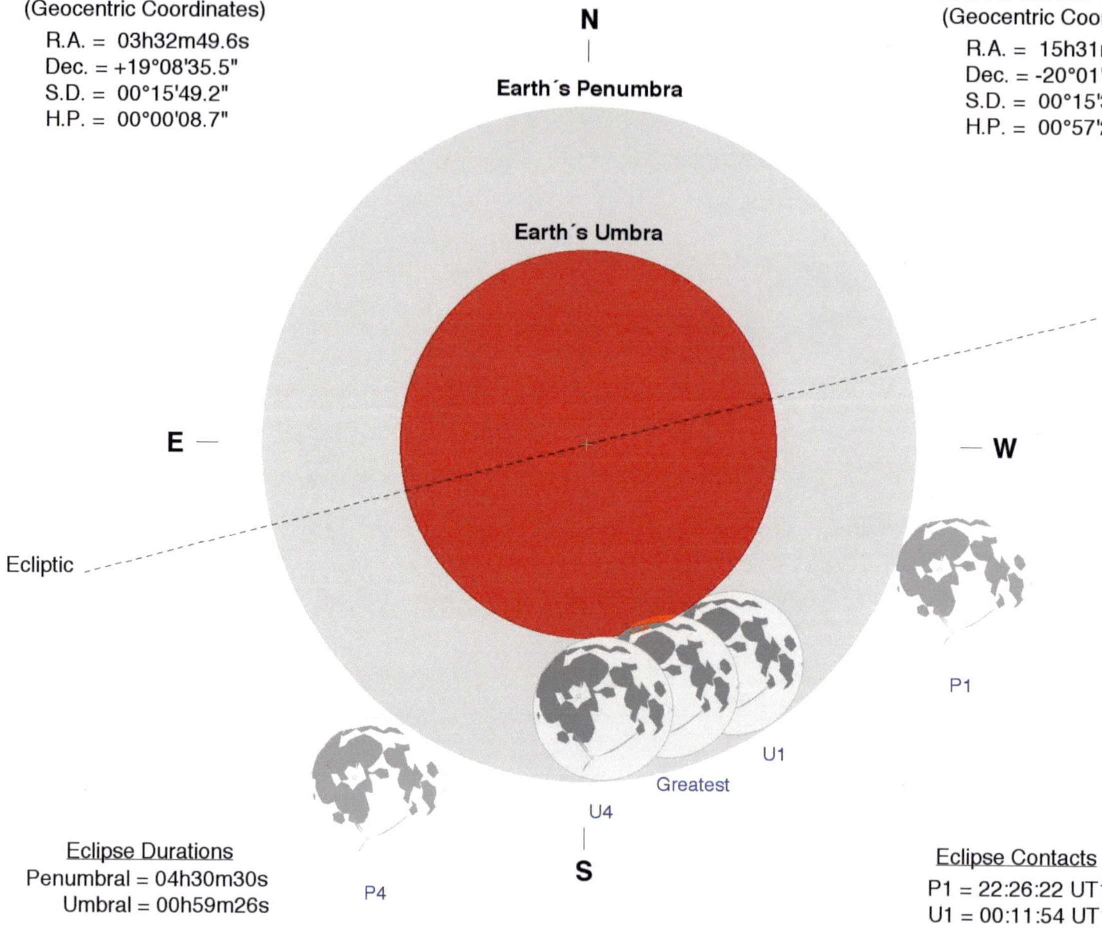

Eclipse Durations
Penumbral = 04h30m30s
Umbral = 00h59m26s

Eclipse Contacts
P1 = 22:26:22 UT1
U1 = 00:11:54 UT1
U4 = 01:11:20 UT1
P4 = 02:56:51 UT1

Eph. = JPL DE430
Rule = Herald-Sinnott
ΔT = 80 s

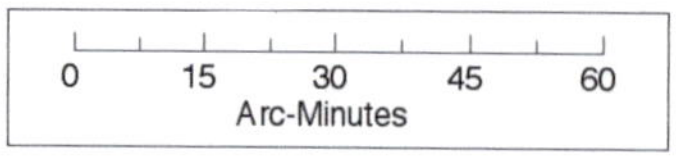

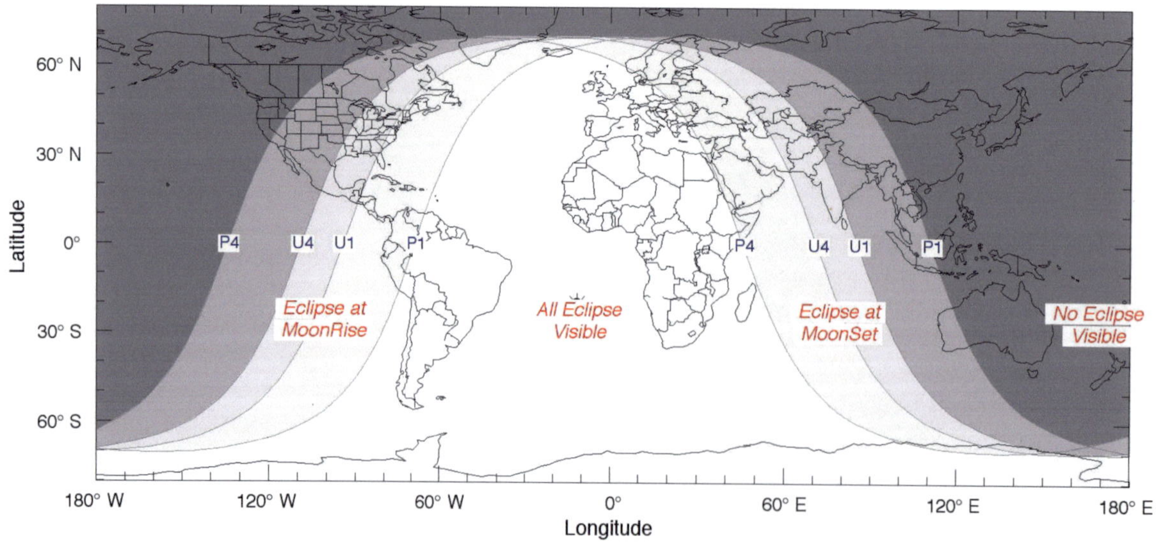

Partial Lunar Eclipse of 2041 Nov 08

Greatest Eclipse = 04:35:04.2 TD (= 04:33:44.0 UT1)

Penumbral Magnitude = 1.1675 Gamma = 0.9212 Saros Series = 146
Umbral Magnitude = 0.1714 Axis = 0.9131° Saros Member = 12 of 72

Sun at Greatest Eclipse
(Geocentric Coordinates)
R.A. = 14h54m42.6s
Dec. = -16°39'56.0"
S.D. = 00°16'08.5"
H.P. = 00°00'08.9"

Moon at Greatest Eclipse
(Geocentric Coordinates)
R.A. = 02h53m15.3s
Dec. = +17°30'36.2"
S.D. = 00°16'12.4"
H.P. = 00°59'28.8"

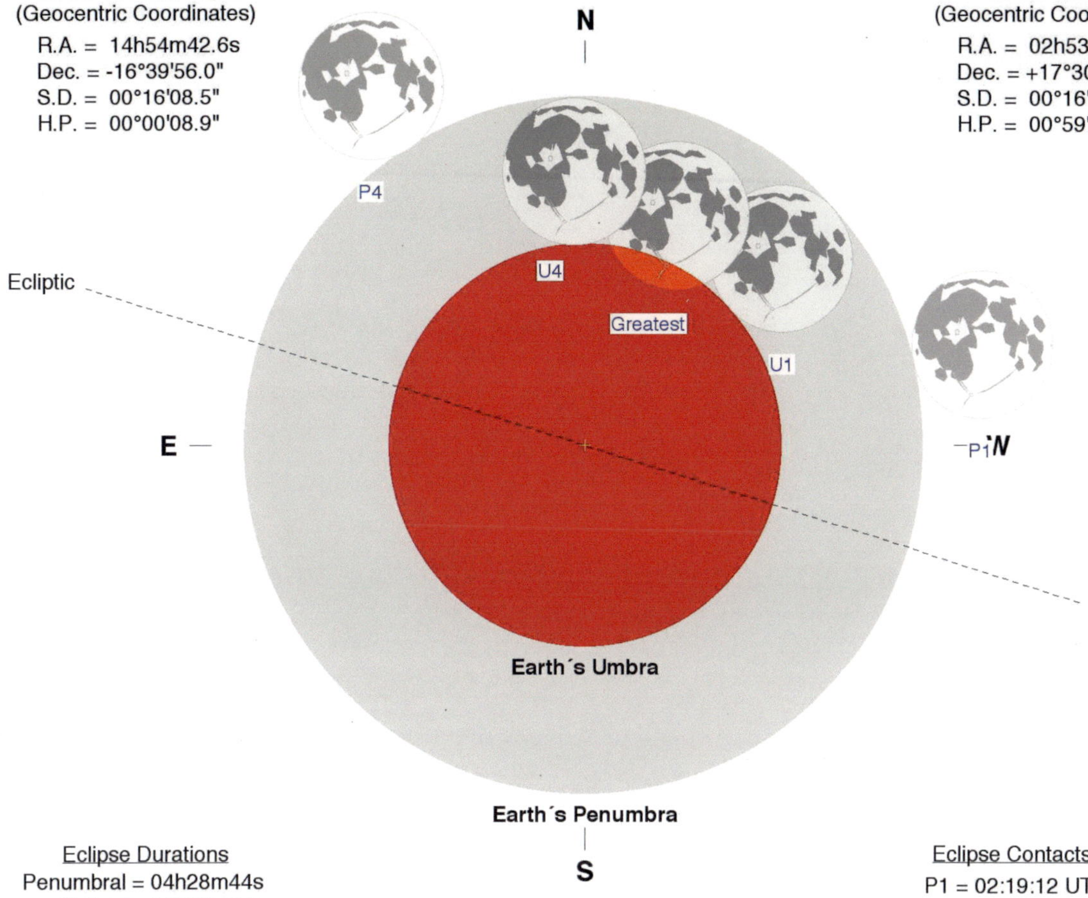

Eclipse Durations
Penumbral = 04h28m44s
Umbral = 01h31m04s

Eclipse Contacts
P1 = 02:19:12 UT1
U1 = 03:47:54 UT1
U4 = 05:18:59 UT1
P4 = 06:47:57 UT1

Eph. = JPL DE430
Rule = Herald-Sinnott
ΔT = 80 s

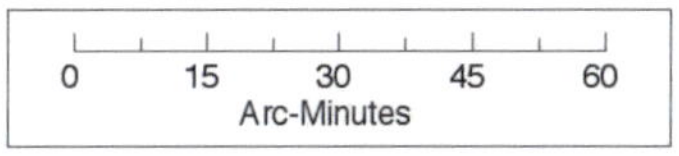

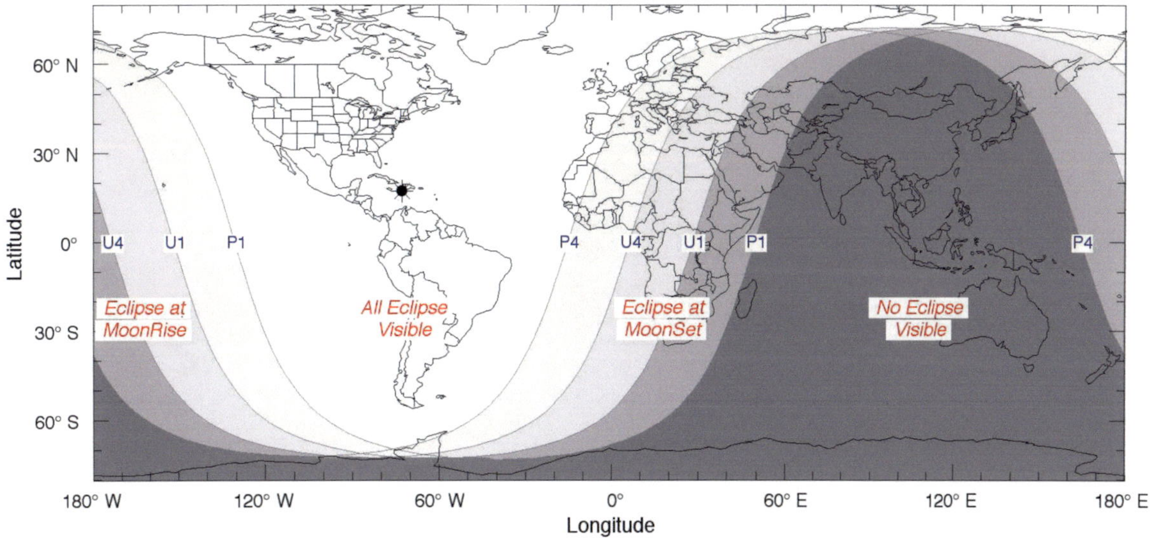

Penumbral Lunar Eclipse of 2042 Apr 05

Greatest Eclipse = 14:30:12.0 TD (= 14:28:51.6 UT1)

Penumbral Magnitude = 0.8700
Umbral Magnitude = -0.2156

Gamma = 1.1080
Axis = 0.9981°

Saros Series = 113
Saros Member = 65 of 71

Sun at Greatest Eclipse
(Geocentric Coordinates)
R.A. = 00h58m43.2s
Dec. = +06°16'08.8"
S.D. = 00°15'59.3"
H.P. = 00°00'08.8"

Moon at Greatest Eclipse
(Geocentric Coordinates)
R.A. = 13h00m37.2s
Dec. = -05°23'23.8"
S.D. = 00°14'43.6"
H.P. = 00°54'03.0"

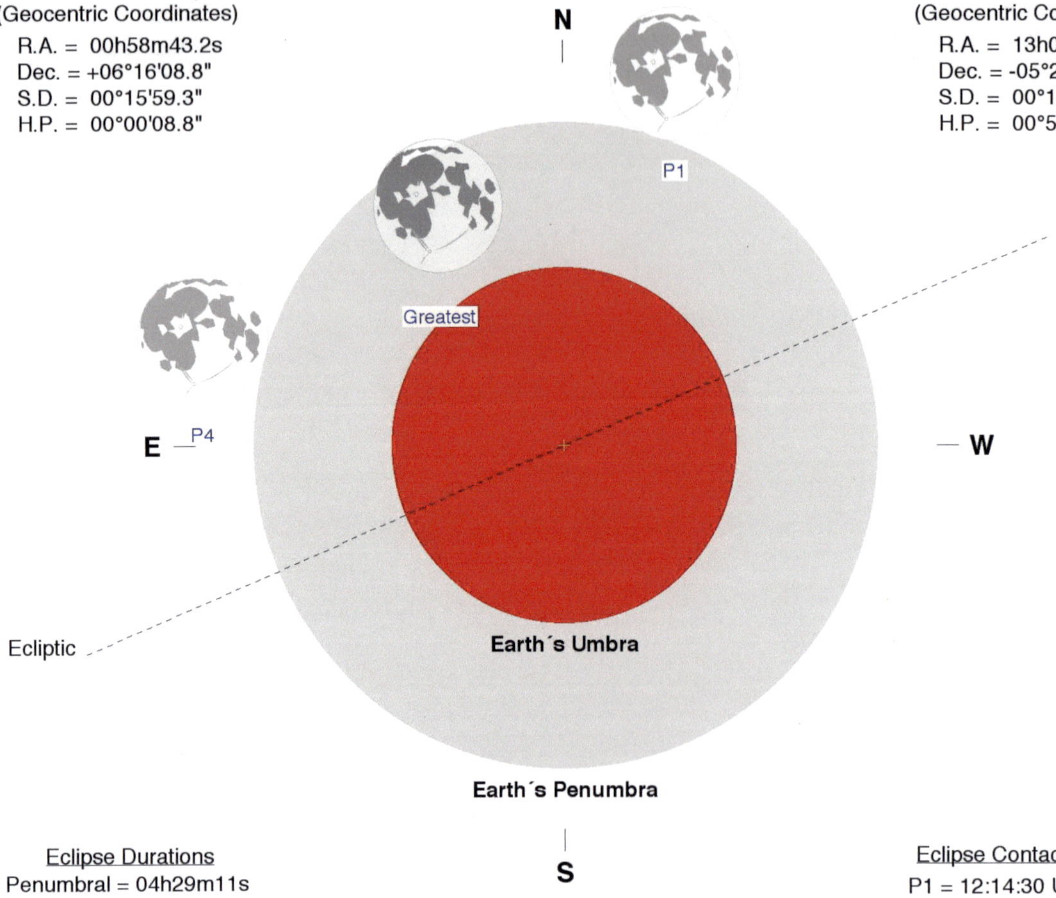

Eclipse Durations
Penumbral = 04h29m11s

Eclipse Contacts
P1 = 12:14:30 UT1
P4 = 16:43:42 UT1

Eph. = JPL DE430
Rule = Herald-Sinnott
ΔT = 80 s

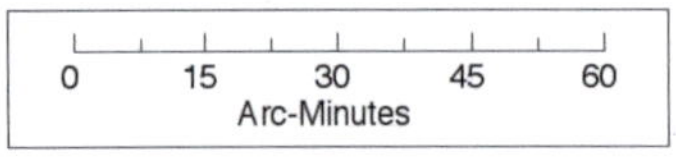

©2020 F. Espenak, www.EclipseWise.com

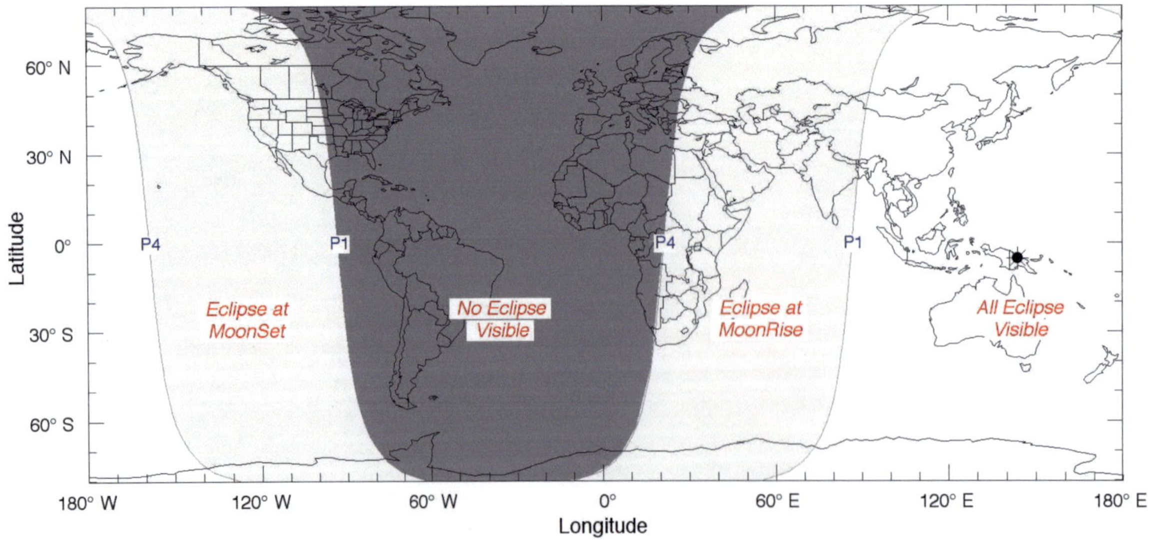

Penumbral Lunar Eclipse of 2042 Sep 29

Greatest Eclipse = 10:45:47.4 TD (= 10:44:26.8 UT1)

Penumbral Magnitude = 0.9548
Umbral Magnitude = -0.0010

Gamma = -1.0262
Axis = 1.0483°

Saros Series = 118
Saros Member = 53 of 73

Sun at Greatest Eclipse
(Geocentric Coordinates)
R.A. = 12h23m37.3s
Dec. = -02°33'13.4"
S.D. = 00°15'57.9"
H.P. = 00°00'08.8"

Moon at Greatest Eclipse
(Geocentric Coordinates)
R.A. = 00h25m38.7s
Dec. = +01°38'07.3"
S.D. = 00°16'42.1"
H.P. = 01°01'18.0"

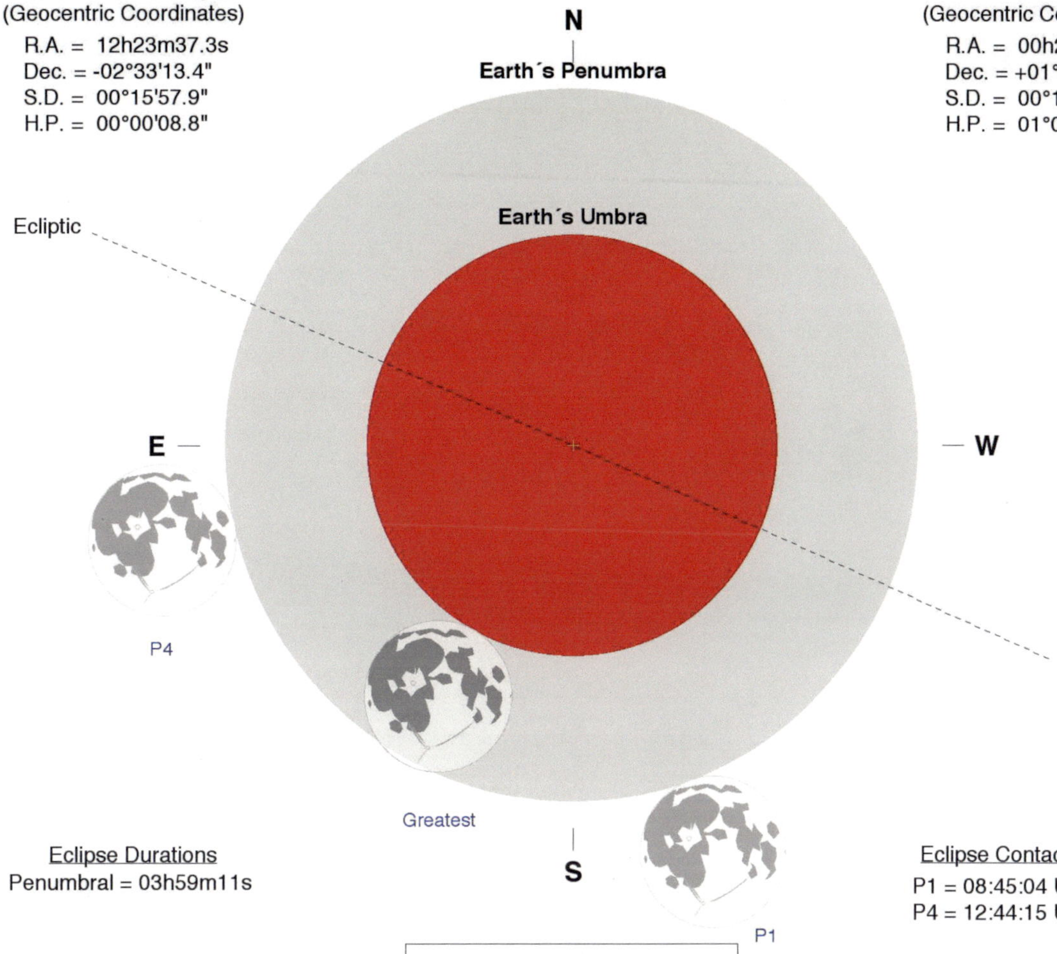

Eclipse Durations
Penumbral = 03h59m11s

Eclipse Contacts
P1 = 08:45:04 UT1
P4 = 12:44:15 UT1

Eph. = JPL DE430
Rule = Herald-Sinnott
ΔT = 81 s

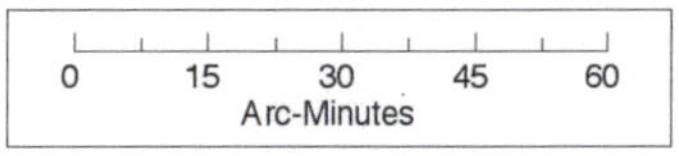

©2020 F. Espenak, www.EclipseWise.com

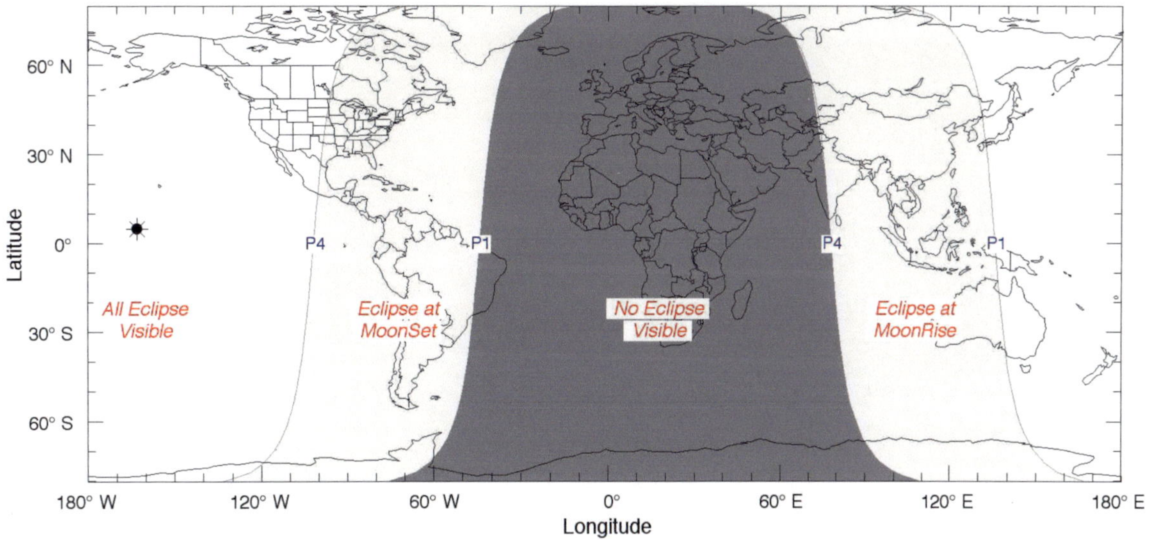

Total Lunar Eclipse of 2043 Mar 25

Greatest Eclipse = 14:32:03.6 TD (= 14:30:42.6 UT1)

Penumbral Magnitude = 2.1920 Gamma = 0.3849 Saros Series = 123
Umbral Magnitude = 1.1161 Axis = 0.3510° Saros Member = 54 of 72

Sun at Greatest Eclipse
(Geocentric Coordinates)
R.A. = 00h17m45.9s
Dec. = +01°55'21.5"
S.D. = 00°16'02.4"
H.P. = 00°00'08.8"

Moon at Greatest Eclipse
(Geocentric Coordinates)
R.A. = 12h18m26.9s
Dec. = -01°36'57.6"
S.D. = 00°14'54.5"
H.P. = 00°54'42.9"

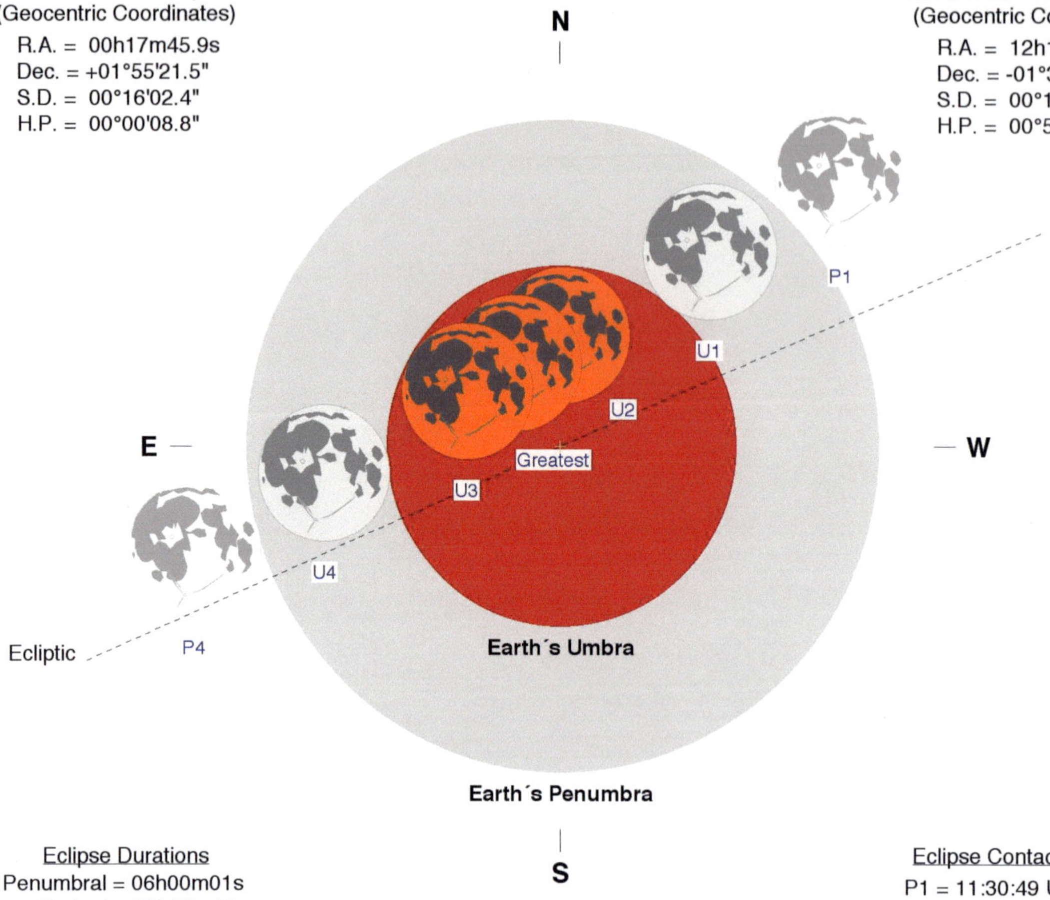

Eclipse Durations
Penumbral = 06h00m01s
Umbral = 03h35m18s
Total = 00h54m06s

Eph. = JPL DE430
Rule = Herald-Sinnott
ΔT = 81 s

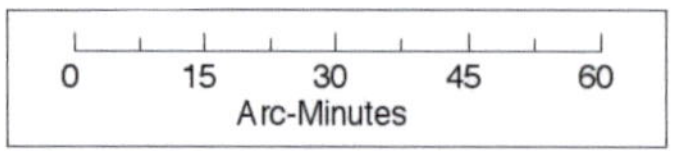

Eclipse Contacts
P1 = 11:30:49 UT1
U1 = 12:43:10 UT1
U2 = 14:03:56 UT1
U3 = 14:58:02 UT1
U4 = 16:18:28 UT1
P4 = 17:30:50 UT1

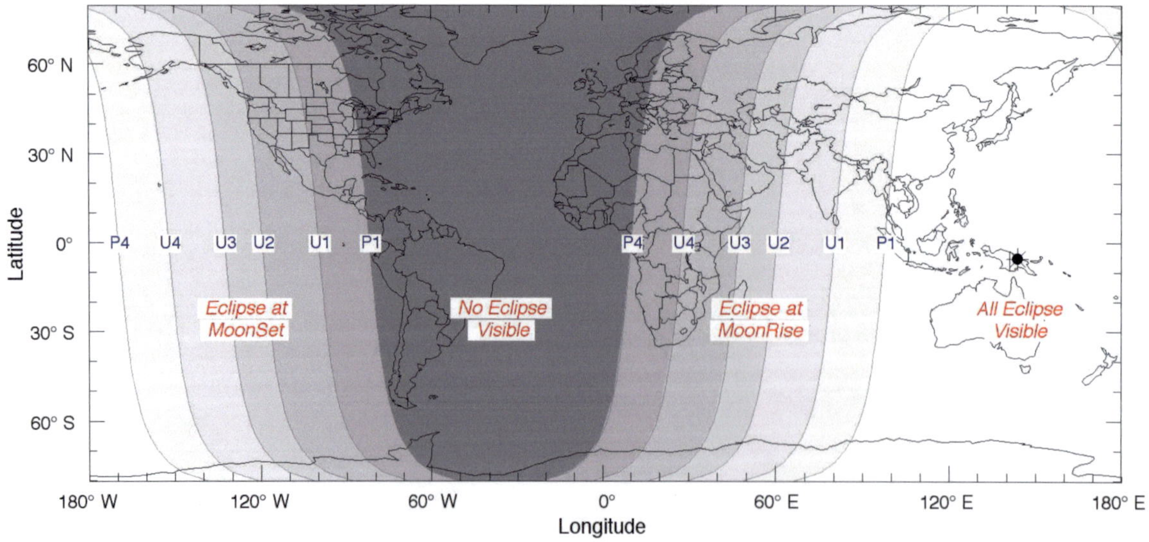

Total Lunar Eclipse of 2043 Sep 19

Greatest Eclipse = 01:51:49.8 TD (= 01:50:28.6 UT1)

Penumbral Magnitude = 2.2452 Gamma = -0.3316 Saros Series = 128
Umbral Magnitude = 1.2575 Axis = 0.3269° Saros Member = 42 of 71

Sun at Greatest Eclipse
(Geocentric Coordinates)
R.A. = 11h45m28.0s
Dec. = +01°34'24.4"
S.D. = 00°15'55.1"
H.P. = 00°00'08.8"

Moon at Greatest Eclipse
(Geocentric Coordinates)
R.A. = 23h46m06.1s
Dec. = -01°51'33.2"
S.D. = 00°16'07.0"
H.P. = 00°59'08.8"

N

Earth´s Penumbra

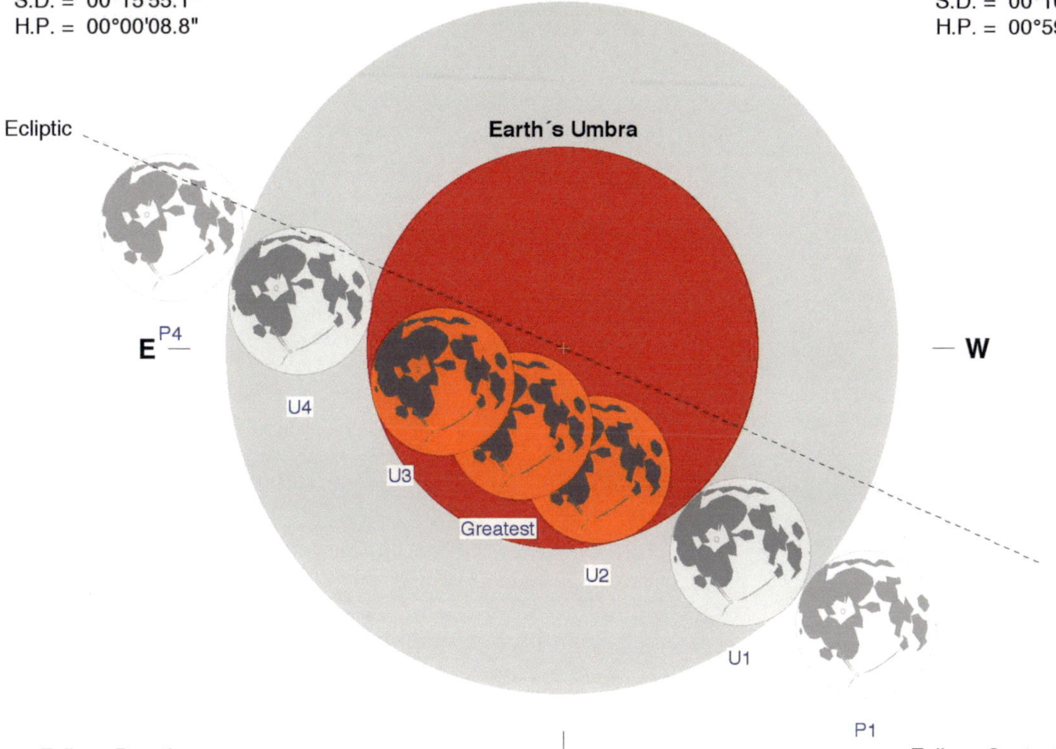

S

Eclipse Durations
Penumbral = 05h26m27s
Umbral = 03h26m39s
Total = 01h12m20s

Eph. = JPL DE430
Rule = Herald-Sinnott
ΔT = 81 s

0 15 30 45 60
Arc-Minutes

Eclipse Contacts
P1 = 23:07:17 UT1
U1 = 00:07:16 UT1
U2 = 01:14:33 UT1
U3 = 02:26:52 UT1
U4 = 03:33:56 UT1
P4 = 04:33:44 UT1

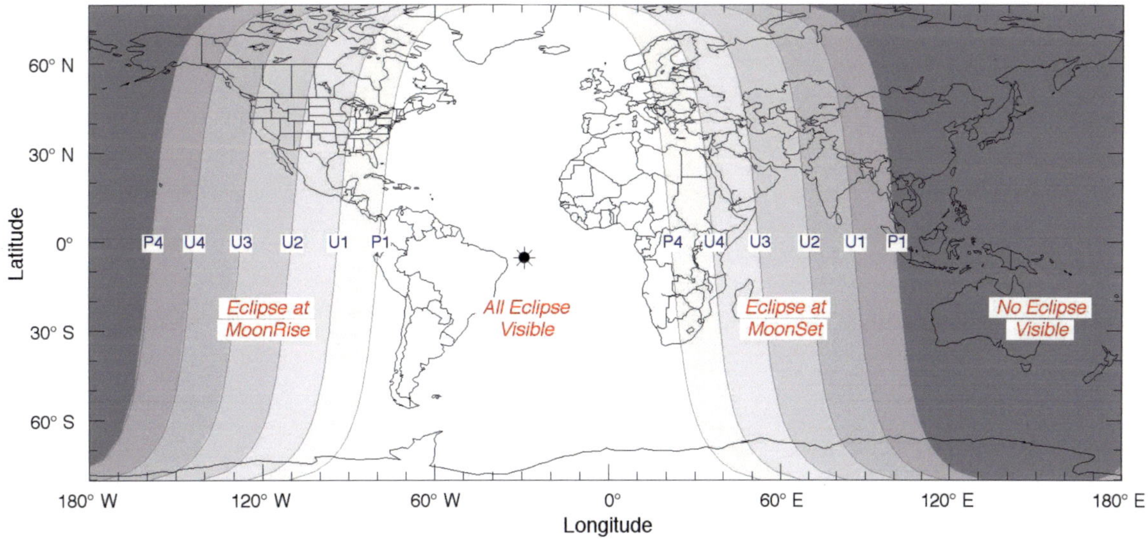

Total Lunar Eclipse of 2044 Mar 13

Greatest Eclipse = 19:38:32.5 TD (= 19:37:11.0 UT1)

Penumbral Magnitude = 2.2322 Gamma = -0.3496 Saros Series = 133
Umbral Magnitude = 1.2050 Axis = 0.3349° Saros Member = 28 of 71

Sun at Greatest Eclipse
(Geocentric Coordinates)
R.A. = 23h37m30.3s
Dec. = -02°25'56.9"
S.D. = 00°16'05.4"
H.P. = 00°00'08.8"

Moon at Greatest Eclipse
(Geocentric Coordinates)
R.A. = 11h36m51.3s
Dec. = +02°08'22.5"
S.D. = 00°15'39.8"
H.P. = 00°57'29.1"

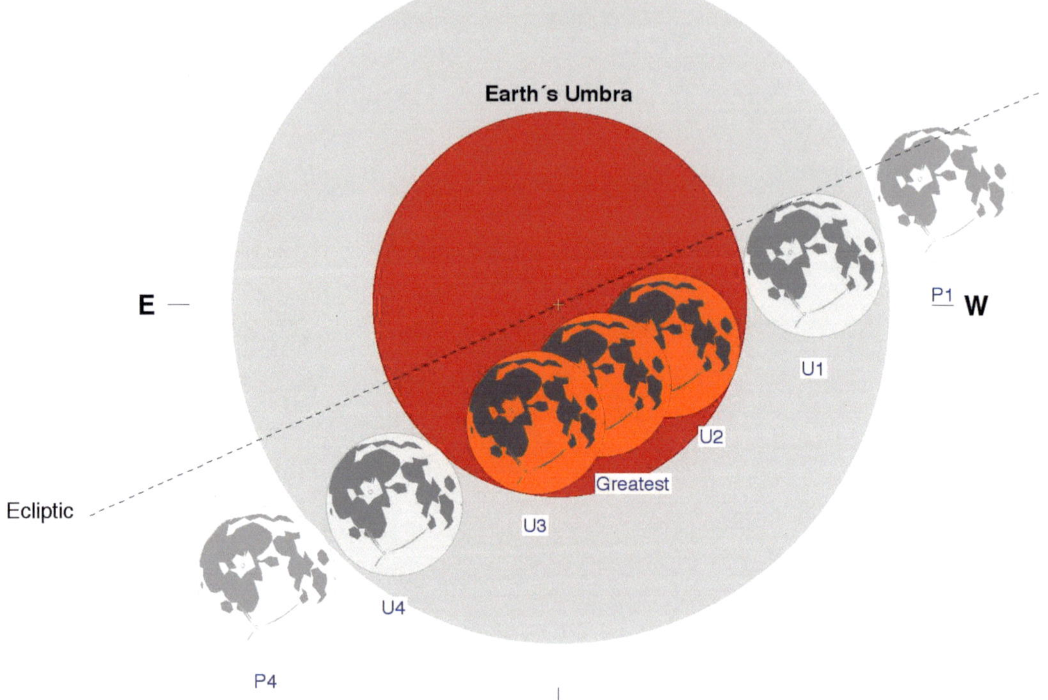

Eclipse Durations
Penumbral = 05h39m06s
Umbral = 03h29m43s
Total = 01h07m02s

Eph. = JPL DE430
Rule = Herald-Sinnott
ΔT = 82 s

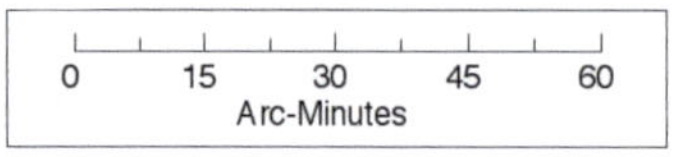

©2020 F. Espenak, www.EclipseWise.com

Eclipse Contacts
P1 = 16:47:36 UT1
U1 = 17:52:11 UT1
U2 = 19:03:24 UT1
U3 = 20:10:26 UT1
U4 = 21:21:54 UT1
P4 = 22:26:42 UT1

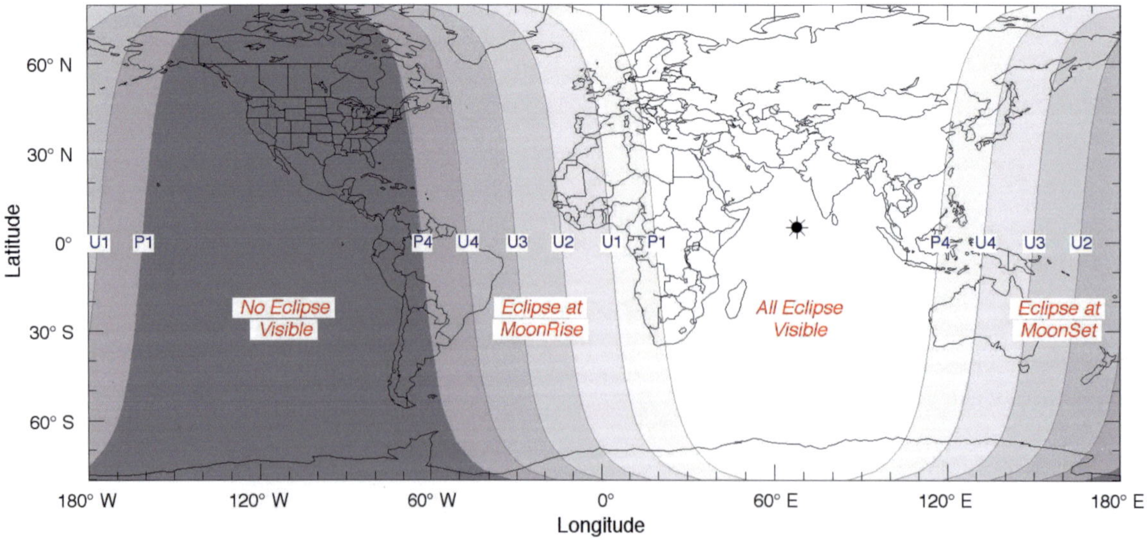

Total Lunar Eclipse of 2044 Sep 07

Greatest Eclipse = 11:20:44.4 TD (= 11:19:22.6 UT1)

Penumbral Magnitude = 2.0879	Gamma = 0.4318	Saros Series = 138
Umbral Magnitude = 1.0476	Axis = 0.4030°	Saros Member = 30 of 82

Sun at Greatest Eclipse
(Geocentric Coordinates)
R.A. = 11h06m33.5s
Dec. = +05°43'12.4"
S.D. = 00°15'52.4"
H.P. = 00°00'08.7"

Moon at Greatest Eclipse
(Geocentric Coordinates)
R.A. = 23h05m47.2s
Dec. = -05°21'56.9"
S.D. = 00°15'15.4"
H.P. = 00°55'59.6"

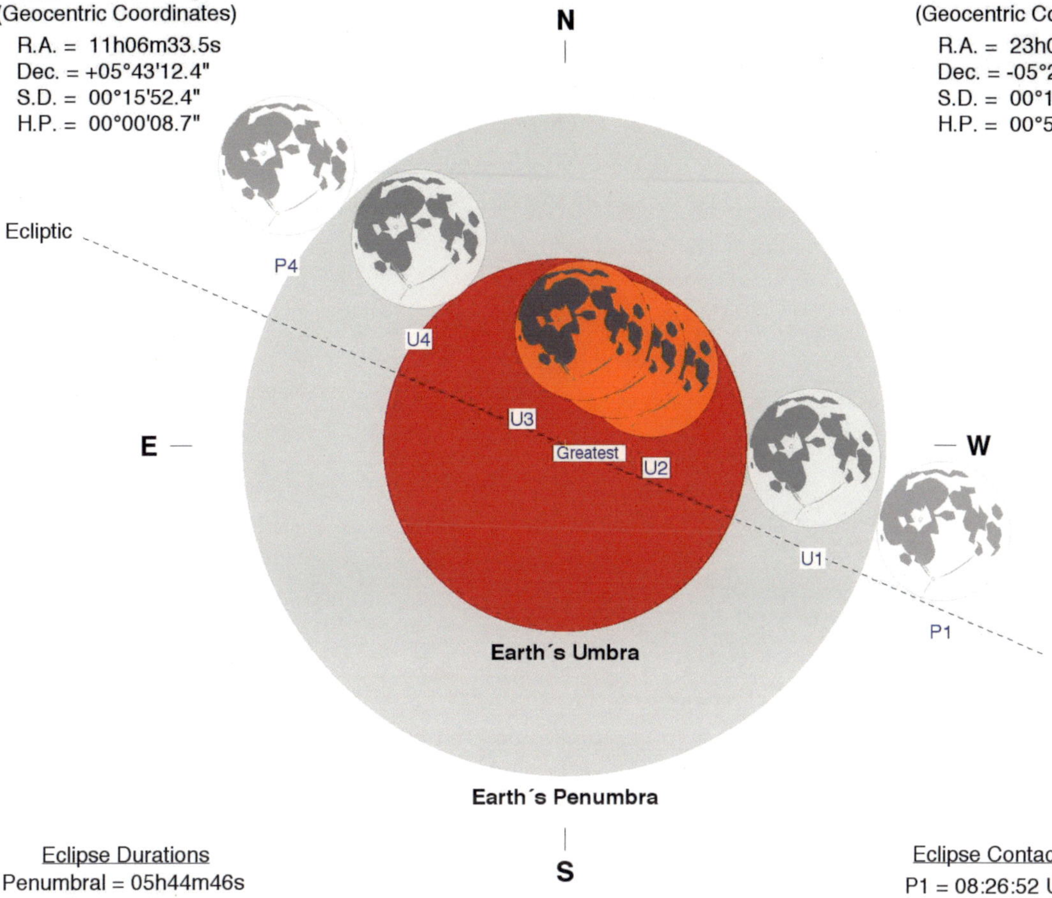

Eclipse Durations
Penumbral = 05h44m46s
Umbral = 03h26m52s
Total = 00h34m47s

Eph. = JPL DE430
Rule = Herald-Sinnott
ΔT = 82 s

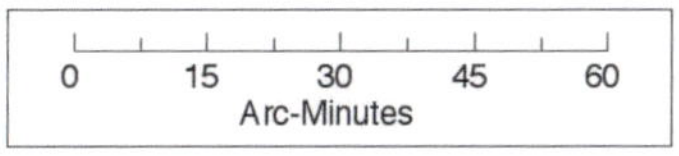

©2020 F. Espenak, www.EclipseWise.com

Eclipse Contacts
P1 = 08:26:52 UT1
U1 = 09:35:50 UT1
U2 = 11:01:42 UT1
U3 = 11:36:30 UT1
U4 = 13:02:42 UT1
P4 = 14:11:38 UT1

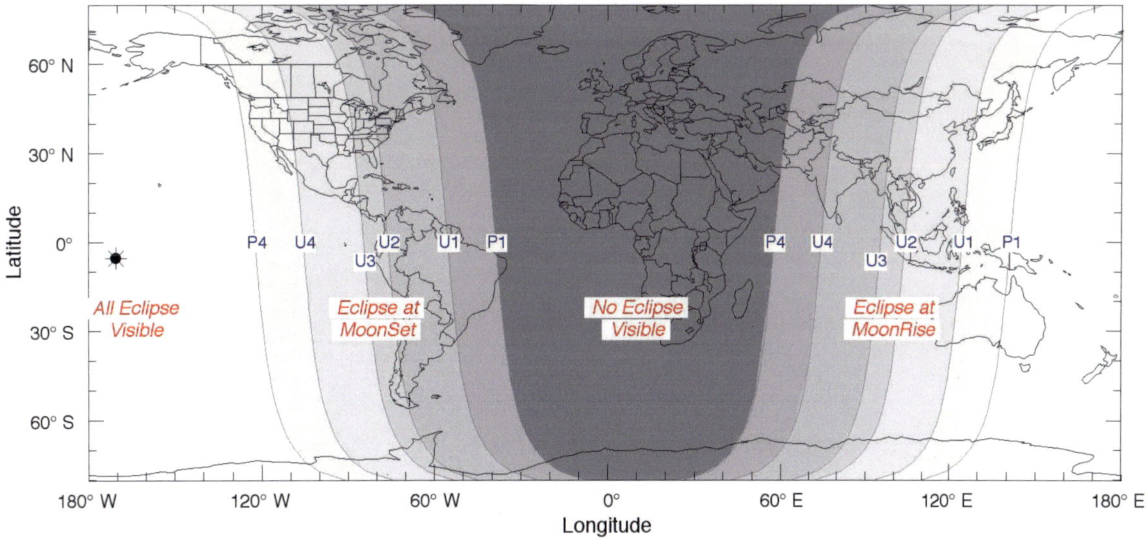

Penumbral Lunar Eclipse of 2045 Mar 03

Greatest Eclipse = 07:43:24.9 TD (= 07:42:02.8 UT1)

Penumbral Magnitude = 0.9643 Gamma = -1.0274 Saros Series = 143
Umbral Magnitude = -0.0148 Axis = 1.0354° Saros Member = 19 of 72

Sun at Greatest Eclipse
(Geocentric Coordinates)
R.A. = 22h57m49.1s
Dec. = -06°37'35.6"
S.D. = 00°16'08.1"
H.P. = 00°00'08.9"

Moon at Greatest Eclipse
(Geocentric Coordinates)
R.A. = 10h55m51.5s
Dec. = +05°42'46.0"
S.D. = 00°16'28.7"
H.P. = 01°00'28.6"

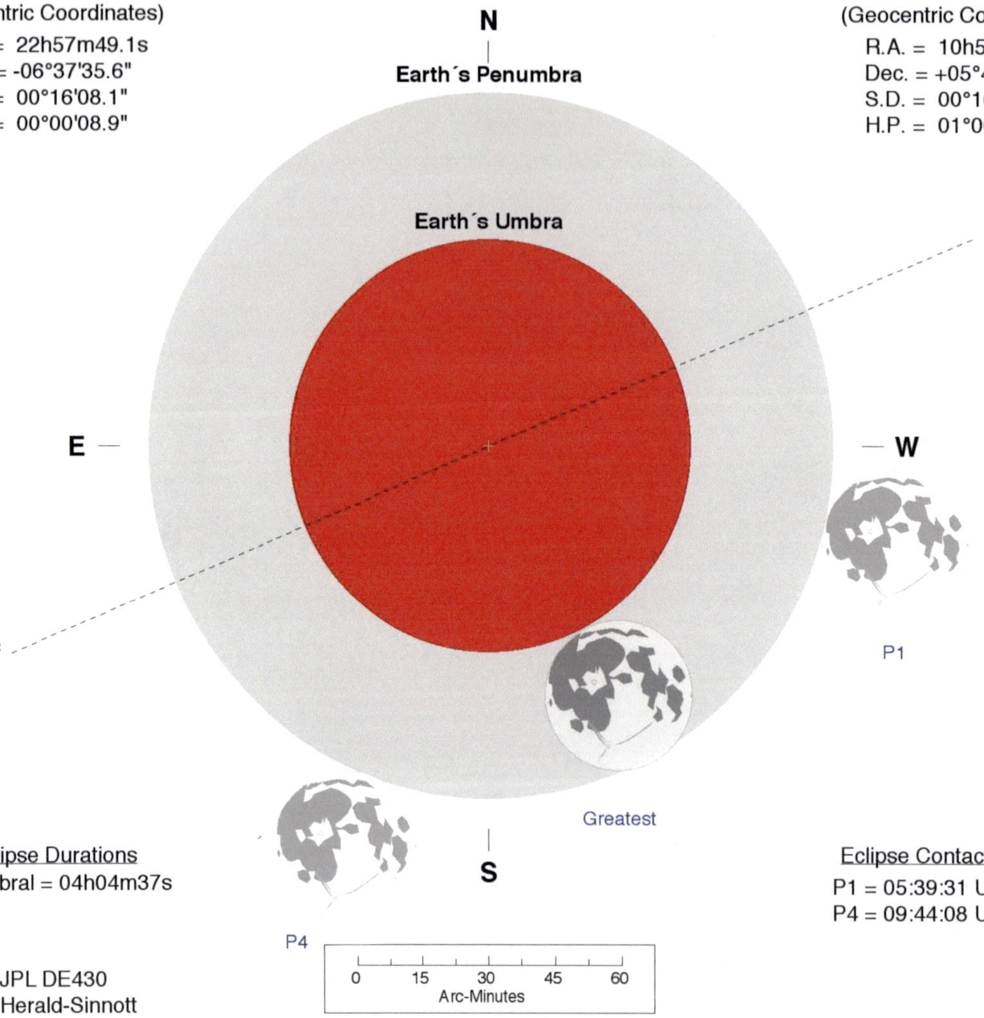

Eclipse Durations
Penumbral = 04h04m37s

Eclipse Contacts
P1 = 05:39:31 UT1
P4 = 09:44:08 UT1

Eph. = JPL DE430
Rule = Herald-Sinnott
ΔT = 82 s

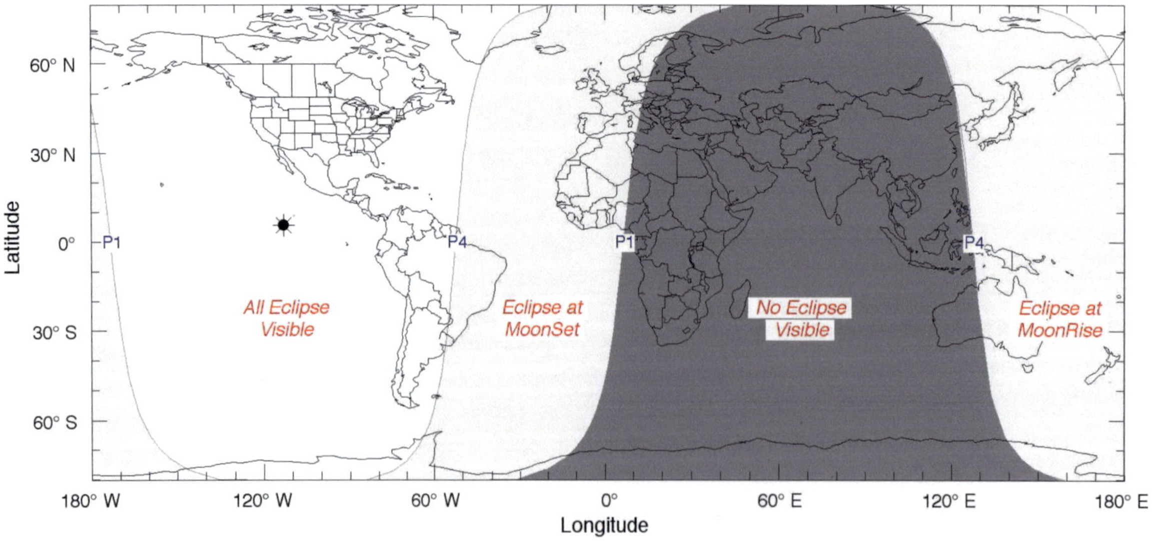

Penumbral Lunar Eclipse of 2045 Aug 27

Greatest Eclipse = 13:54:48.8 TD (= 13:53:26.4 UT1)

Penumbral Magnitude = 0.6845 Gamma = 1.2061 Saros Series = 148
Umbral Magnitude = -0.3899 Axis = 1.0869° Saros Member = 5 of 70

Sun at Greatest Eclipse
(Geocentric Coordinates)
R.A. = 10h26m15.1s
Dec. = +09°46'56.3"
S.D. = 00°15'49.9"
H.P. = 00°00'08.7"

Moon at Greatest Eclipse
(Geocentric Coordinates)
R.A. = 22h24m15.1s
Dec. = -08°48'49.2"
S.D. = 00°14'44.1"
H.P. = 00°54'04.7"

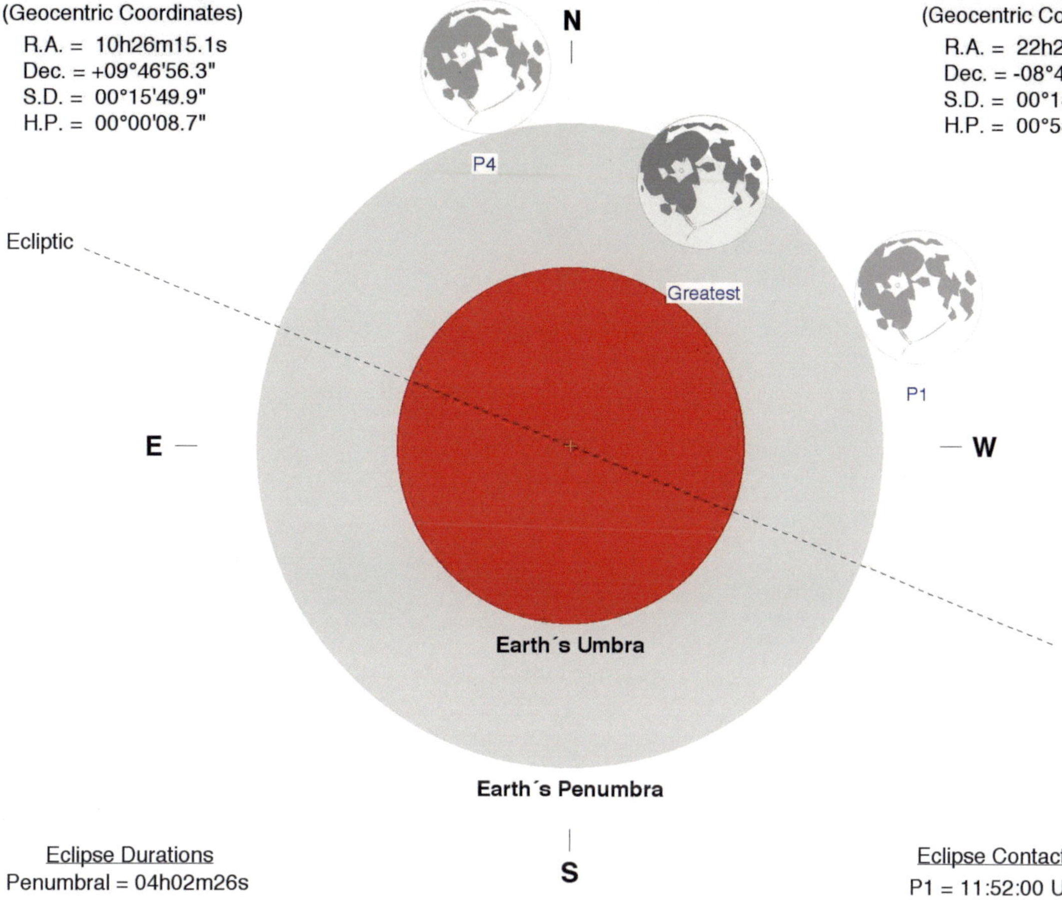

Eclipse Durations
Penumbral = 04h02m26s

Eclipse Contacts
P1 = 11:52:00 UT1
P4 = 15:54:26 UT1

Eph. = JPL DE430
Rule = Herald-Sinnott
ΔT = 82 s

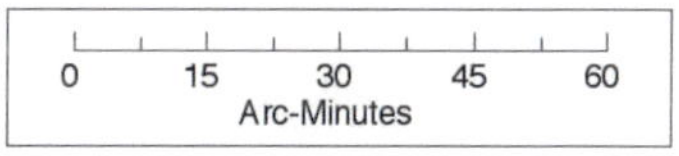

©2020 F. Espenak, www.EclipseWise.com

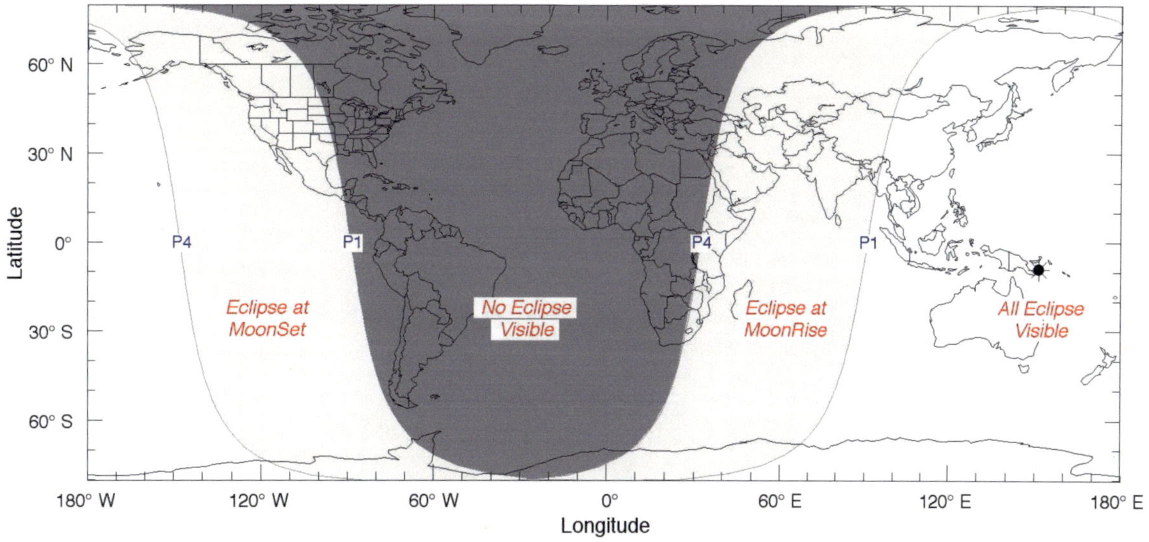

Partial Lunar Eclipse of 2046 Jan 22

Greatest Eclipse = 13:02:37.2 TD (= 13:01:14.5 UT1)

Penumbral Magnitude = 1.0365　Gamma = 0.9886　Saros Series = 115
Umbral Magnitude = 0.0550　Axis = 1.0011°　Saros Member = 59 of 72

Sun at Greatest Eclipse
(Geocentric Coordinates)
R.A. = 20h19m45.5s
Dec. = -19°33'42.8"
S.D. = 00°16'15.1"
H.P. = 00°00'08.9"

Moon at Greatest Eclipse
(Geocentric Coordinates)
R.A. = 08h21m07.9s
Dec. = +20°30'34.8"
S.D. = 00°16'33.4"
H.P. = 01°00'46.0"

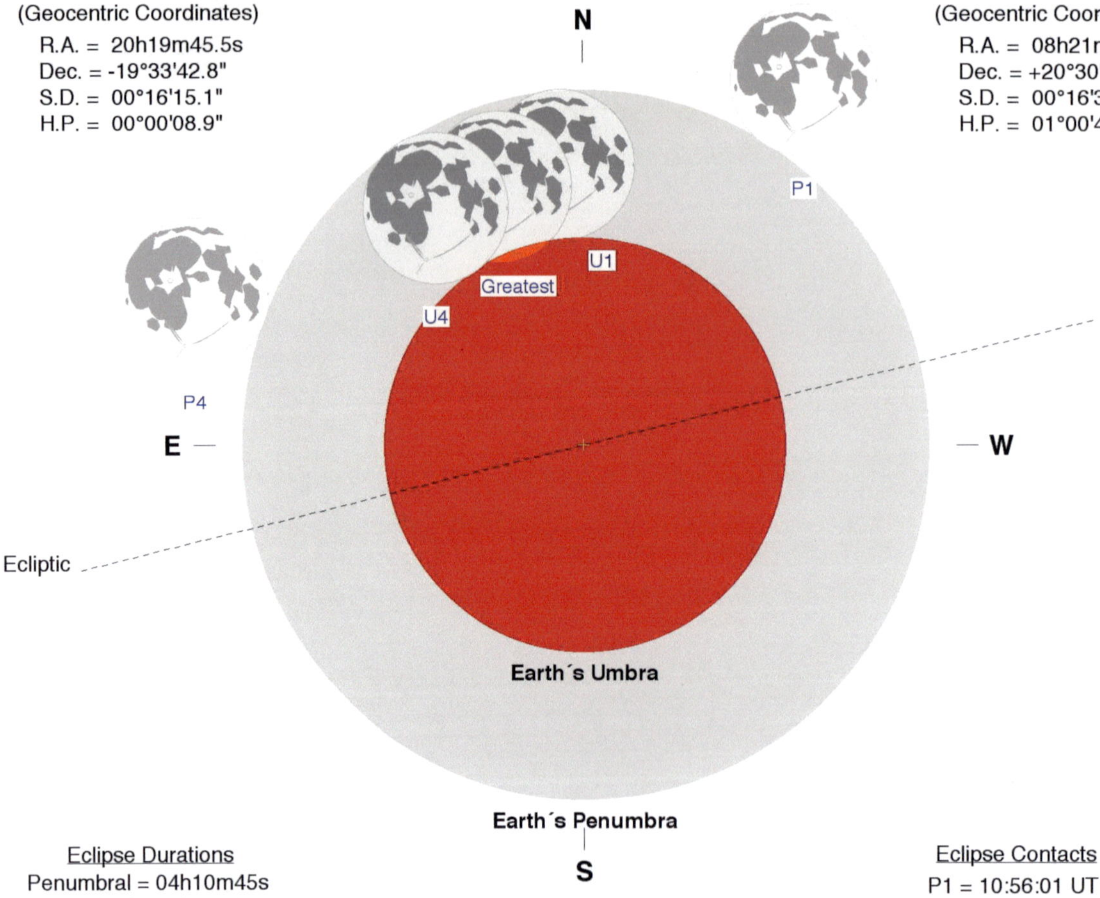

Eclipse Durations
Penumbral = 04h10m45s
Umbral = 00h51m22s

Eclipse Contacts
P1 = 10:56:01 UT1
U1 = 12:35:48 UT1
U4 = 13:27:10 UT1
P4 = 15:06:45 UT1

Eph. = JPL DE430
Rule = Herald-Sinnott
ΔT = 83 s

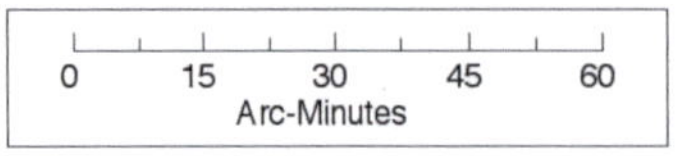

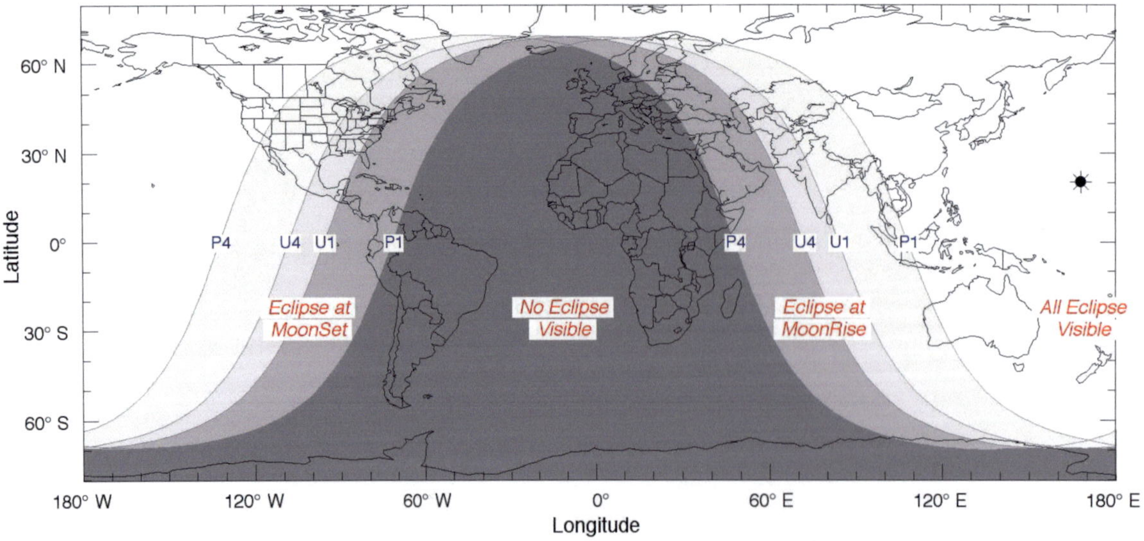

Partial Lunar Eclipse of 2046 Jul 18

Greatest Eclipse = 01:06:05.5 TD (= 01:04:42.5 UT1)

Penumbral Magnitude = 1.2824 Gamma = -0.8692 Saros Series = 120
Umbral Magnitude = 0.2478 Axis = 0.8086° Saros Member = 59 of 83

<u>Sun at Greatest Eclipse</u>
(Geocentric Coordinates)
R.A. = 07h50m23.8s
Dec. = +21°00'48.3"
S.D. = 00°15'44.2"
H.P. = 00°00'08.7"

<u>Moon at Greatest Eclipse</u>
(Geocentric Coordinates)
R.A. = 19h51m22.3s
Dec. = -21°47'22.3"
S.D. = 00°15'12.7"
H.P. = 00°55'49.5"

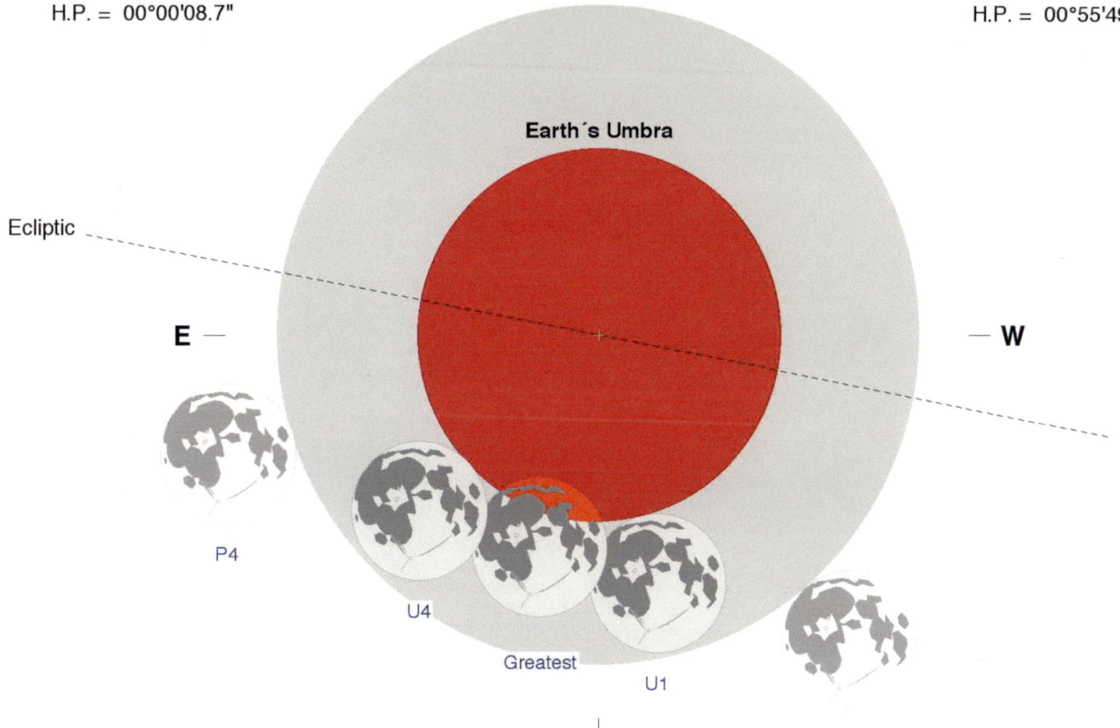

<u>Eclipse Durations</u>
Penumbral = 04h58m57s
Umbral = 01h55m18s

<u>Eclipse Contacts</u>
P1 = 22:35:19 UT1
U1 = 00:07:08 UT1
U4 = 02:02:26 UT1
P4 = 03:34:16 UT1

Eph. = JPL DE430
Rule = Herald-Sinnott
ΔT = 83 s

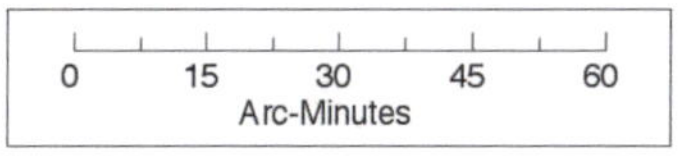

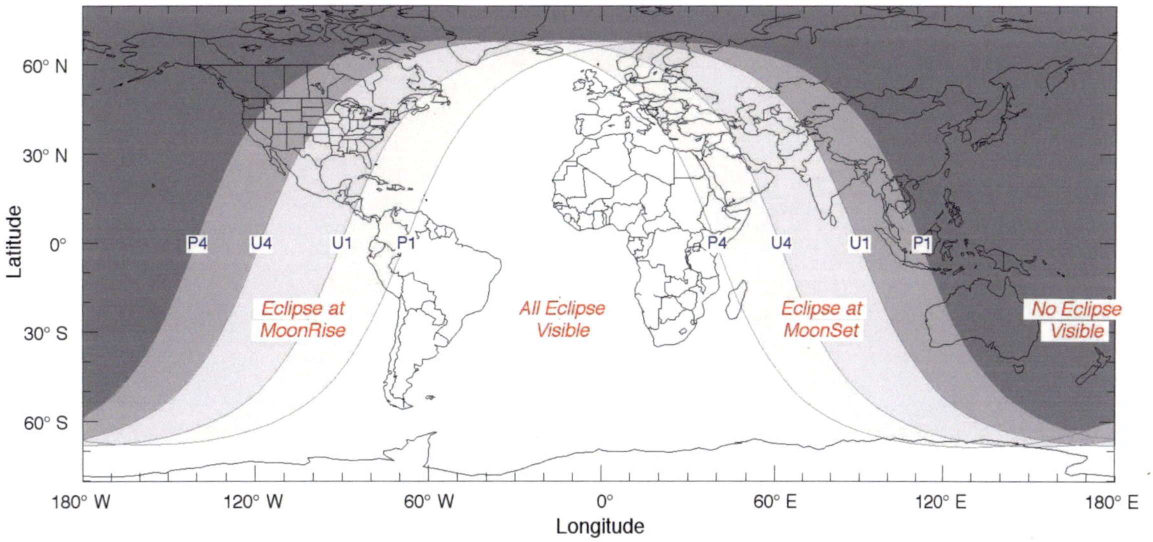

Total Lunar Eclipse of 2047 Jan 12

Greatest Eclipse = 01:26:14.4 TD (= 01:24:51.2 UT1)

Penumbral Magnitude = 2.2665 Gamma = 0.3317 Saros Series = 125
Umbral Magnitude = 1.2358 Axis = 0.3201° Saros Member = 50 of 72

Sun at Greatest Eclipse
(Geocentric Coordinates)
R.A. = 19h33m56.9s
Dec. = -21°40'46.3"
S.D. = 00°16'15.8"
H.P. = 00°00'08.9"

Moon at Greatest Eclipse
(Geocentric Coordinates)
R.A. = 07h34m18.1s
Dec. = +21°59'20.2"
S.D. = 00°15'46.6"
H.P. = 00°57'54.2"

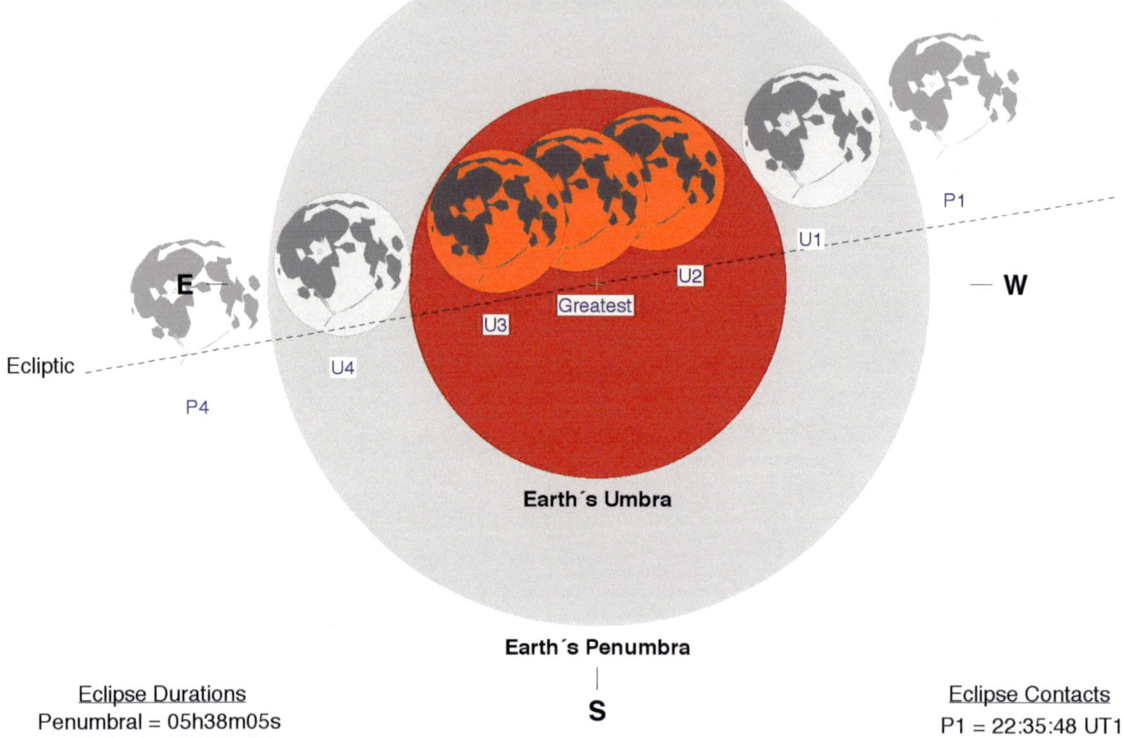

Eclipse Durations
Penumbral = 05h38m05s
Umbral = 03h29m39s
Total = 01h10m41s

Eph. = JPL DE430
Rule = Herald-Sinnott
ΔT = 83 s

0 15 30 45 60
Arc-Minutes

Eclipse Contacts
P1 = 22:35:48 UT1
U1 = 23:40:07 UT1
U2 = 00:49:39 UT1
U3 = 02:00:20 UT1
U4 = 03:09:45 UT1
P4 = 04:13:53 UT1

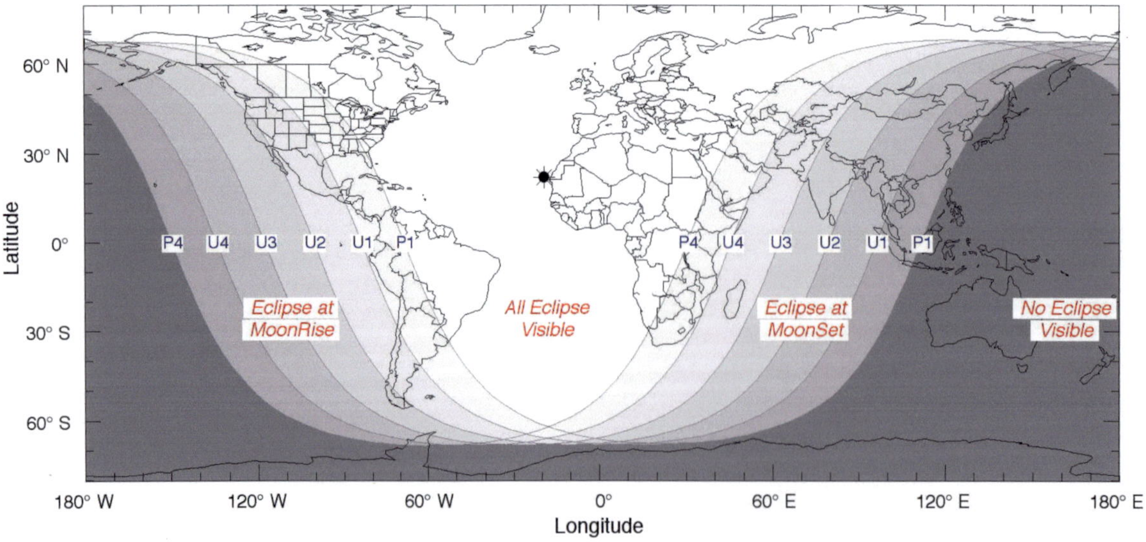

Total Lunar Eclipse of 2047 Jul 07

Greatest Eclipse = 10:35:45.4 TD (= 10:34:21.9 UT1)

Penumbral Magnitude = 2.7326 Gamma = -0.0636 Saros Series = 130
Umbral Magnitude = 1.7529 Axis = 0.0625° Saros Member = 36 of 71

Sun at Greatest Eclipse
(Geocentric Coordinates)
R.A. = 07h06m19.6s
Dec. = +22°33'30.9"
S.D. = 00°15'43.9"
H.P. = 00°00'08.7"

Moon at Greatest Eclipse
(Geocentric Coordinates)
R.A. = 19h06m23.0s
Dec. = -22°37'10.8"
S.D. = 00°16'03.5"
H.P. = 00°58'56.1"

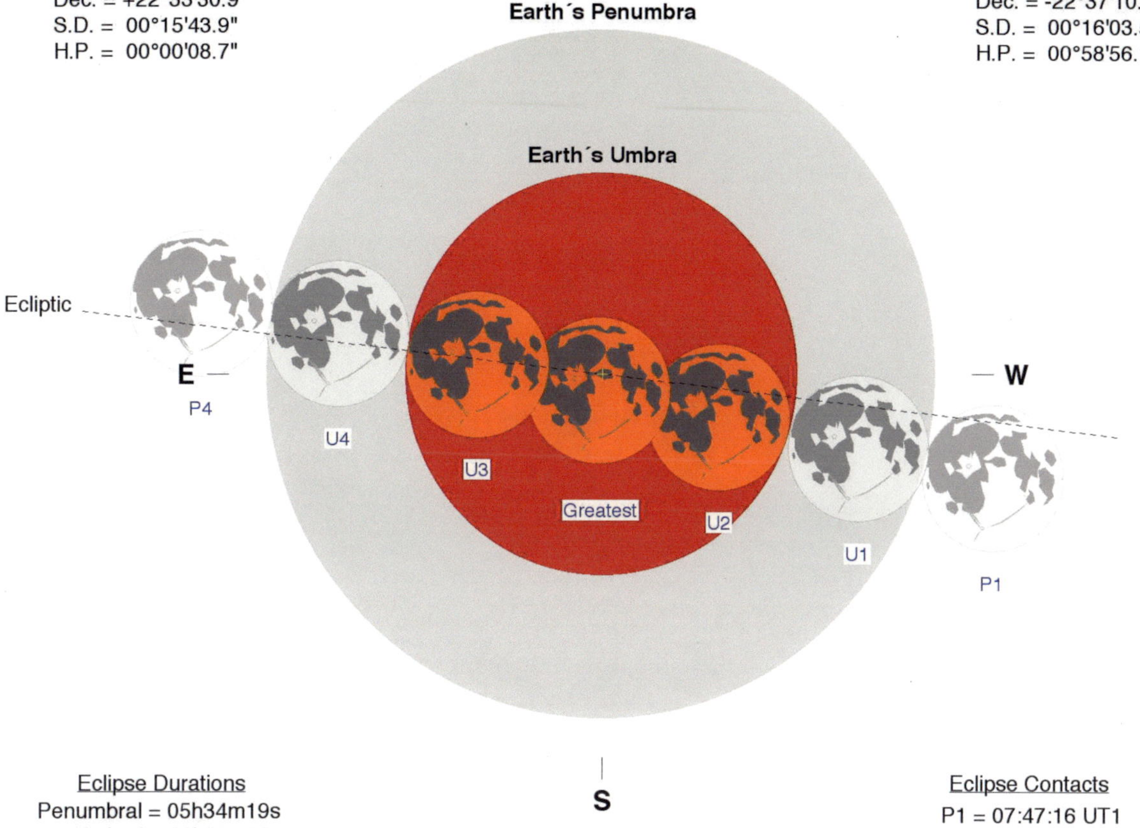

Eclipse Durations
Penumbral = 05h34m19s
Umbral = 03h39m18s
Total = 01h41m35s

Eph. = JPL DE430
Rule = Herald-Sinnott
ΔT = 84 s

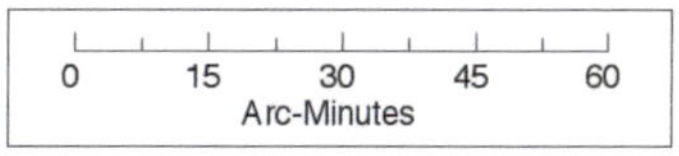

Eclipse Contacts
P1 = 07:47:16 UT1
U1 = 08:44:42 UT1
U2 = 09:43:35 UT1
U3 = 11:25:10 UT1
U4 = 12:24:01 UT1
P4 = 13:21:34 UT1

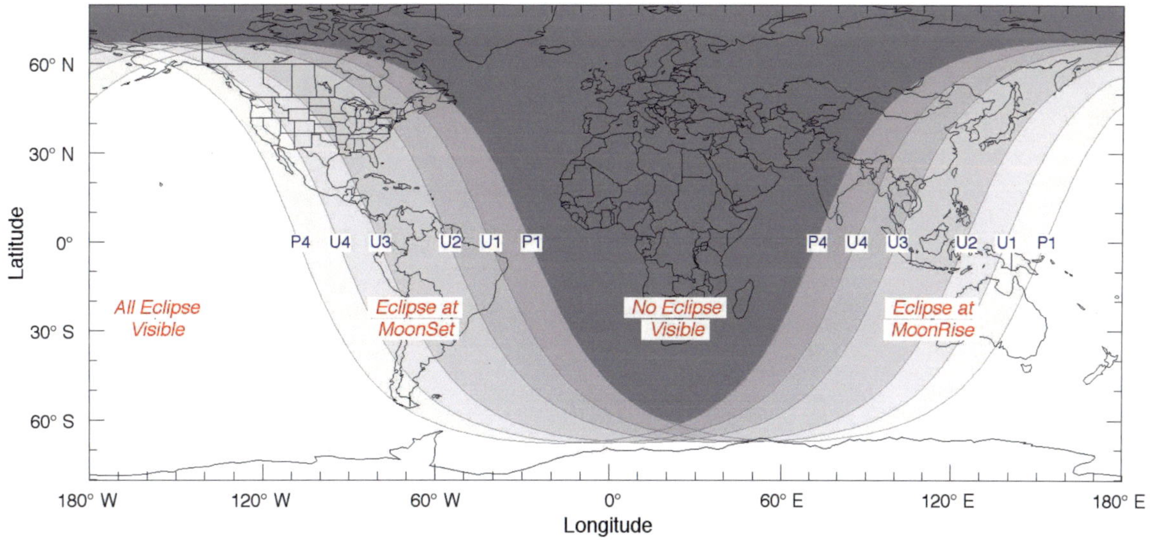

Total Lunar Eclipse of 2048 Jan 01

Greatest Eclipse = 06:53:54.8 TD (= 06:52:30.9 UT1)

Penumbral Magnitude = 2.2158 Gamma = -0.3746 Saros Series = 135
Umbral Magnitude = 1.1297 Axis = 0.3431° Saros Member = 25 of 71

Sun at Greatest Eclipse
(Geocentric Coordinates)
R.A. = 18h45m45.0s
Dec. = -23°01'00.1"
S.D. = 00°16'15.9"
H.P. = 00°00'08.9"

Moon at Greatest Eclipse
(Geocentric Coordinates)
R.A. = 06h45m29.1s
Dec. = +22°40'44.8"
S.D. = 00°14'58.6"
H.P. = 00°54'57.7"

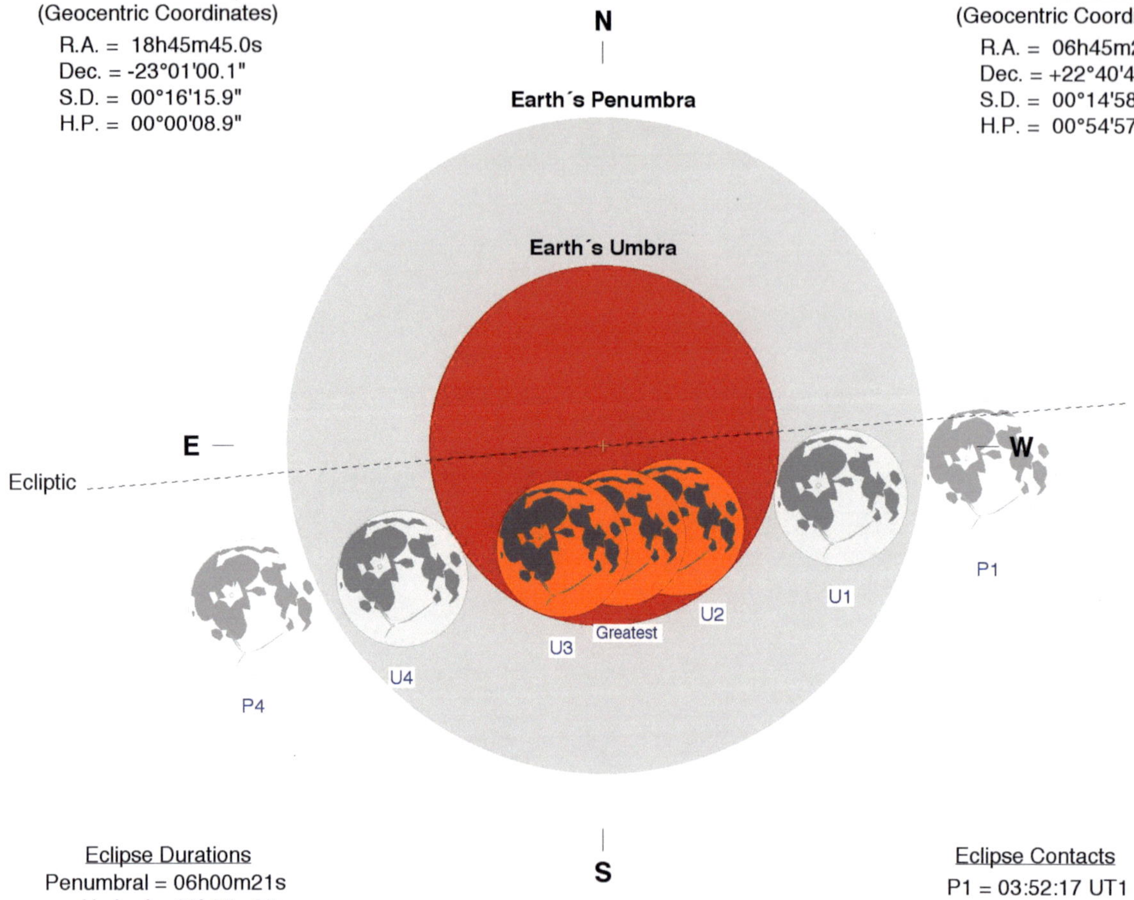

Eclipse Durations
Penumbral = 06h00m21s
Umbral = 03h35m05s
Total = 00h56m40s

Eph. = JPL DE430
Rule = Herald-Sinnott
ΔT = 84 s

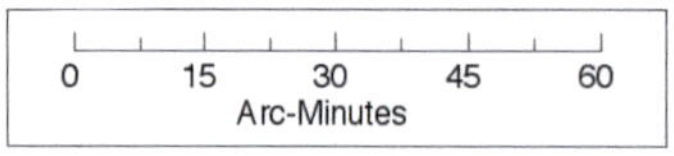

Eclipse Contacts
P1 = 03:52:17 UT1
U1 = 05:04:57 UT1
U2 = 06:24:06 UT1
U3 = 07:20:46 UT1
U4 = 08:40:02 UT1
P4 = 09:52:38 UT1

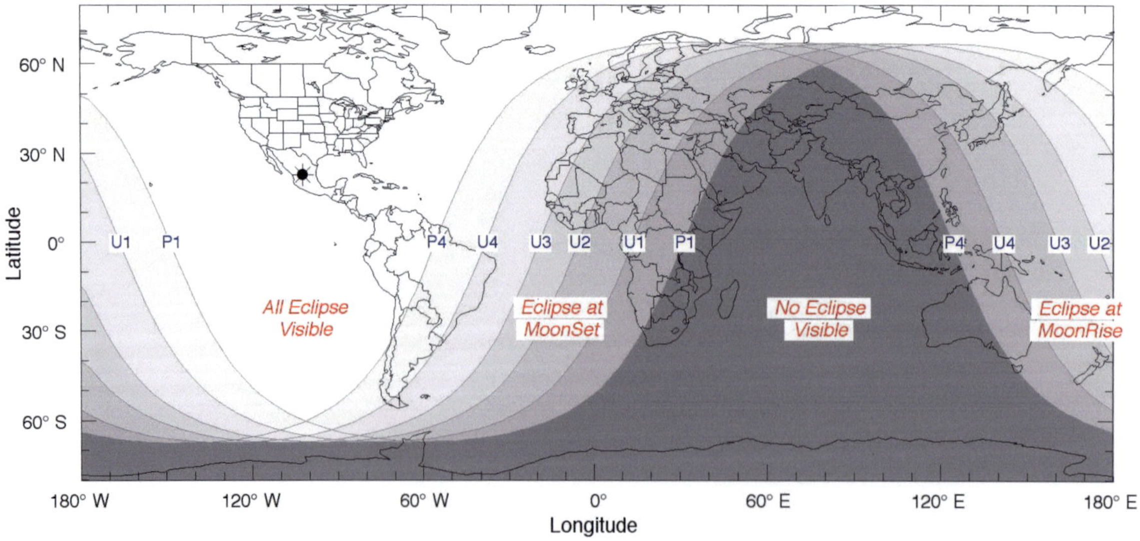

Partial Lunar Eclipse of 2048 Jun 26

Greatest Eclipse = 02:02:28.0 TD (= 02:01:03.9 UT1)

Penumbral Magnitude = 1.5841 Gamma = 0.6797 Saros Series = 140
Umbral Magnitude = 0.6404 Axis = 0.6931° Saros Member = 26 of 77

Sun at Greatest Eclipse
(Geocentric Coordinates)
R.A. = 06h22m31.9s
Dec. = +23°19'54.0"
S.D. = 00°15'44.1"
H.P. = 00°00'08.7"

Moon at Greatest Eclipse
(Geocentric Coordinates)
R.A. = 18h22m07.4s
Dec. = -22°38'42.2"
S.D. = 00°16'40.4"
H.P. = 01°01'11.5"

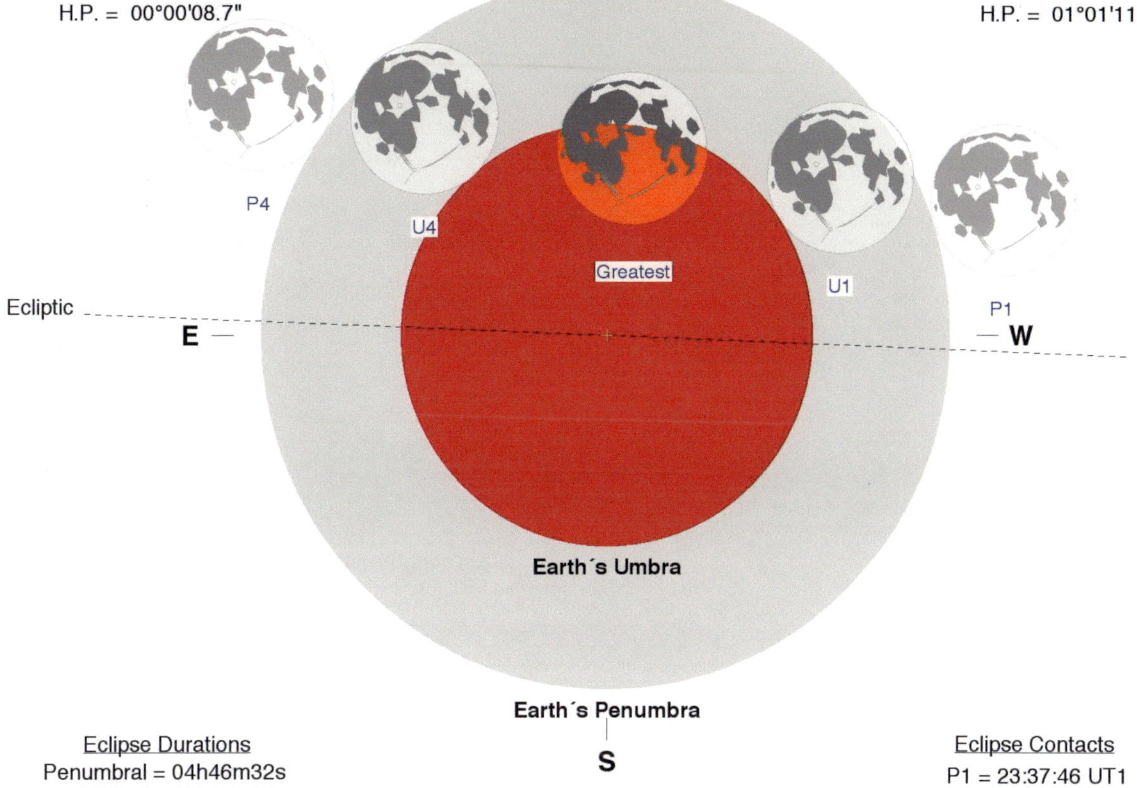

Eclipse Durations
Penumbral = 04h46m32s
Umbral = 02h39m51s

Eclipse Contacts
P1 = 23:37:46 UT1
U1 = 00:41:04 UT1
U4 = 03:20:55 UT1
P4 = 04:24:18 UT1

Eph. = JPL DE430
Rule = Herald-Sinnott
ΔT = 84 s

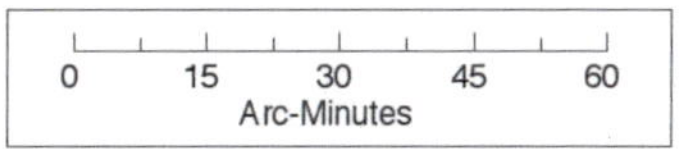

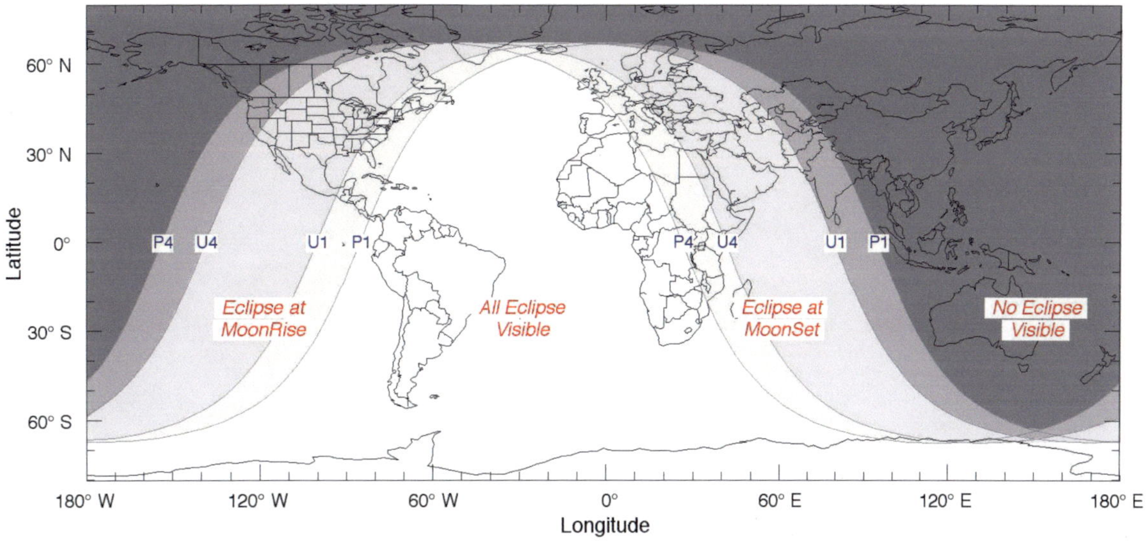

Penumbral Lunar Eclipse of 2048 Dec 20

Greatest Eclipse = 06:27:47.7 TD (= 06:26:23.3 UT1)

Penumbral Magnitude = 0.9632 Gamma = -1.0624 Saros Series = 145
Umbral Magnitude = -0.1420 Axis = 0.9558° Saros Member = 13 of 71

Sun at Greatest Eclipse
(Geocentric Coordinates)
R.A. = 17h55m49.3s
Dec. = -23°25'43.8"
S.D. = 00°16'15.4"
H.P. = 00°00'08.9"

Moon at Greatest Eclipse
(Geocentric Coordinates)
R.A. = 05h55m26.5s
Dec. = +22°28'37.2"
S.D. = 00°14'42.6"
H.P. = 00°53'59.0"

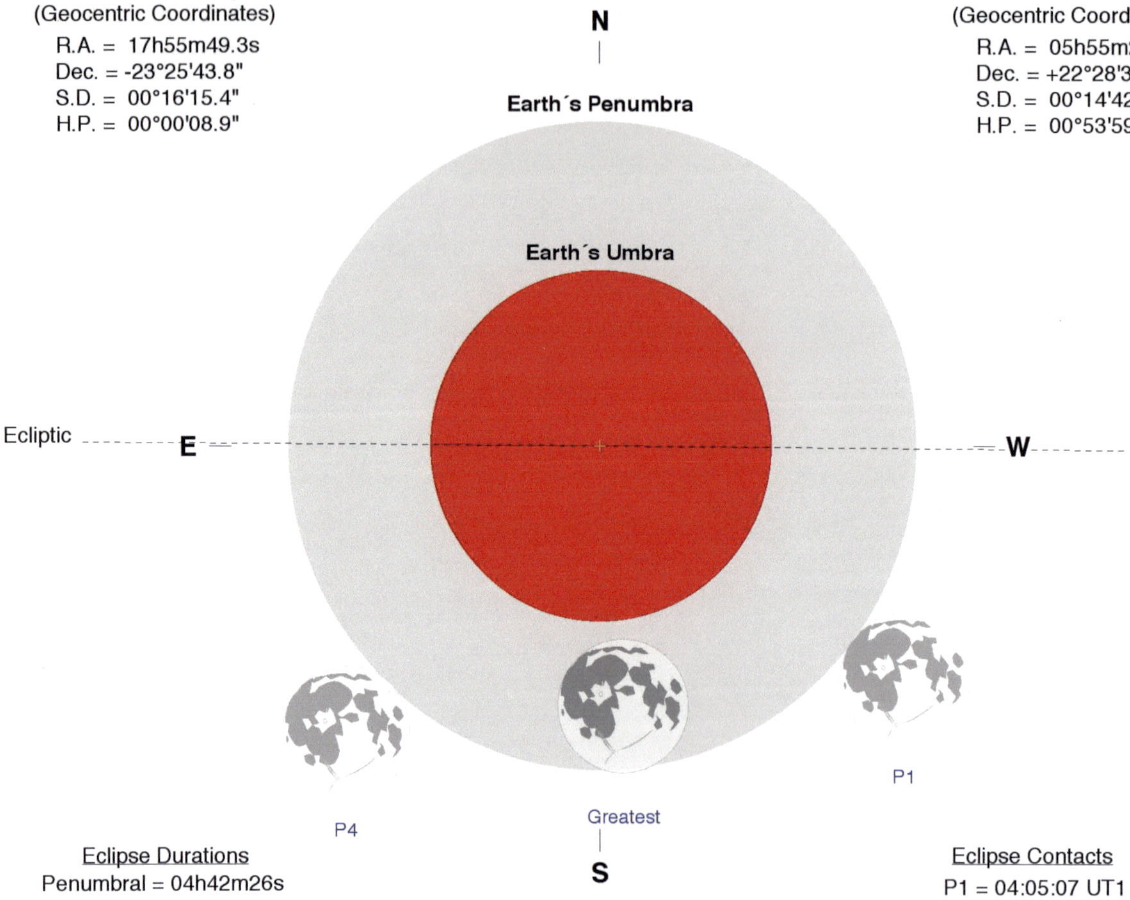

Eclipse Durations
Penumbral = 04h42m26s

Eclipse Contacts
P1 = 04:05:07 UT1
P4 = 08:47:34 UT1

Eph. = JPL DE430
Rule = Herald-Sinnott
ΔT = 84 s

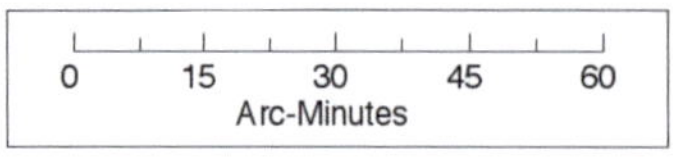

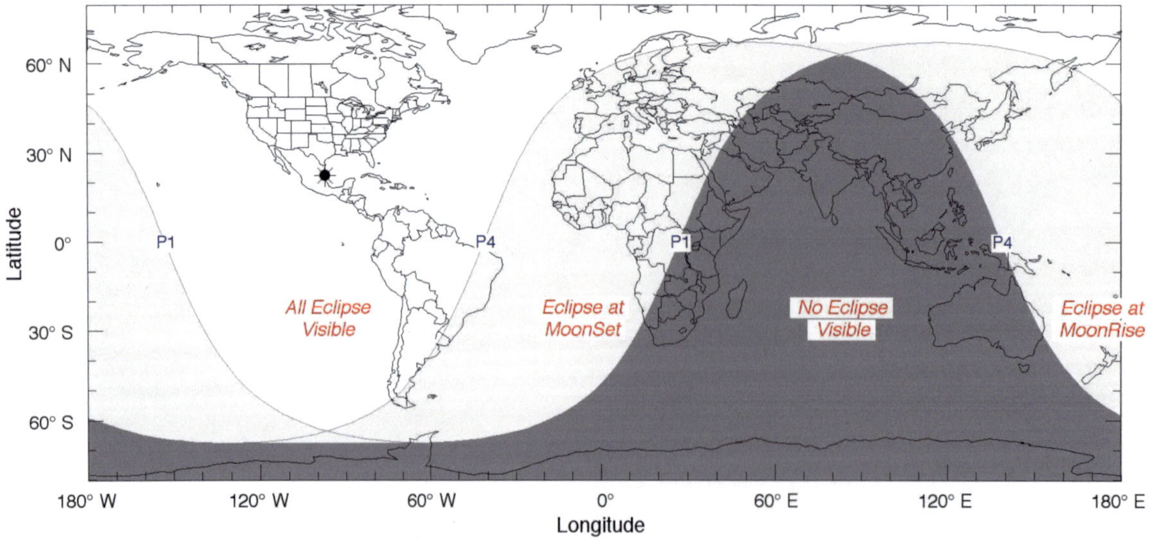

Penumbral Lunar Eclipse of 2049 May 17

Greatest Eclipse = 11:26:38.4 TD (= 11:25:13.7 UT1)

Penumbral Magnitude = 0.7650 Gamma = -1.1337 Saros Series = 112
Umbral Magnitude = -0.2073 Axis = 1.1279° Saros Member = 67 of 72

Sun at Greatest Eclipse
(Geocentric Coordinates)
R.A. = 03h38m51.9s
Dec. = +19°28'58.4"
S.D. = 00°15'49.0"
H.P. = 00°00'08.7"

Moon at Greatest Eclipse
(Geocentric Coordinates)
R.A. = 15h38m12.8s
Dec. = -20°36'01.8"
S.D. = 00°16'16.0"
H.P. = 00°59'41.9"

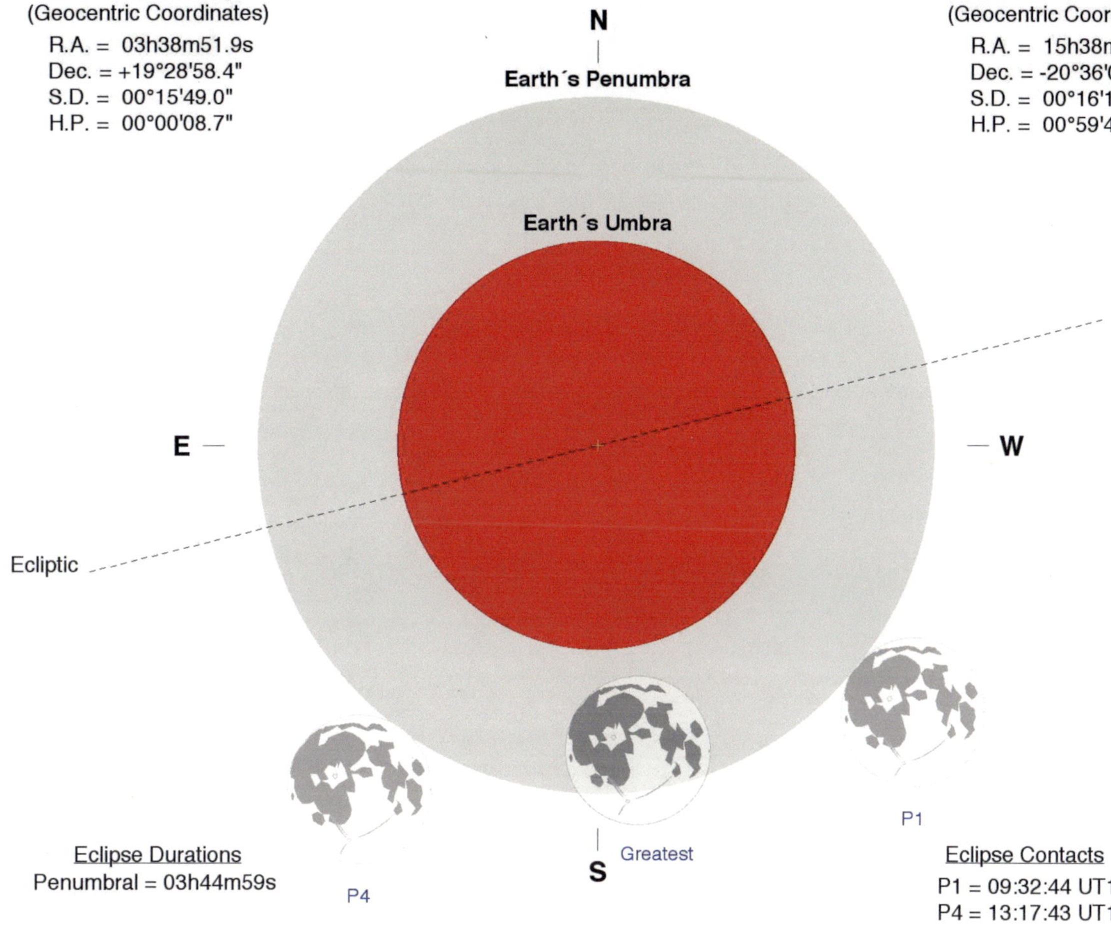

Eclipse Durations
Penumbral = 03h44m59s

Eclipse Contacts
P1 = 09:32:44 UT1
P4 = 13:17:43 UT1

Eph. = JPL DE430
Rule = Herald-Sinnott
ΔT = 85 s

0 15 30 45 60
Arc-Minutes

©2020 F. Espenak, www.EclipseWise.com

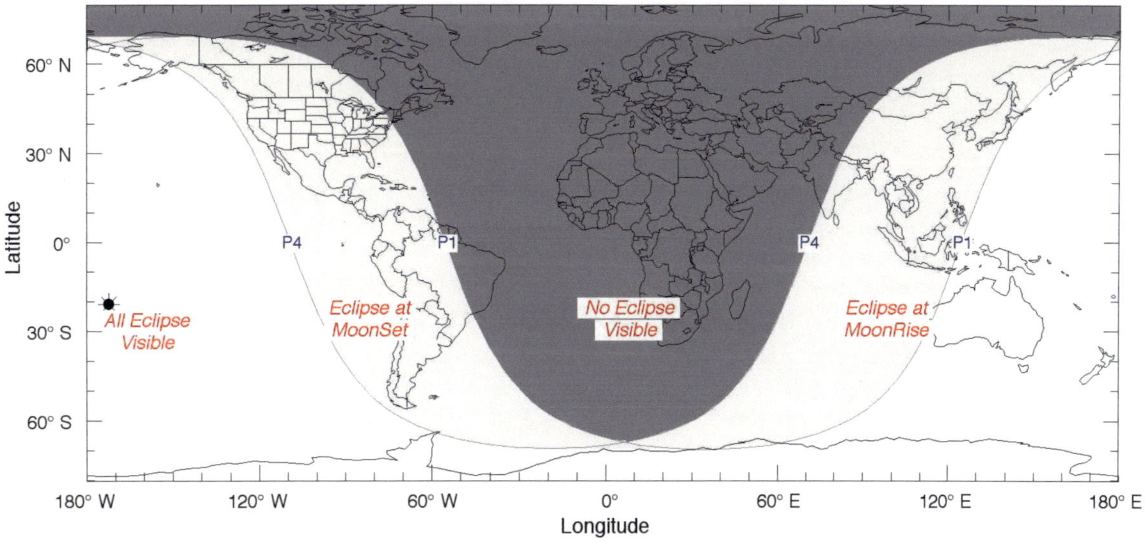

Penumbral Lunar Eclipse of 2049 Jun 15

Greatest Eclipse = 19:14:11.5 TD (= 19:12:46.8 UT1)

Penumbral Magnitude = 0.2526 Gamma = 1.4069 Saros Series = 150
Umbral Magnitude = -0.6970 Axis = 1.4268° Saros Member = 3 of 71

Sun at Greatest Eclipse
(Geocentric Coordinates)
R.A. = 05h38m45.5s
Dec. = +23°20'31.0"
S.D. = 00°15'44.8"
H.P. = 00°00'08.7"

Moon at Greatest Eclipse
(Geocentric Coordinates)
R.A. = 17h38m24.2s
Dec. = -21°55'02.3"
S.D. = 00°16'34.9"
H.P. = 01°00'51.4"

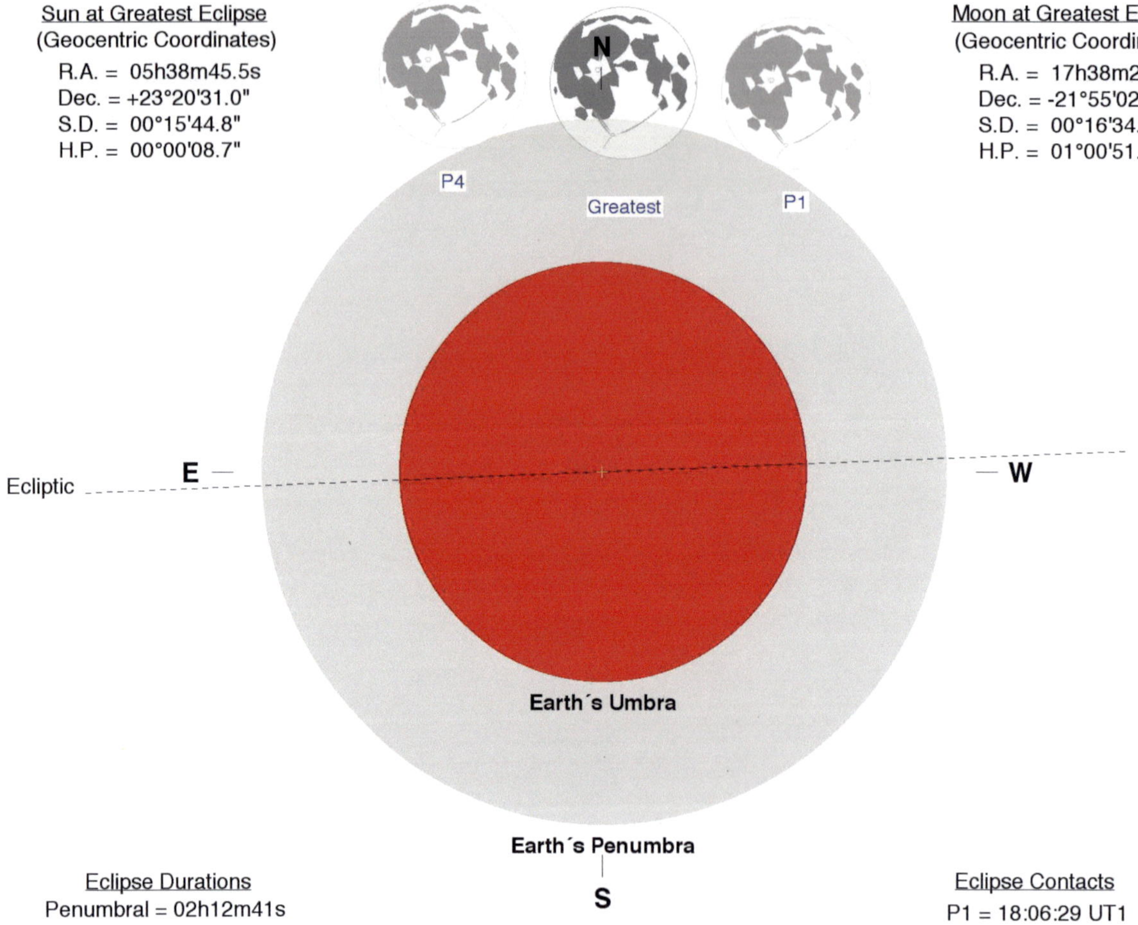

Eclipse Durations
Penumbral = 02h12m41s

Eclipse Contacts
P1 = 18:06:29 UT1
P4 = 20:19:09 UT1

Eph. = JPL DE430
Rule = Herald-Sinnott
ΔT = 85 s

0 15 30 45 60
Arc-Minutes

©2020 F. Espenak, www.EclipseWise.com

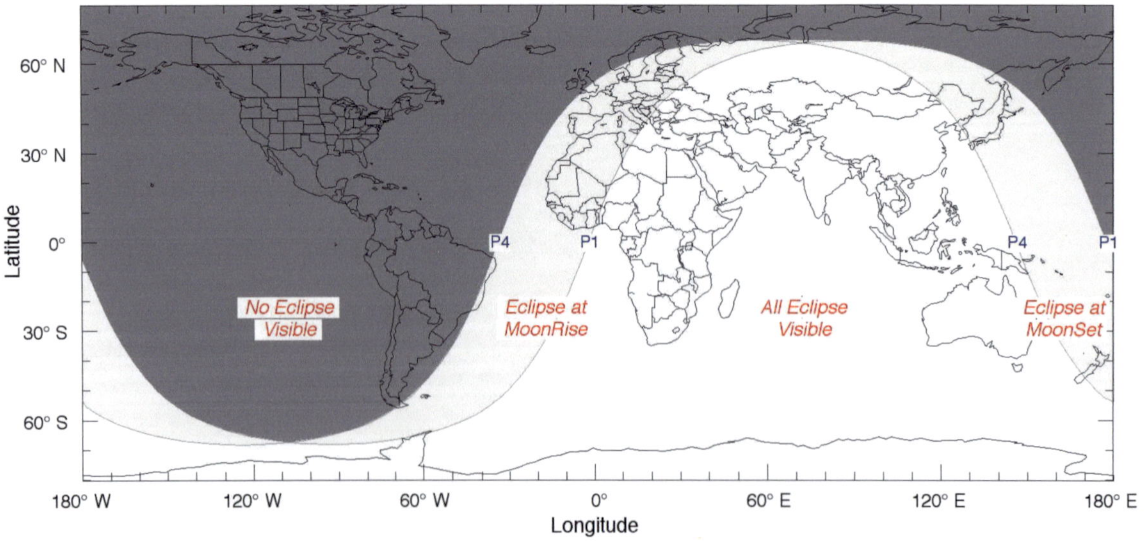

Penumbral Lunar Eclipse of 2049 Nov 09

Greatest Eclipse = 15:52:10.7 TD (= 15:50:45.7 UT1)

Penumbral Magnitude = 0.6821 Gamma = 1.1965 Saros Series = 117
Umbral Magnitude = -0.3541 Axis = 1.1404° Saros Member = 54 of 71

Sun at Greatest Eclipse
(Geocentric Coordinates)
R.A. = 15h00m53.5s
Dec. = -17°06'00.6"
S.D. = 00°16'08.8"
H.P. = 00°00'08.9"

Moon at Greatest Eclipse
(Geocentric Coordinates)
R.A. = 03h00m00.0s
Dec. = +18°13'14.6"
S.D. = 00°15'35.1"
H.P. = 00°57'11.8"

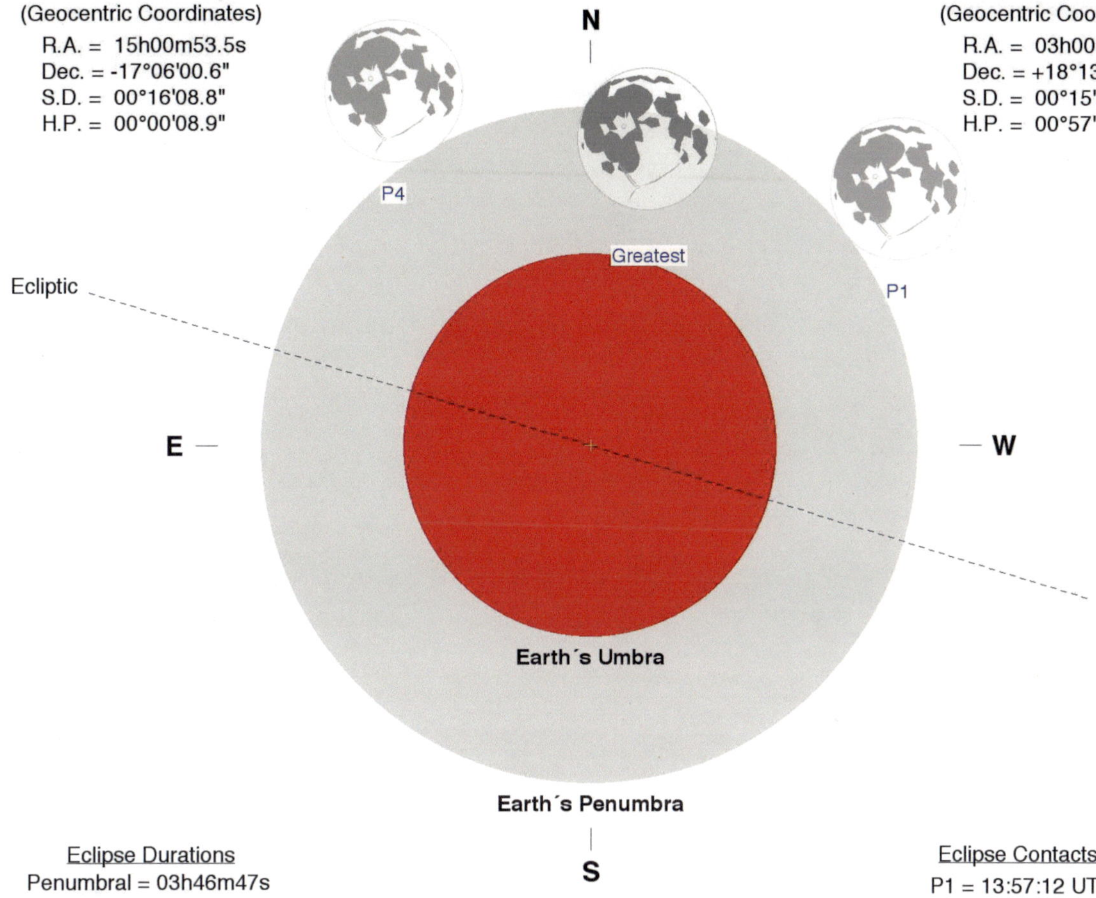

Eclipse Durations
Penumbral = 03h46m47s

Eclipse Contacts
P1 = 13:57:12 UT1
P4 = 17:43:59 UT1

Eph. = JPL DE430
Rule = Herald-Sinnott
ΔT = 85 s

0 15 30 45 60
Arc-Minutes

©2020 F. Espenak, www.EclipseWise.com

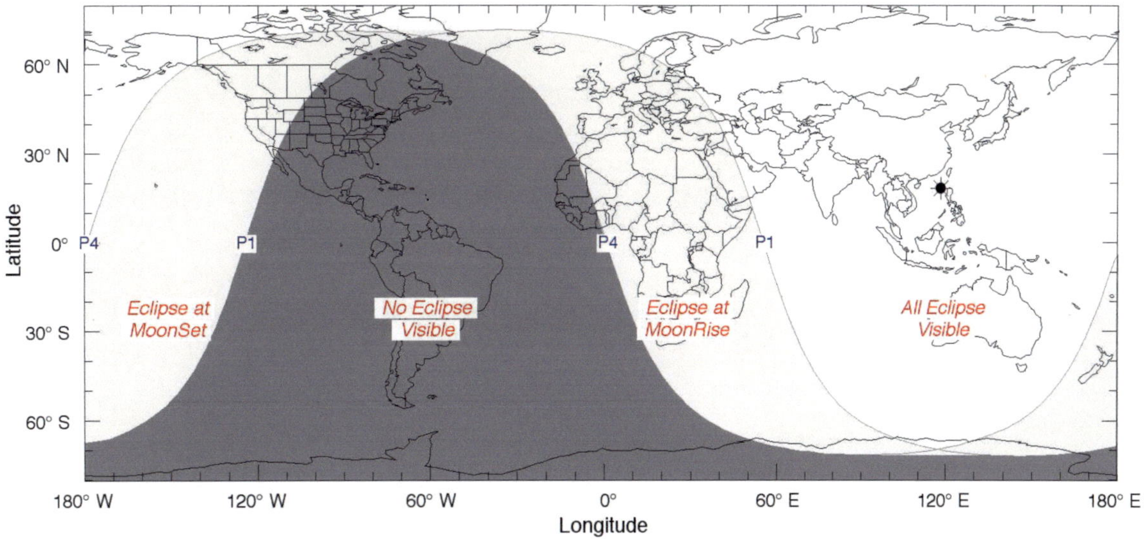

Total Lunar Eclipse of 2050 May 06

Greatest Eclipse = 22:32:01.6 TD (= 22:30:36.3 UT1)

Penumbral Magnitude = 2.1064 Gamma = -0.4181 Saros Series = 122
Umbral Magnitude = 1.0779 Axis = 0.3942° Saros Member = 58 of 74

Sun at Greatest Eclipse
(Geocentric Coordinates)
R.A. = 02h56m30.8s
Dec. = +16°47'28.5"
S.D. = 00°15'51.3"
H.P. = 00°00'08.7"

Moon at Greatest Eclipse
(Geocentric Coordinates)
R.A. = 14h56m12.1s
Dec. = -17°10'41.9"
S.D. = 00°15'24.9"
H.P. = 00°56'34.4"

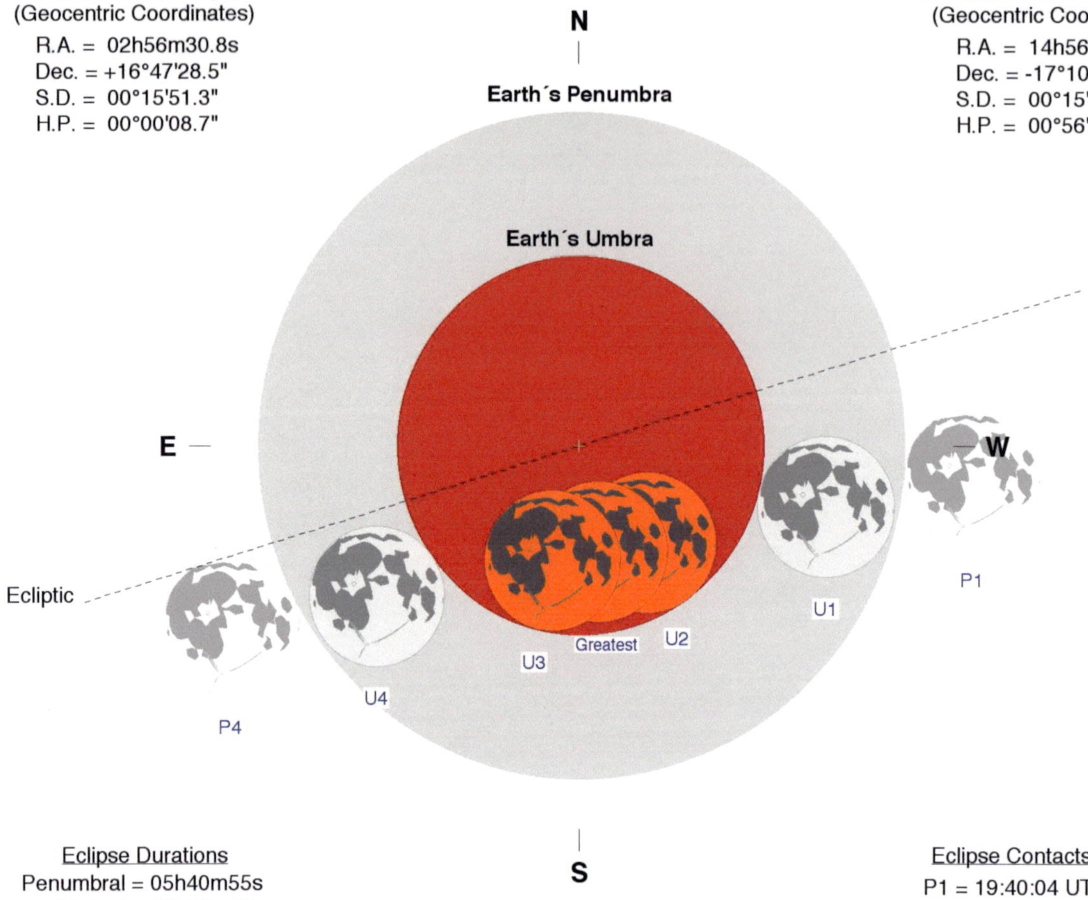

Eclipse Durations
Penumbral = 05h40m55s
Umbral = 03h26m46s
Total = 00h43m49s

Eph. = JPL DE430
Rule = Herald-Sinnott
ΔT = 85 s

0 15 30 45 60
Arc-Minutes

©2020 F. Espenak, www.EclipseWise.com

Eclipse Contacts
P1 = 19:40:04 UT1
U1 = 20:47:12 UT1
U2 = 22:08:36 UT1
U3 = 22:52:25 UT1
U4 = 00:13:58 UT1
P4 = 01:20:59 UT1

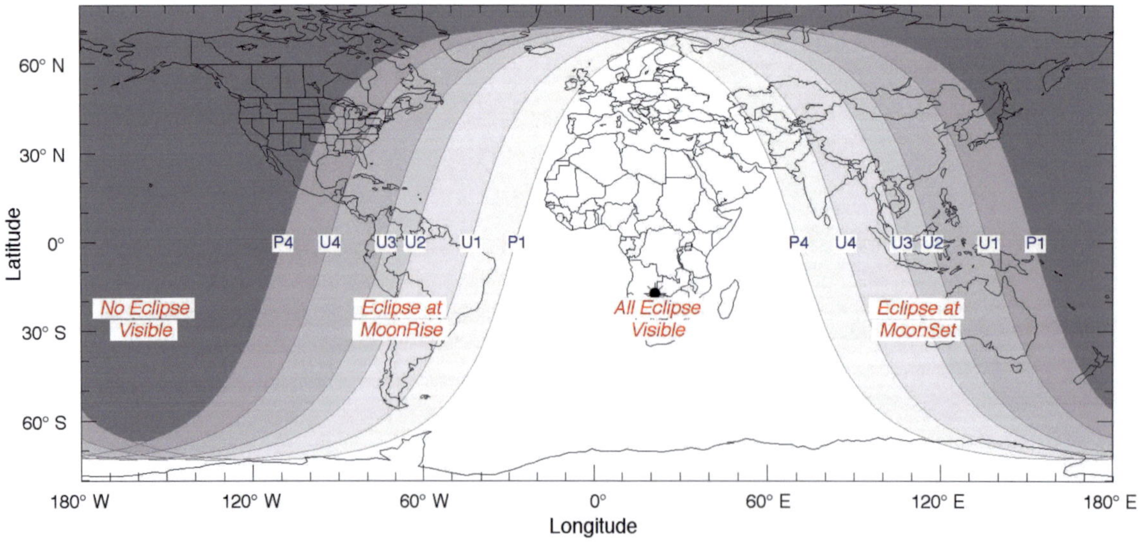

Total Lunar Eclipse of 2050 Oct 30

Greatest Eclipse = 03:21:46.8 TD (= 03:20:21.2 UT1)

Penumbral Magnitude = 2.0356 Gamma = 0.4435 Saros Series = 127
Umbral Magnitude = 1.0549 Axis = 0.4454° Saros Member = 44 of 72

Sun at Greatest Eclipse
(Geocentric Coordinates)
R.A. = 14h18m15.4s
Dec. = -13°48'46.9"
S.D. = 00°16'06.2"
H.P. = 00°00'08.9"

Moon at Greatest Eclipse
(Geocentric Coordinates)
R.A. = 02h17m49.7s
Dec. = +14°14'46.2"
S.D. = 00°16'25.2"
H.P. = 01°00'15.6"

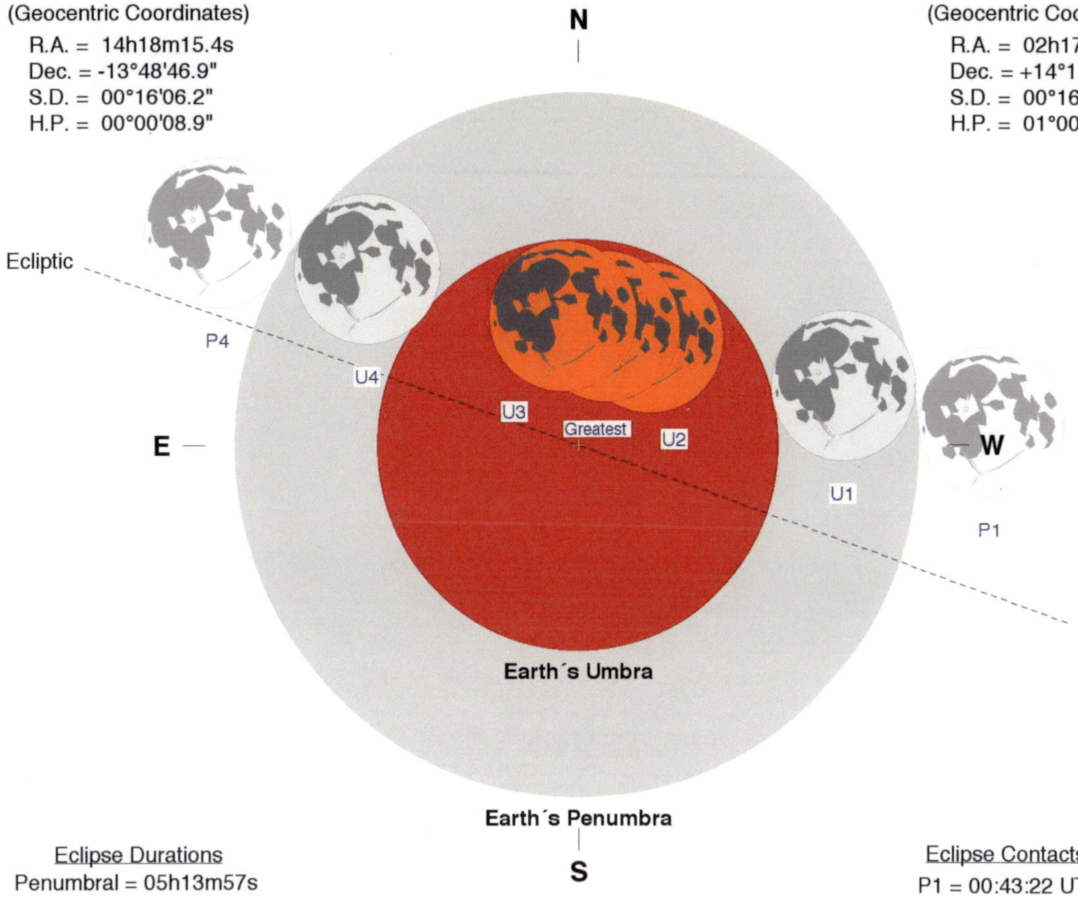

Eclipse Durations
Penumbral = 05h13m57s
Umbral = 03h13m34s
Total = 00h35m05s

Eph. = JPL DE430
Rule = Herald-Sinnott
ΔT = 86 s

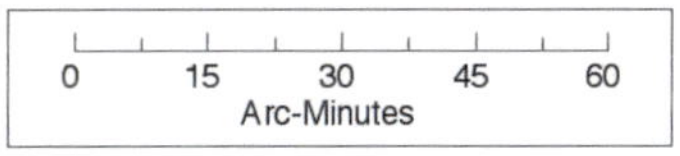

Eclipse Contacts
P1 = 00:43:22 UT1
U1 = 01:43:29 UT1
U2 = 03:02:39 UT1
U3 = 03:37:44 UT1
U4 = 04:57:03 UT1
P4 = 05:57:18 UT1

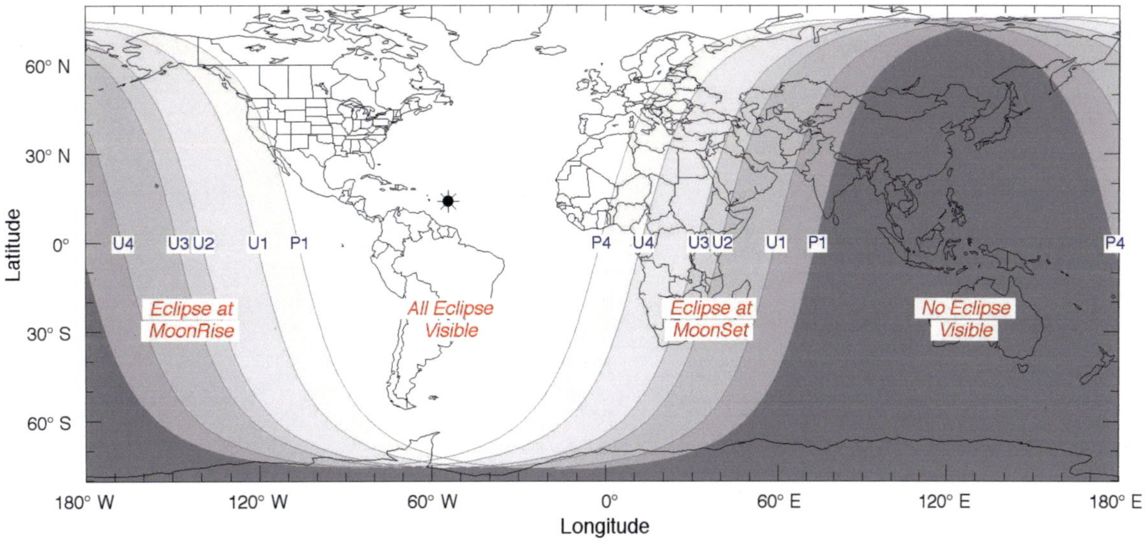

Total Lunar Eclipse of 2051 Apr 26

Greatest Eclipse = 02:16:27.8 TD (= 02:15:01.8 UT1)

Penumbral Magnitude = 2.2785 | Gamma = 0.3371 | Saros Series = 132
Umbral Magnitude = 1.2034 | Axis = 0.3049° | Saros Member = 32 of 71

Sun at Greatest Eclipse
(Geocentric Coordinates)
R.A. = 02h14m06.4s
Dec. = +13°27'39.8"
S.D. = 00°15'53.9"
H.P. = 00°00'08.7"

Moon at Greatest Eclipse
(Geocentric Coordinates)
R.A. = 14h14m24.0s
Dec. = -13°09'52.9"
S.D. = 00°14'47.3"
H.P. = 00°54'16.4"

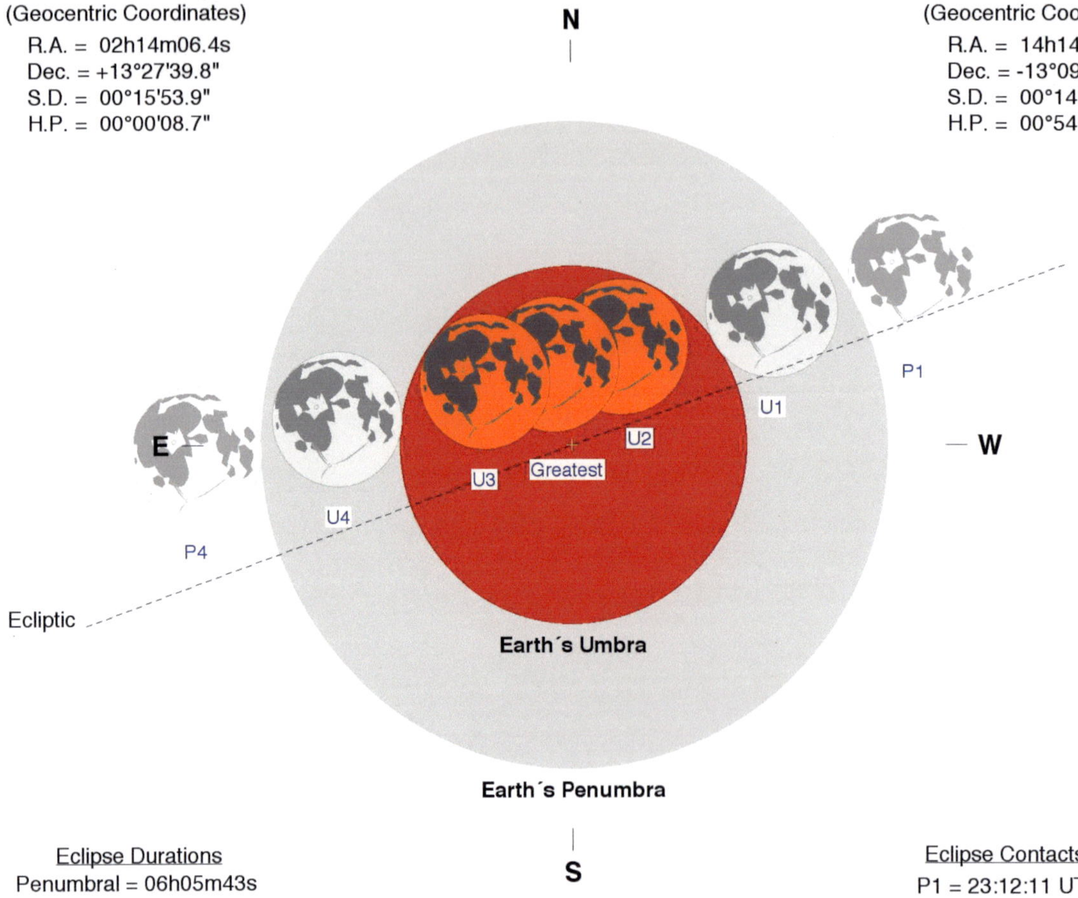

Eclipse Durations
Penumbral = 06h05m43s
Umbral = 03h41m40s
Total = 01h10m15s

Eph. = JPL DE430
Rule = Herald-Sinnott
ΔT = 86 s

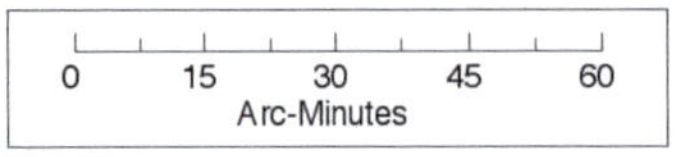

Eclipse Contacts
P1 = 23:12:11 UT1
U1 = 00:24:16 UT1
U2 = 01:40:03 UT1
U3 = 02:50:18 UT1
U4 = 04:05:56 UT1
P4 = 05:17:54 UT1

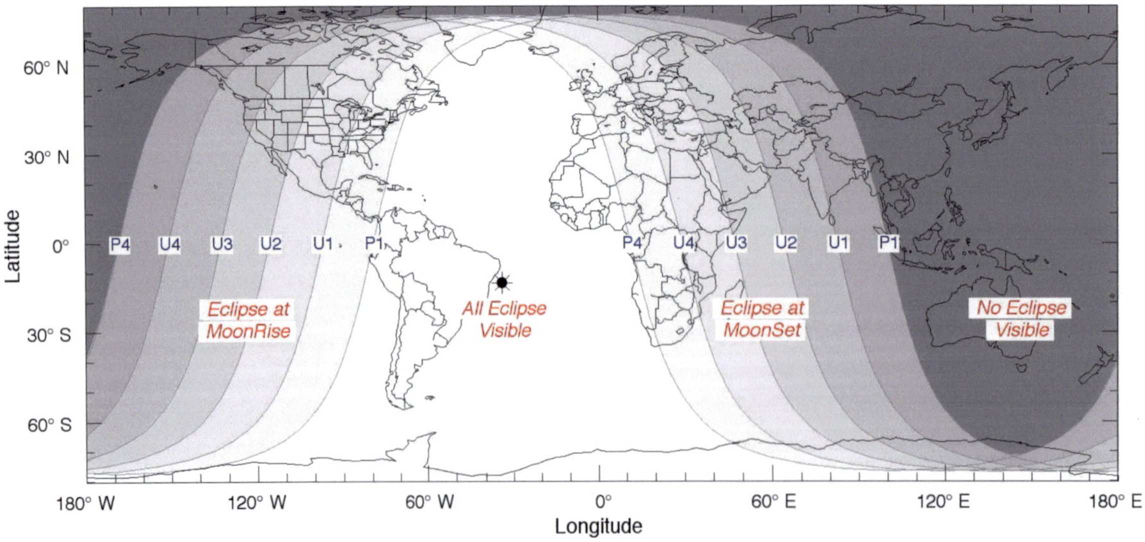

Total Lunar Eclipse of 2051 Oct 19

Greatest Eclipse = 19:11:49.6 TD (= 19:10:23.3 UT1)

Penumbral Magnitude = 2.3719 | Gamma = -0.2542 | Saros Series = 137
Umbral Magnitude = 1.4130 | Axis = 0.2603° | Saros Member = 28 of 78

Sun at Greatest Eclipse
(Geocentric Coordinates)
R.A. = 13h37m47.1s
Dec. = -10°10'03.3"
S.D. = 00°16'03.4"
H.P. = 00°00'08.8"

Moon at Greatest Eclipse
(Geocentric Coordinates)
R.A. = 01h38m04.1s
Dec. = +09°55'00.4"
S.D. = 00°16'44.6"
H.P. = 01°01'27.1"

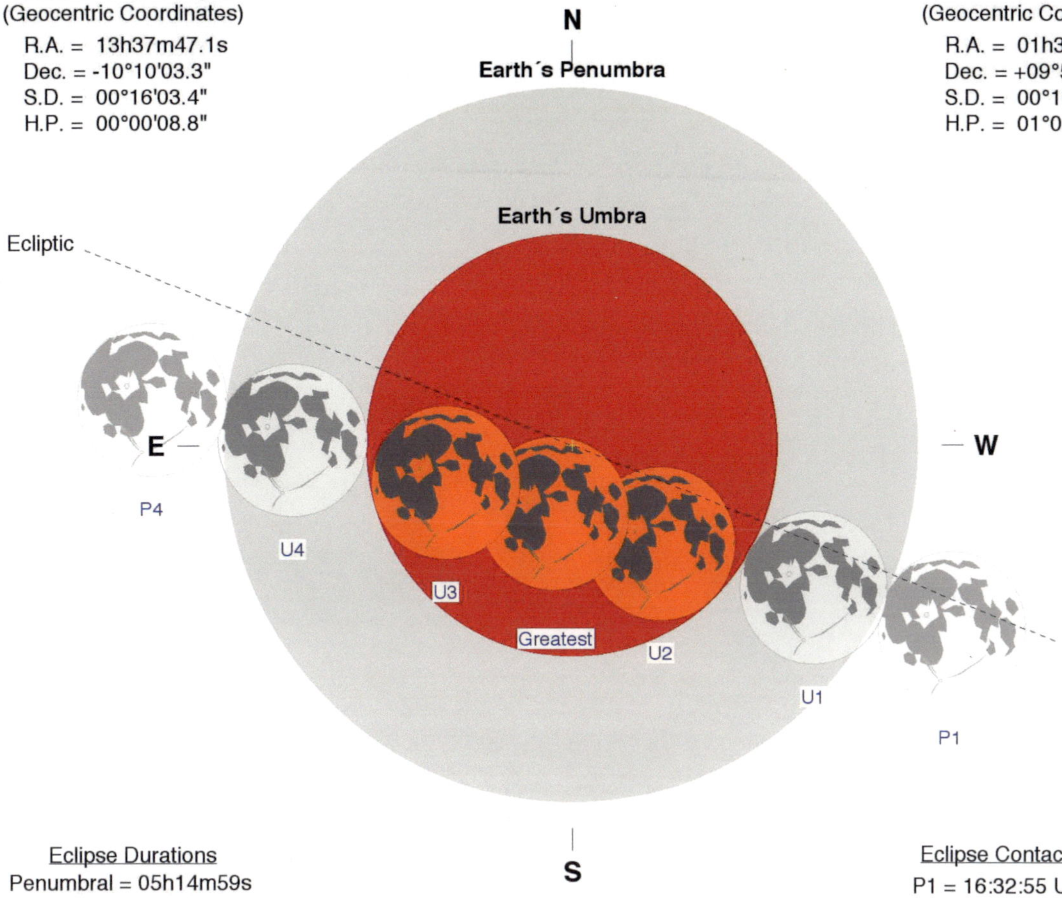

Eclipse Durations
Penumbral = 05h14m59s
Umbral = 03h25m00s
Total = 01h24m12s

Eph. = JPL DE430
Rule = Herald-Sinnott
ΔT = 86 s

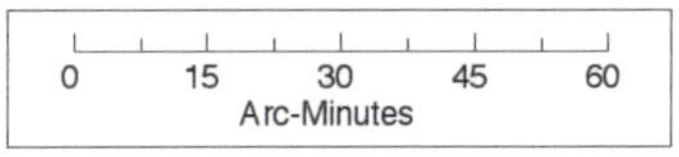

©2020 F. Espenak, www.EclipseWise.com

Eclipse Contacts
P1 = 16:32:55 UT1
U1 = 17:27:56 UT1
U2 = 18:28:23 UT1
U3 = 19:52:35 UT1
U4 = 20:52:56 UT1
P4 = 21:47:55 UT1

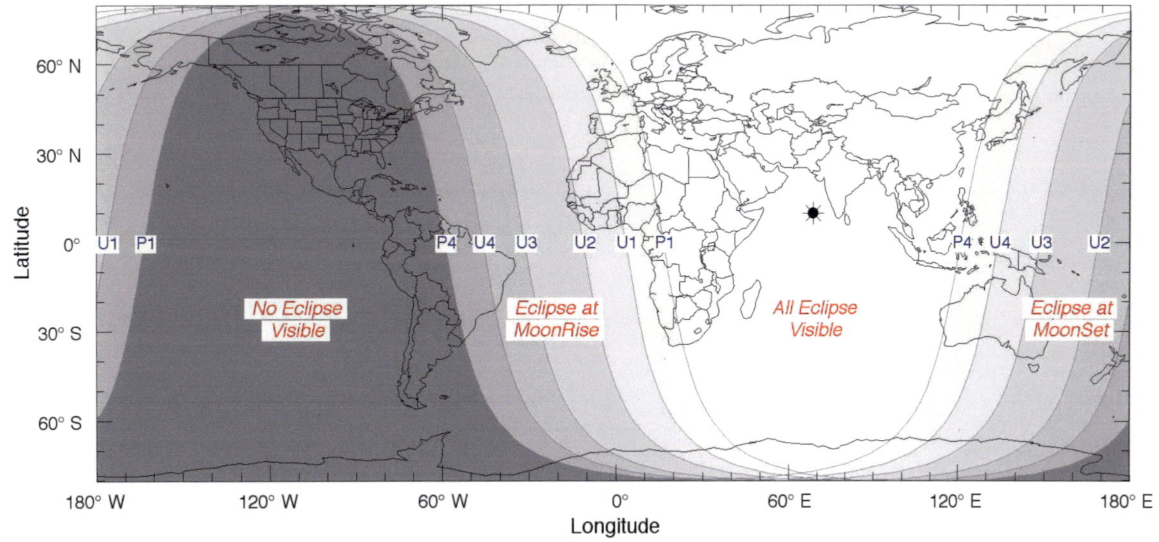

Penumbral Lunar Eclipse of 2052 Apr 14

Greatest Eclipse = 02:18:06.3 TD (= 02:16:39.7 UT1)

Penumbral Magnitude = 0.9478 Gamma = 1.0629 Saros Series = 142
Umbral Magnitude = -0.1294 Axis = 0.9625° Saros Member = 20 of 73

Sun at Greatest Eclipse
(Geocentric Coordinates)
R.A. = 01h32m05.6s
Dec. = +09°37'10.9"
S.D. = 00°15'56.9"
H.P. = 00°00'08.8"

Moon at Greatest Eclipse
(Geocentric Coordinates)
R.A. = 13h33m09.3s
Dec. = -08°41'36.6"
S.D. = 00°14'48.3"
H.P. = 00°54'20.2"

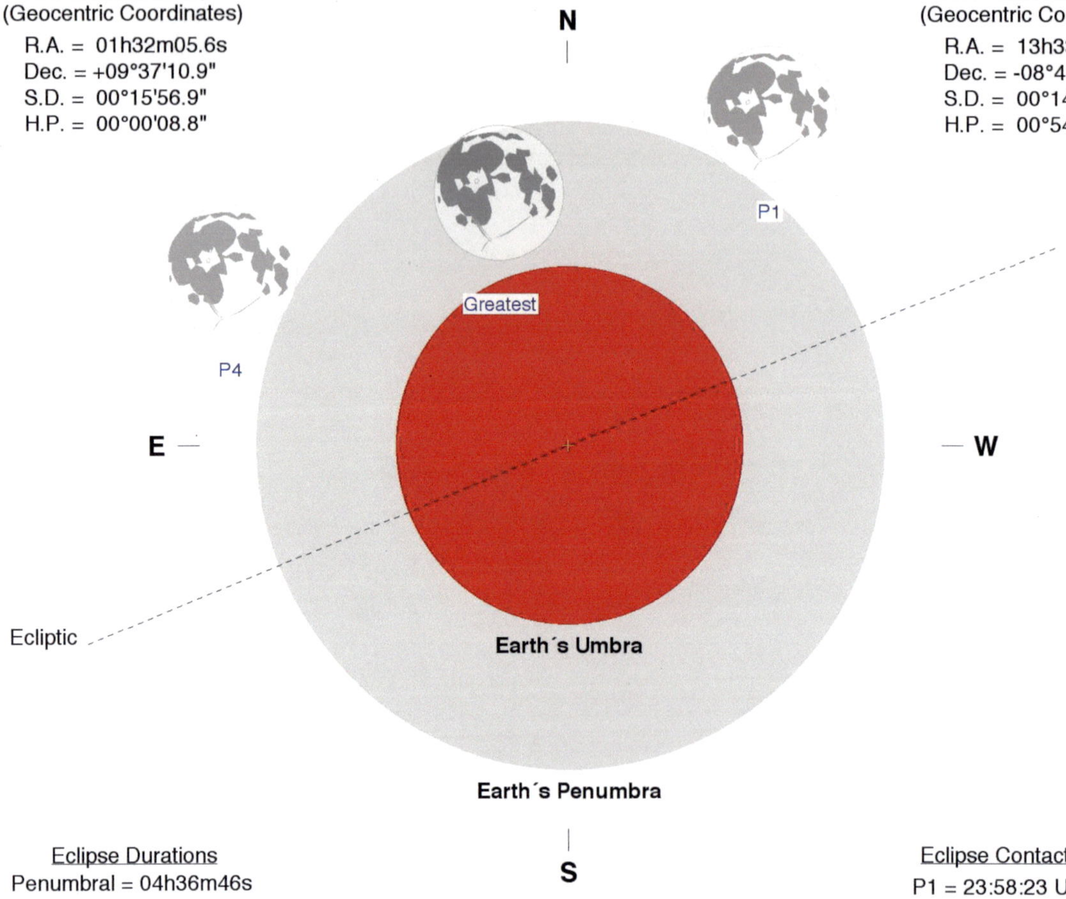

Eclipse Durations
Penumbral = 04h36m46s

Eclipse Contacts
P1 = 23:58:23 UT1
P4 = 04:35:09 UT1

Eph. = JPL DE430
Rule = Herald-Sinnott
ΔT = 87 s

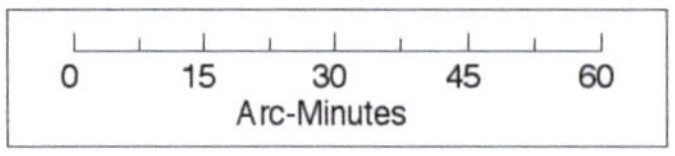

©2020 F. Espenak, www.EclipseWise.com

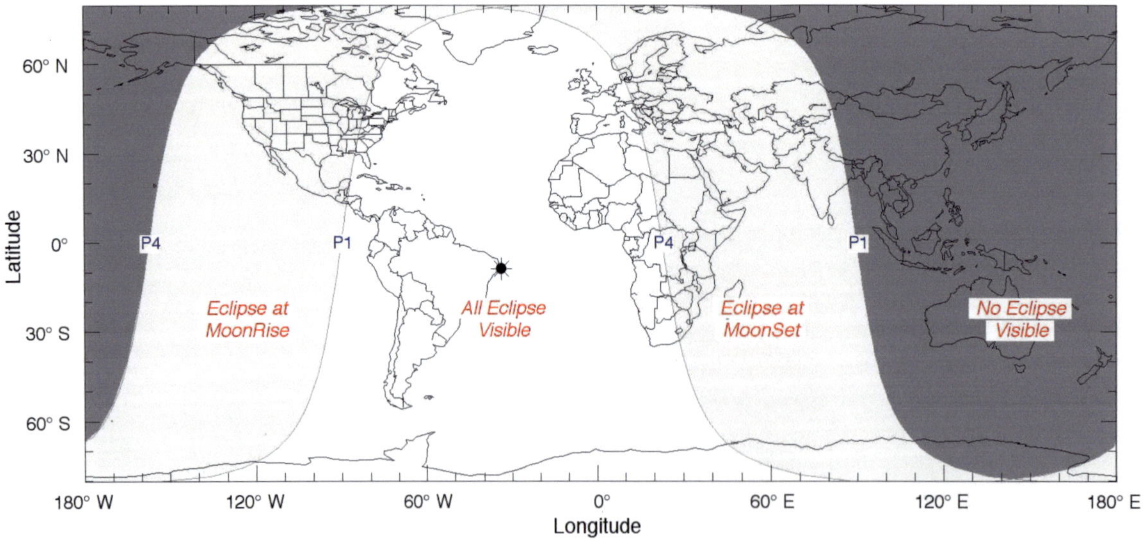

Partial Lunar Eclipse of 2052 Oct 08

Greatest Eclipse = 10:45:58.0 TD (= 10:44:31.1 UT1)

Penumbral Magnitude = 1.0653 Gamma = -0.9727 Saros Series = 147
Umbral Magnitude = 0.0832 Axis = 0.9697° Saros Member = 10 of 70

Sun at Greatest Eclipse
(Geocentric Coordinates)
R.A. = 12h58m28.0s
Dec. = -06°14'27.6"
S.D. = 00°16'00.5"
H.P. = 00°00'08.8"

Moon at Greatest Eclipse
(Geocentric Coordinates)
R.A. = 00h59m36.6s
Dec. = +05°18'49.9"
S.D. = 00°16'18.0"
H.P. = 00°59'49.3"

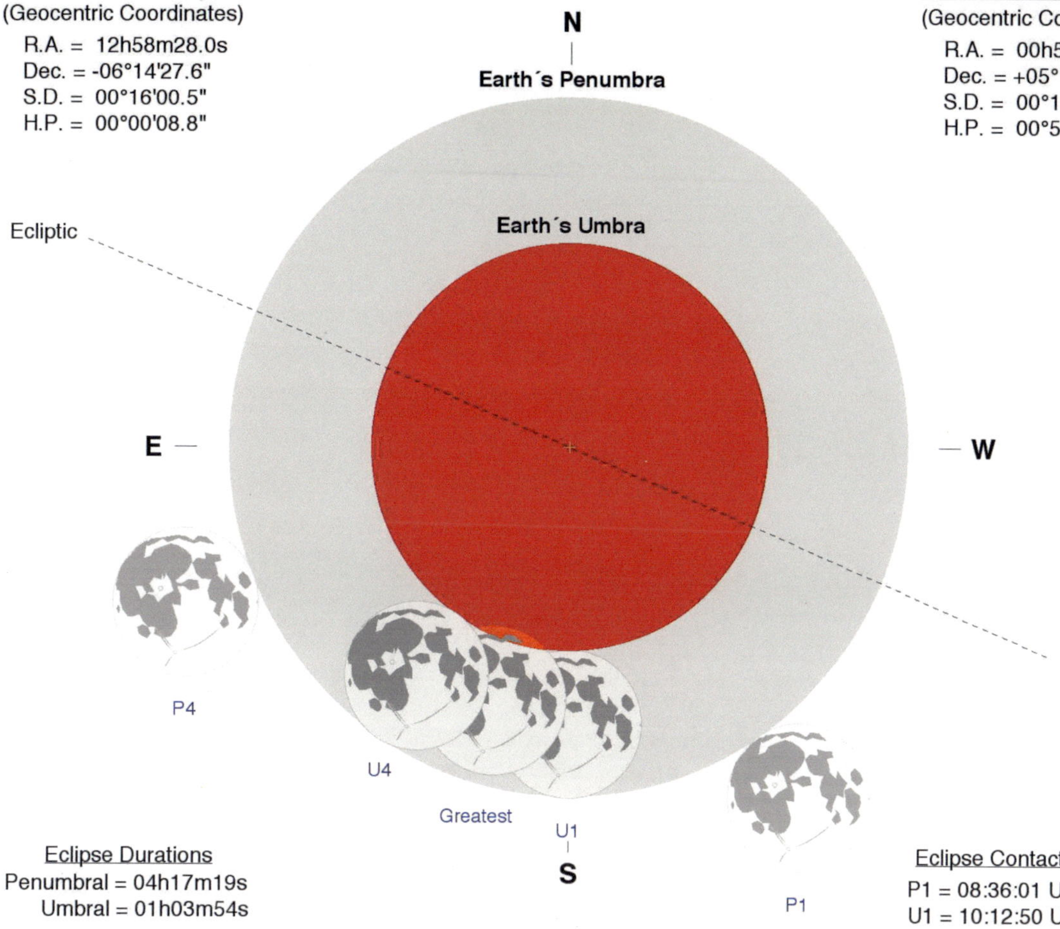

Eclipse Durations
Penumbral = 04h17m19s
Umbral = 01h03m54s

Eclipse Contacts
P1 = 08:36:01 UT1
U1 = 10:12:50 UT1
U4 = 11:16:44 UT1
P4 = 12:53:20 UT1

Eph. = JPL DE430
Rule = Herald-Sinnott
ΔT = 87 s

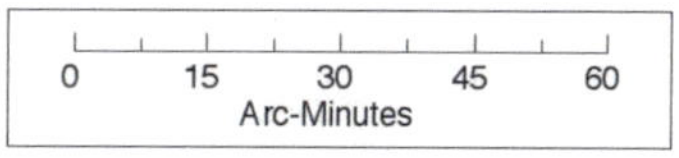

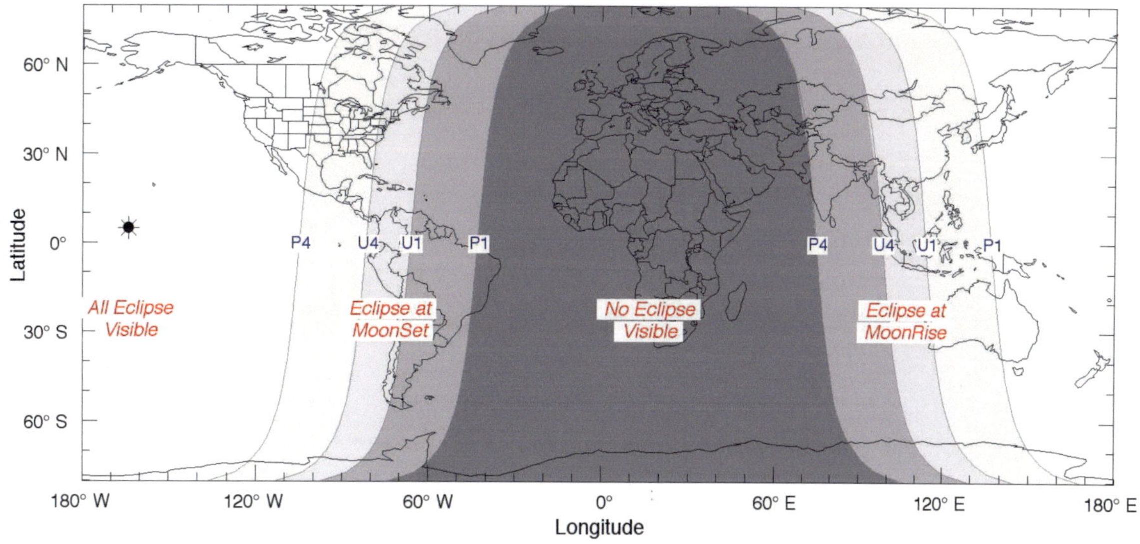

Penumbral Lunar Eclipse of 2053 Mar 04

Greatest Eclipse = 17:22:09.3 TD (= 17:20:42.1 UT1)

Penumbral Magnitude = 0.9334 Gamma = -1.0531 Saros Series = 114
Umbral Magnitude = -0.0796 Axis = 1.0255° Saros Member = 61 of 71

Sun at Greatest Eclipse
(Geocentric Coordinates)
R.A. = 23h03m14.8s
Dec. = -06°03'47.9"
S.D. = 00°16'07.7"
H.P. = 00°00'08.9"

Moon at Greatest Eclipse
(Geocentric Coordinates)
R.A. = 11h02m02.1s
Dec. = +05°04'58.9"
S.D. = 00°15'55.3"
H.P. = 00°58'26.0"

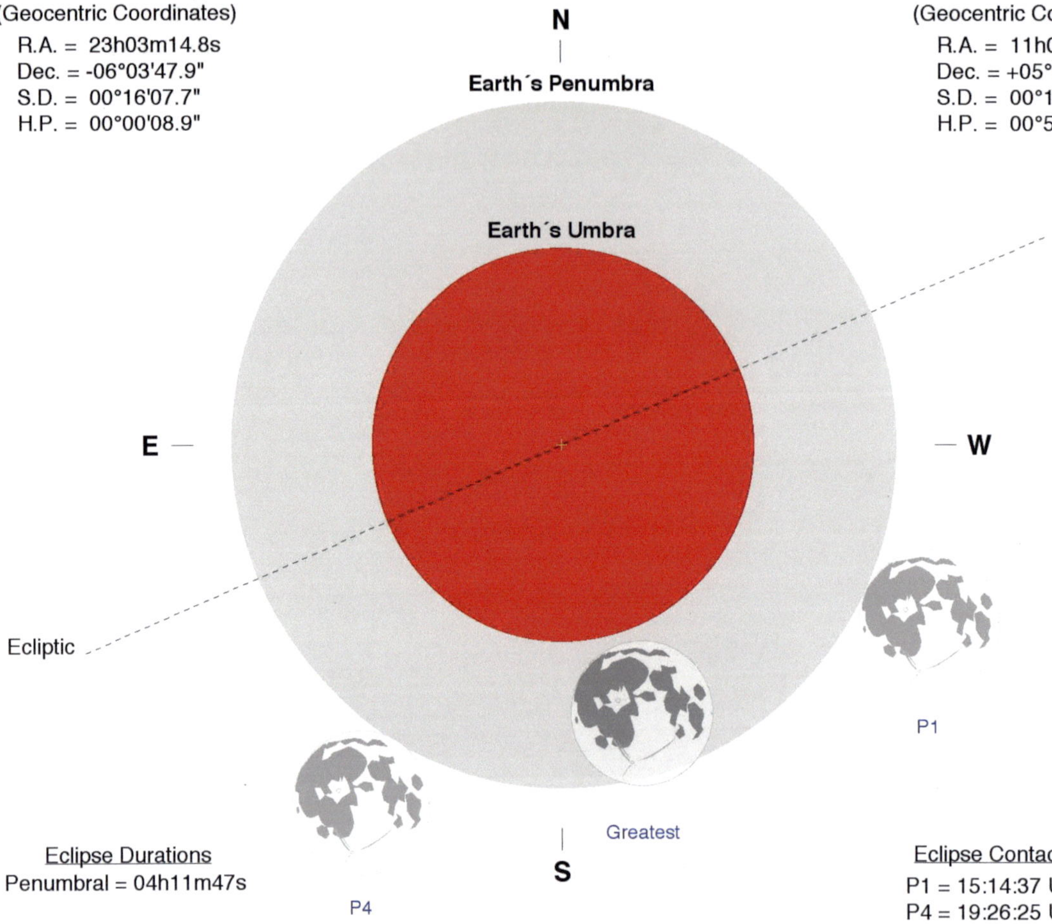

Eclipse Durations
Penumbral = 04h11m47s

Eclipse Contacts
P1 = 15:14:37 UT1
P4 = 19:26:25 UT1

Eph. = JPL DE430
Rule = Herald-Sinnott
ΔT = 87 s

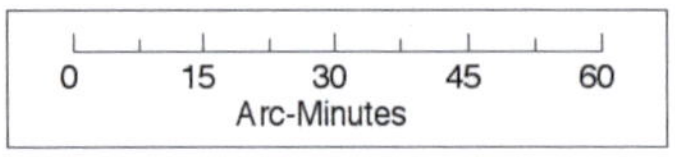

©2020 F. Espenak, www.EclipseWise.com

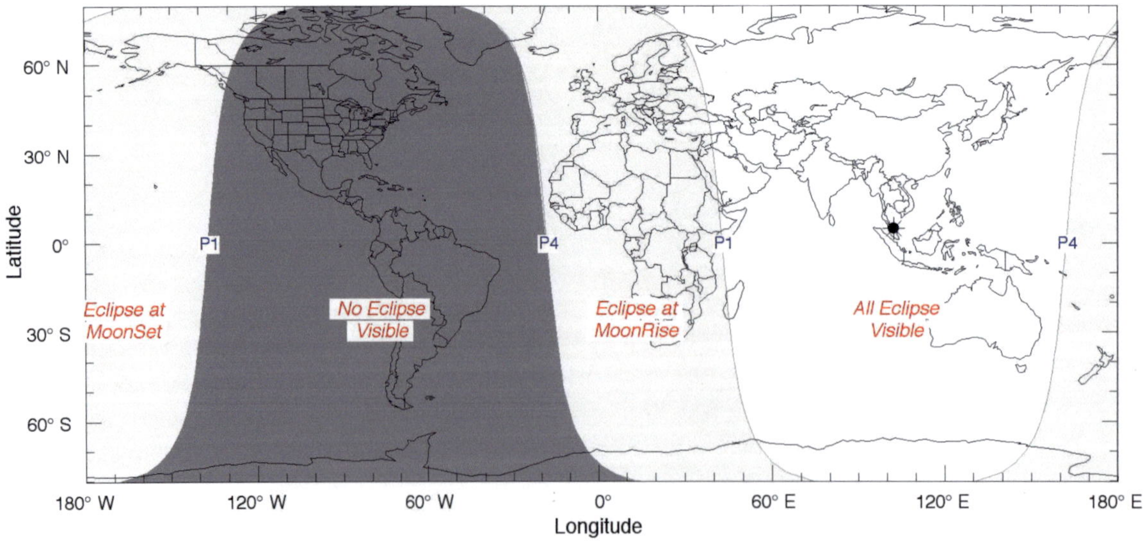

Penumbral Lunar Eclipse of 2053 Aug 29

Greatest Eclipse = 08:05:49.8 TD (= 08:04:22.3 UT1)

Penumbral Magnitude = 1.0203
Umbral Magnitude = -0.0319

Gamma = 1.0165
Axis = 0.9358°

Saros Series = 119
Saros Member = 63 of 82

Sun at Greatest Eclipse
(Geocentric Coordinates)
R.A. = 10h32m52.4s
Dec. = +09°08'07.1"
S.D. = 00°15'50.2"
H.P. = 00°00'08.7"

Moon at Greatest Eclipse
(Geocentric Coordinates)
R.A. = 22h31m49.7s
Dec. = -08°14'09.0"
S.D. = 00°15'03.1"
H.P. = 00°55'14.5"

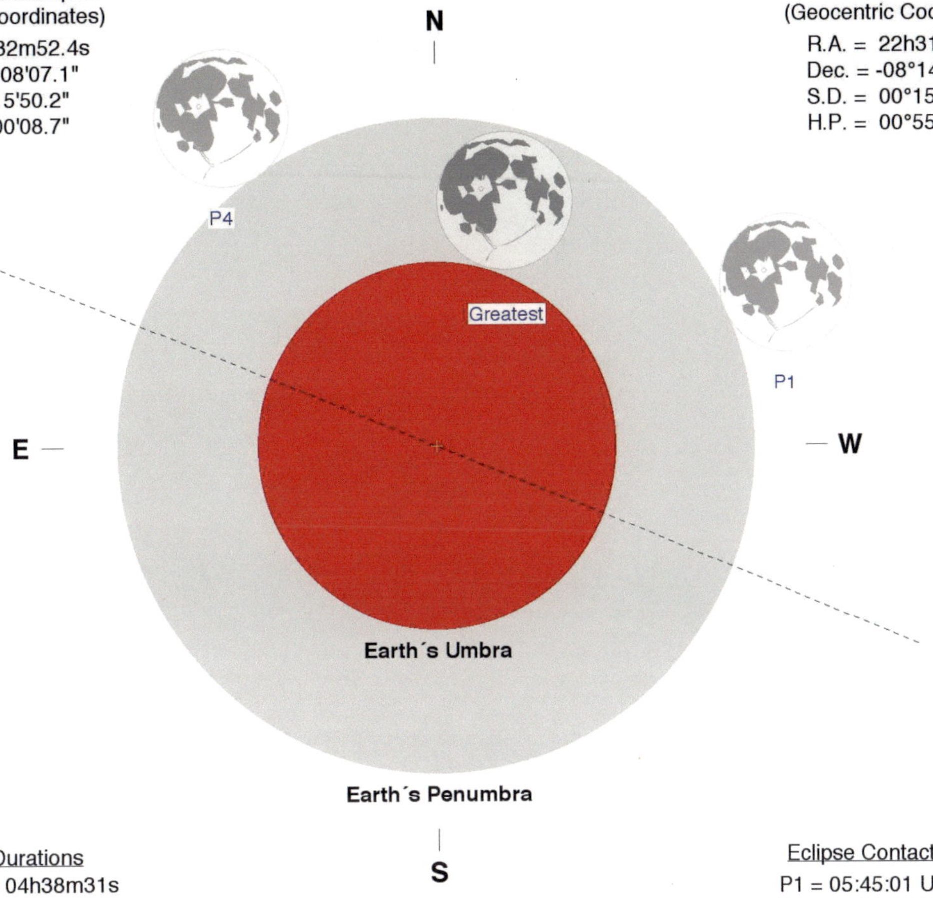

Eclipse Durations
Penumbral = 04h38m31s

Eclipse Contacts
P1 = 05:45:01 UT1
P4 = 10:23:32 UT1

Eph. = JPL DE430
Rule = Herald-Sinnott
ΔT = 87 s

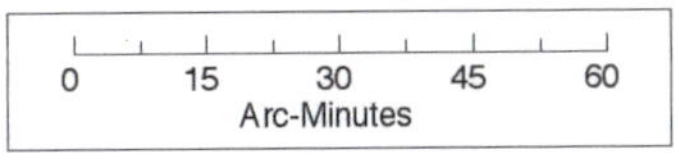

©2020 F. Espenak, www.EclipseWise.com

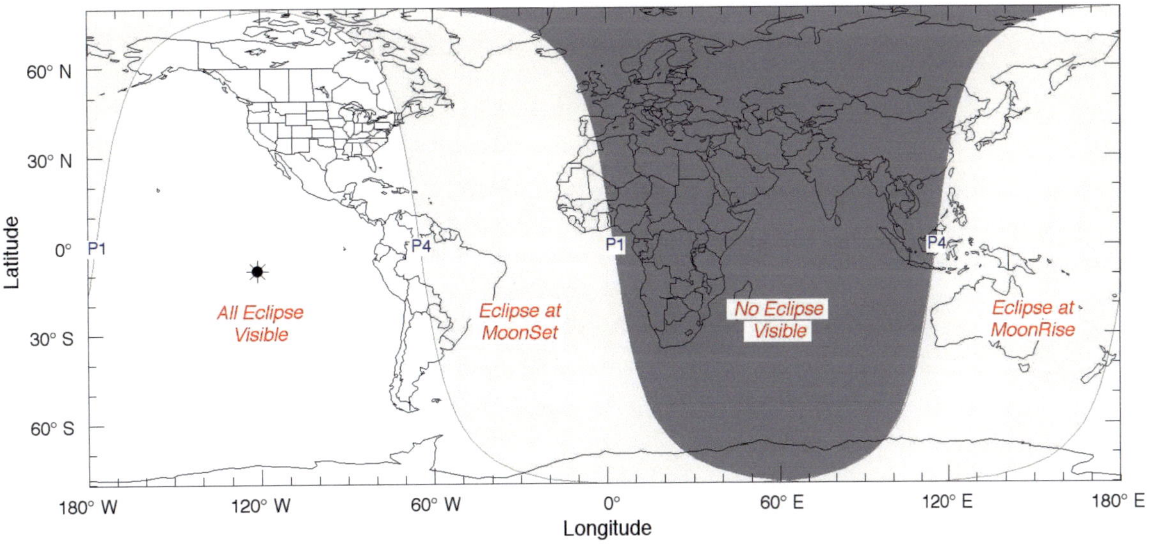

Total Lunar Eclipse of 2054 Feb 22

Greatest Eclipse = 06:51:27.0 TD (= 06:49:59.1 UT1)

Penumbral Magnitude = 2.2502
Umbral Magnitude = 1.2780

Gamma = -0.3242
Axis = 0.3298°

Saros Series = 124
Saros Member = 51 of 73

Sun at Greatest Eclipse
(Geocentric Coordinates)
R.A. = 22h23m02.4s
Dec. = -10°05'18.4"
S.D. = 00°16'10.2"
H.P. = 00°00'08.9"

Moon at Greatest Eclipse
(Geocentric Coordinates)
R.A. = 10h22m40.9s
Dec. = +09°46'14.8"
S.D. = 00°16'38.0"
H.P. = 01°01'02.7"

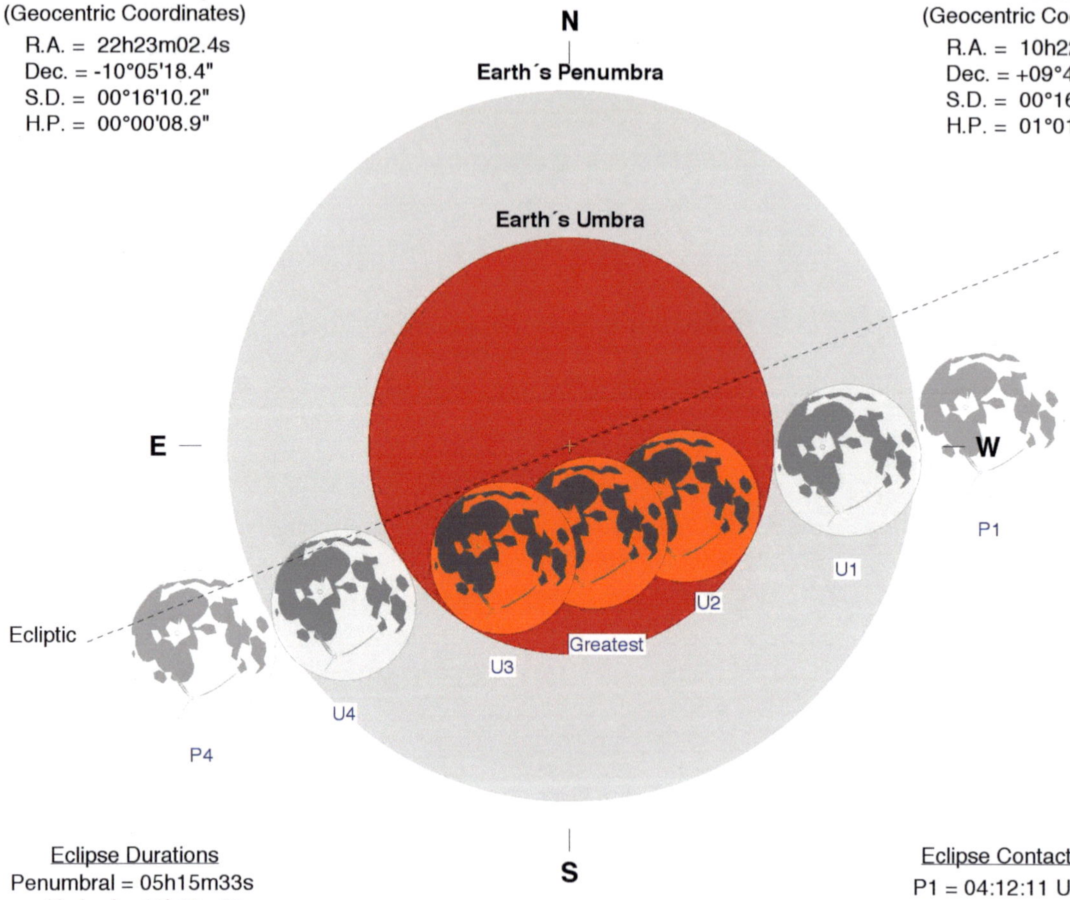

Eclipse Durations
Penumbral = 05h15m33s
Umbral = 03h21m36s
Total = 01h12m44s

Eph. = JPL DE430
Rule = Herald-Sinnott
ΔT = 88 s

0 15 30 45 60
Arc-Minutes

©2020 F. Espenak, www.EclipseWise.com

Eclipse Contacts
P1 = 04:12:11 UT1
U1 = 05:09:07 UT1
U2 = 06:13:29 UT1
U3 = 07:26:14 UT1
U4 = 08:30:43 UT1
P4 = 09:27:44 UT1

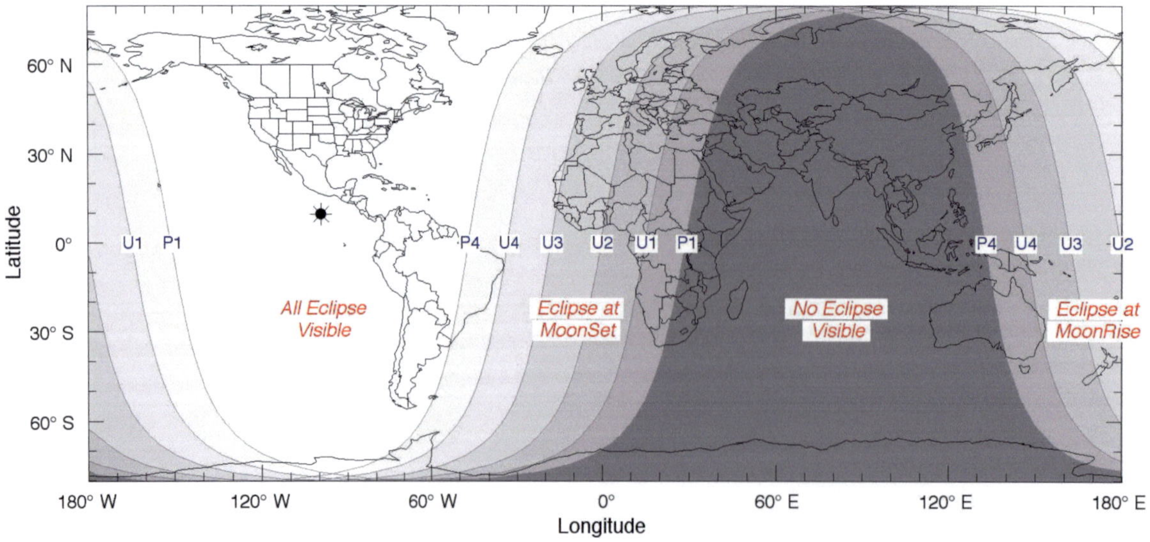

Total Lunar Eclipse of 2054 Aug 18

Greatest Eclipse = 09:26:30.1 TD (= 09:25:01.9 UT1)

Penumbral Magnitude = 2.3817
Umbral Magnitude = 1.3074

Gamma = 0.2806
Axis = 0.2524°

Saros Series = 129
Saros Member = 40 of 71

Sun at Greatest Eclipse
(Geocentric Coordinates)
R.A. = 09h51m47.0s
Dec. = +12°57'08.8"
S.D. = 00°15'48.0"
H.P. = 00°00'08.7"

Moon at Greatest Eclipse
(Geocentric Coordinates)
R.A. = 21h51m32.0s
Dec. = -12°42'26.7"
S.D. = 00°14'42.4"
H.P. = 00°53'58.4"

N

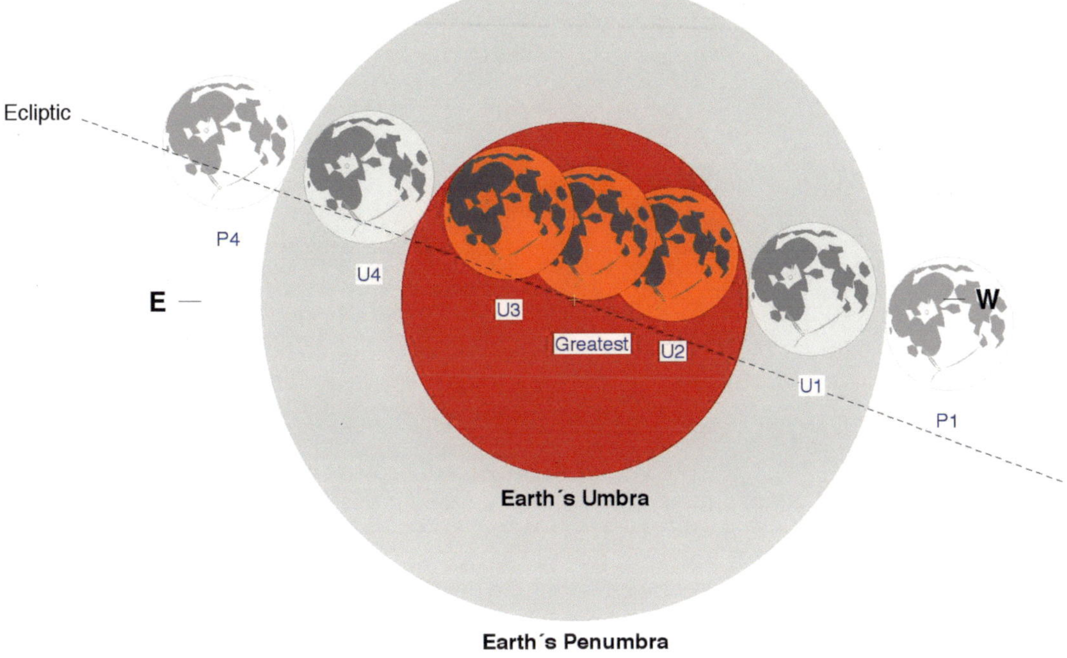

S

Eclipse Durations
Penumbral = 06h10m22s
Umbral = 03h47m21s
Total = 01h23m40s

Eph. = JPL DE430
Rule = Herald-Sinnott
ΔT = 88 s

0 15 30 45 60
Arc-Minutes

Eclipse Contacts
P1 = 06:19:49 UT1
U1 = 07:31:18 UT1
U2 = 08:43:05 UT1
U3 = 10:06:45 UT1
U4 = 11:18:39 UT1
P4 = 12:30:11 UT1

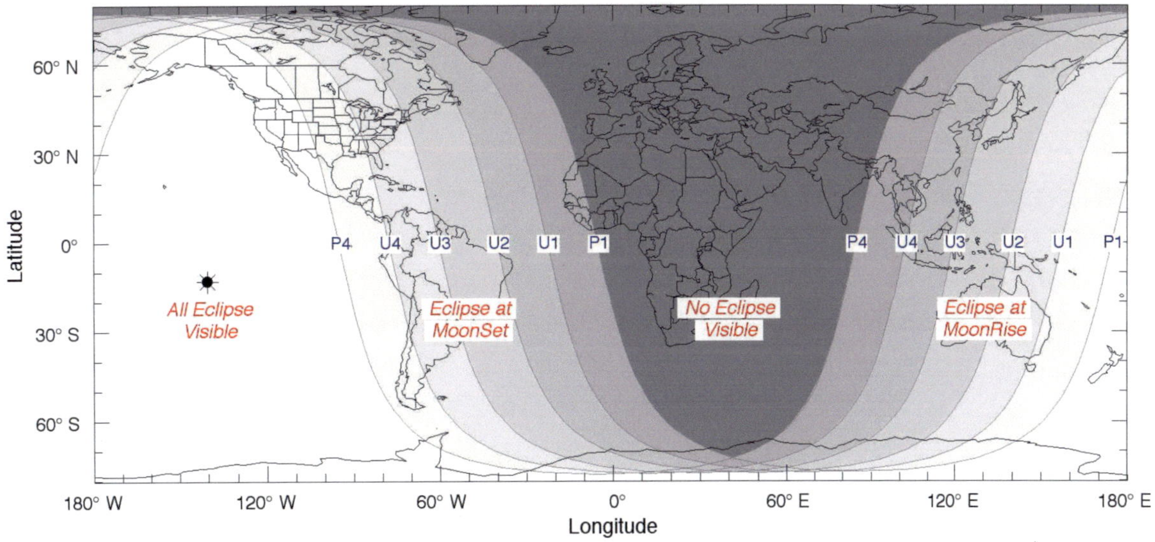

Total Lunar Eclipse of 2055 Feb 11

Greatest Eclipse = 22:46:17.5 TD (= 22:44:49.0 UT1)

Penumbral Magnitude = 2.1982 Gamma = 0.3526 Saros Series = 134
Umbral Magnitude = 1.2258 Axis = 0.3594° Saros Member = 29 of 72

Sun at Greatest Eclipse
(Geocentric Coordinates)
R.A. = 21h42m03.5s
Dec. = -13°47'10.8"
S.D. = 00°16'12.3"
H.P. = 00°00'08.9"

Moon at Greatest Eclipse
(Geocentric Coordinates)
R.A. = 09h42m24.2s
Dec. = +14°08'09.1"
S.D. = 00°16'39.9"
H.P. = 01°01'09.9"

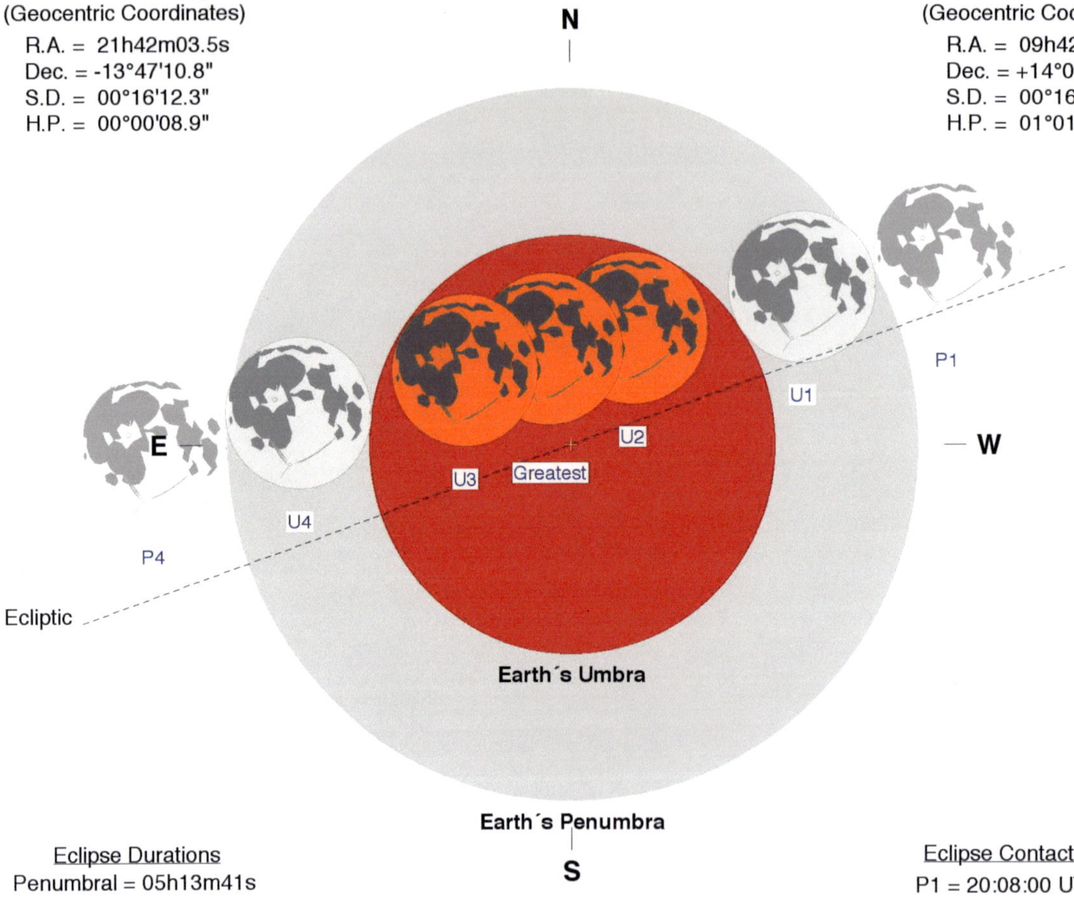

Eclipse Durations
Penumbral = 05h13m41s
Umbral = 03h19m09s
Total = 01h06m35s

Eph. = JPL DE430
Rule = Herald-Sinnott
ΔT = 88 s

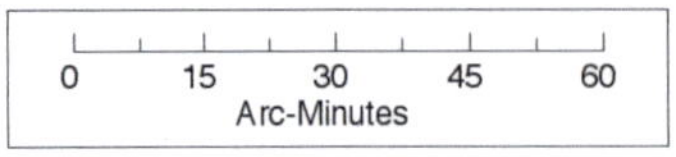

©2020 F. Espenak, www.EclipseWise.com

Eclipse Contacts
P1 = 20:08:00 UT1
U1 = 21:05:18 UT1
U2 = 22:11:39 UT1
U3 = 23:18:14 UT1
U4 = 00:24:27 UT1
P4 = 01:21:40 UT1

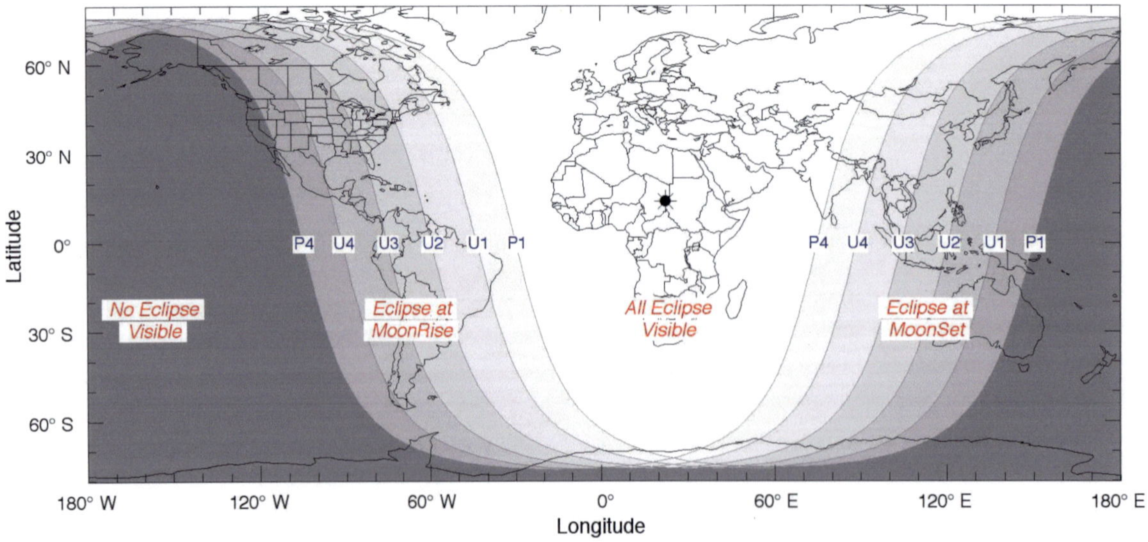

Partial Lunar Eclipse of 2055 Aug 07

Greatest Eclipse = 10:53:18.2 TD (= 10:51:49.4 UT1)

Penumbral Magnitude = 2.0081
Umbral Magnitude = 0.9606

Gamma = -0.4769
Axis = 0.4391°

Saros Series = 139
Saros Member = 23 of 79

Sun at Greatest Eclipse
(Geocentric Coordinates)
R.A. = 09h09m39.9s
Dec. = +16°20'36.9"
S.D. = 00°15'46.2"
H.P. = 00°00'08.7"

Moon at Greatest Eclipse
(Geocentric Coordinates)
R.A. = 21h10m01.5s
Dec. = -16°46'26.9"
S.D. = 00°15'03.3"
H.P. = 00°55'15.1"

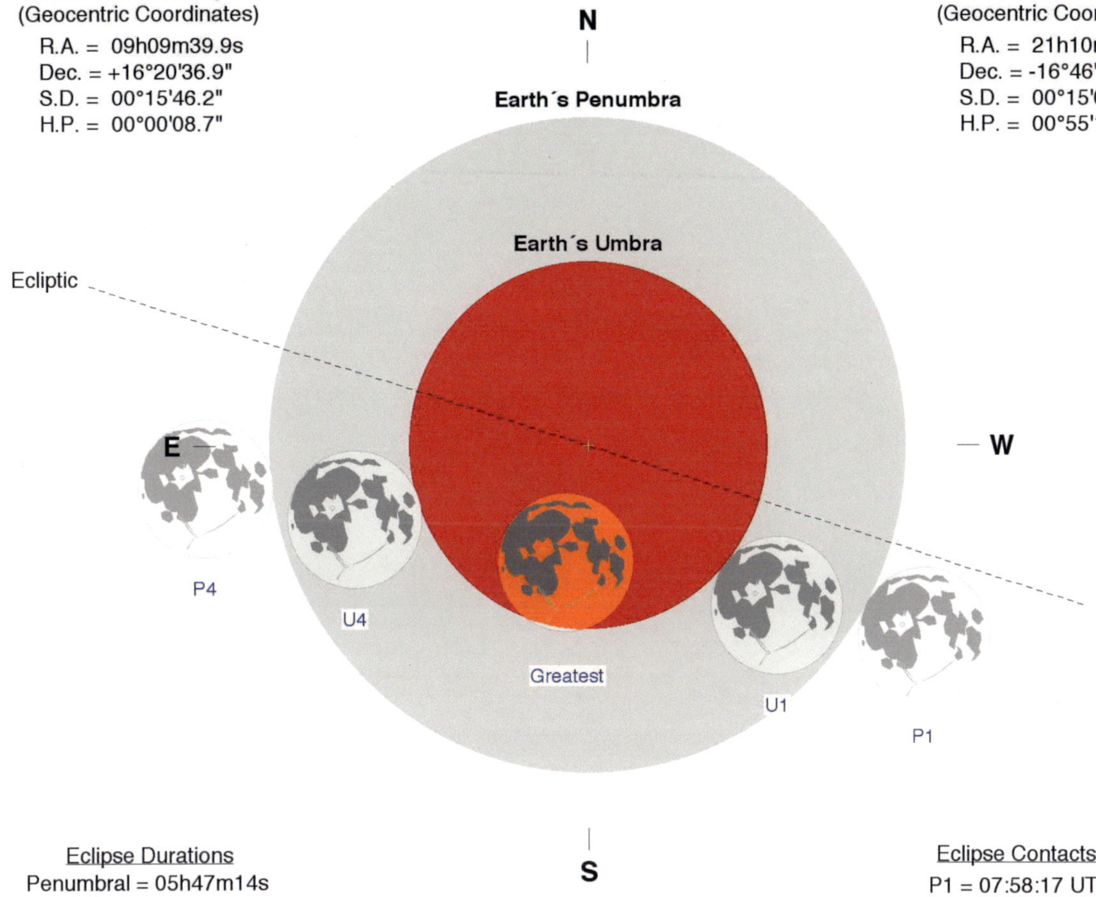

Eclipse Durations
Penumbral = 05h47m14s
Umbral = 03h24m10s

Eclipse Contacts
P1 = 07:58:17 UT1
U1 = 09:09:47 UT1
U4 = 12:33:56 UT1
P4 = 13:45:31 UT1

Eph. = JPL DE430
Rule = Herald-Sinnott
ΔT = 89 s

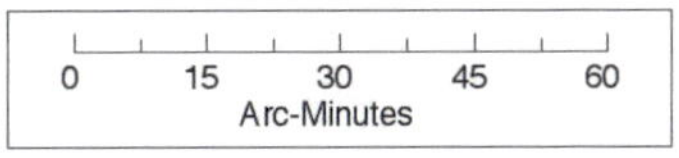

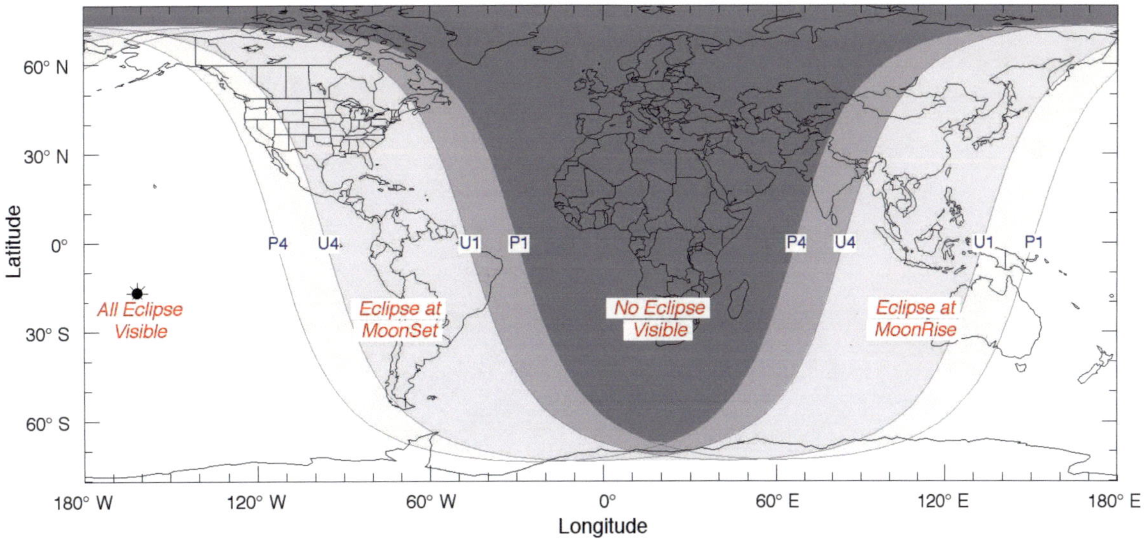

Penumbral Lunar Eclipse of 2056 Feb 01

Greatest Eclipse = 12:26:06.2 TD (= 12:24:37.1 UT1)

Penumbral Magnitude = 0.9069 Gamma = 1.0682 Saros Series = 144
Umbral Magnitude = -0.1084 Axis = 1.0447° Saros Member = 18 of 71

Sun at Greatest Eclipse
(Geocentric Coordinates)
R.A. = 20h59m18.0s
Dec. = -17°05'10.5"
S.D. = 00°16'14.0"
H.P. = 00°00'08.9"

Moon at Greatest Eclipse
(Geocentric Coordinates)
R.A. = 09h00m07.0s
Dec. = +18°06'45.8"
S.D. = 00°15'59.4"
H.P. = 00°58'41.2"

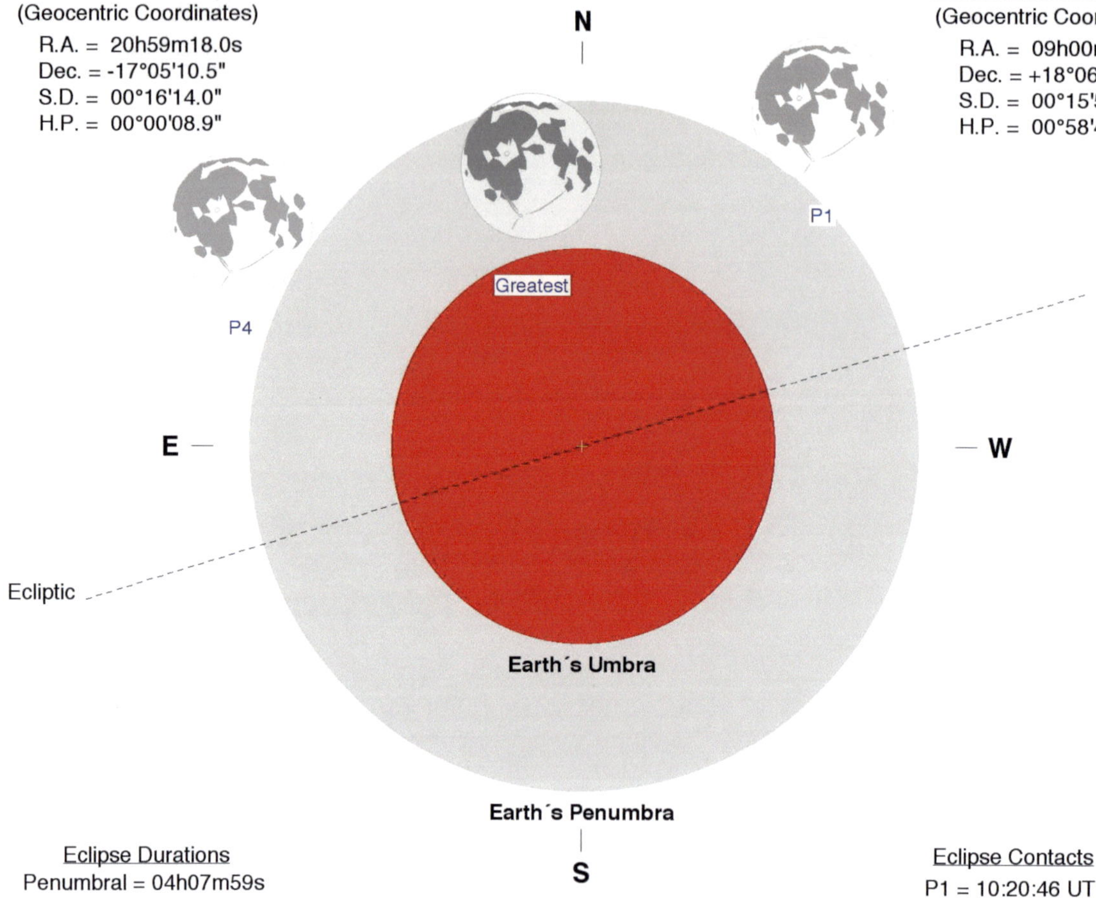

Eclipse Durations
Penumbral = 04h07m59s

Eclipse Contacts
P1 = 10:20:46 UT1
P4 = 14:28:44 UT1

Eph. = JPL DE430
Rule = Herald-Sinnott
ΔT = 89 s

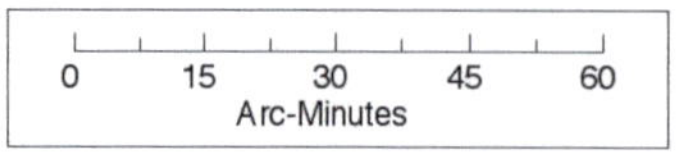

©2020 F. Espenak, www.EclipseWise.com

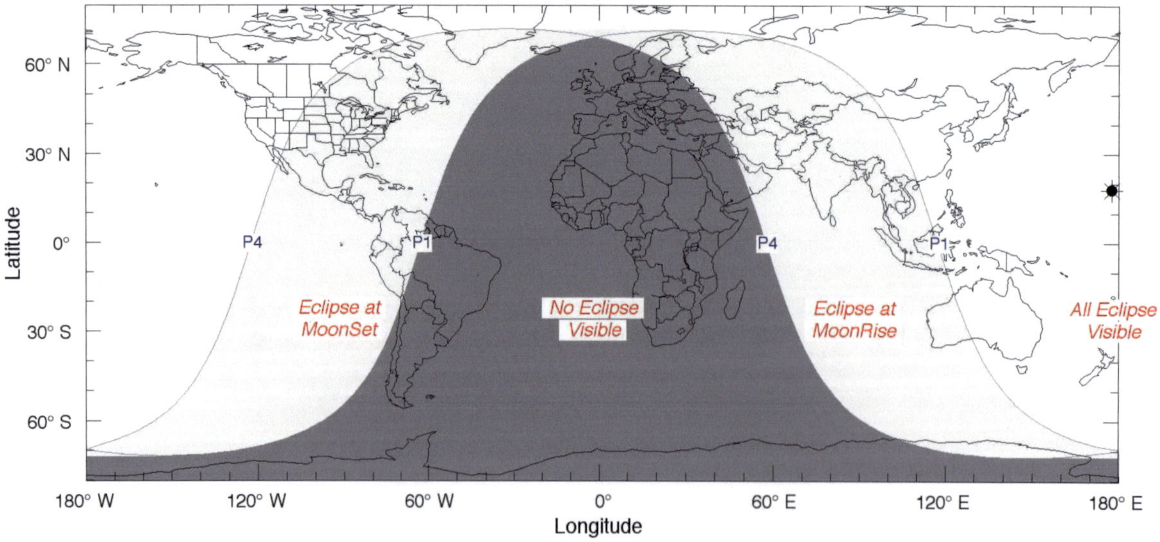

Penumbral Lunar Eclipse of 2056 Jun 27

Greatest Eclipse = 10:03:09.3 TD (= 10:01:39.9 UT1)

Penumbral Magnitude = 0.3158 | Gamma = 1.3770 | Saros Series = 111
Umbral Magnitude = -0.6504 | Axis = 1.3714° | Saros Member = 69 of 71

Sun at Greatest Eclipse
(Geocentric Coordinates)
R.A. = 06h28m18.7s
Dec. = +23°16'17.2"
S.D. = 00°15'44.0"
H.P. = 00°00'08.7"

Moon at Greatest Eclipse
(Geocentric Coordinates)
R.A. = 18h28m35.0s
Dec. = -21°54'04.7"
S.D. = 00°16'17.0"
H.P. = 00°59'45.8"

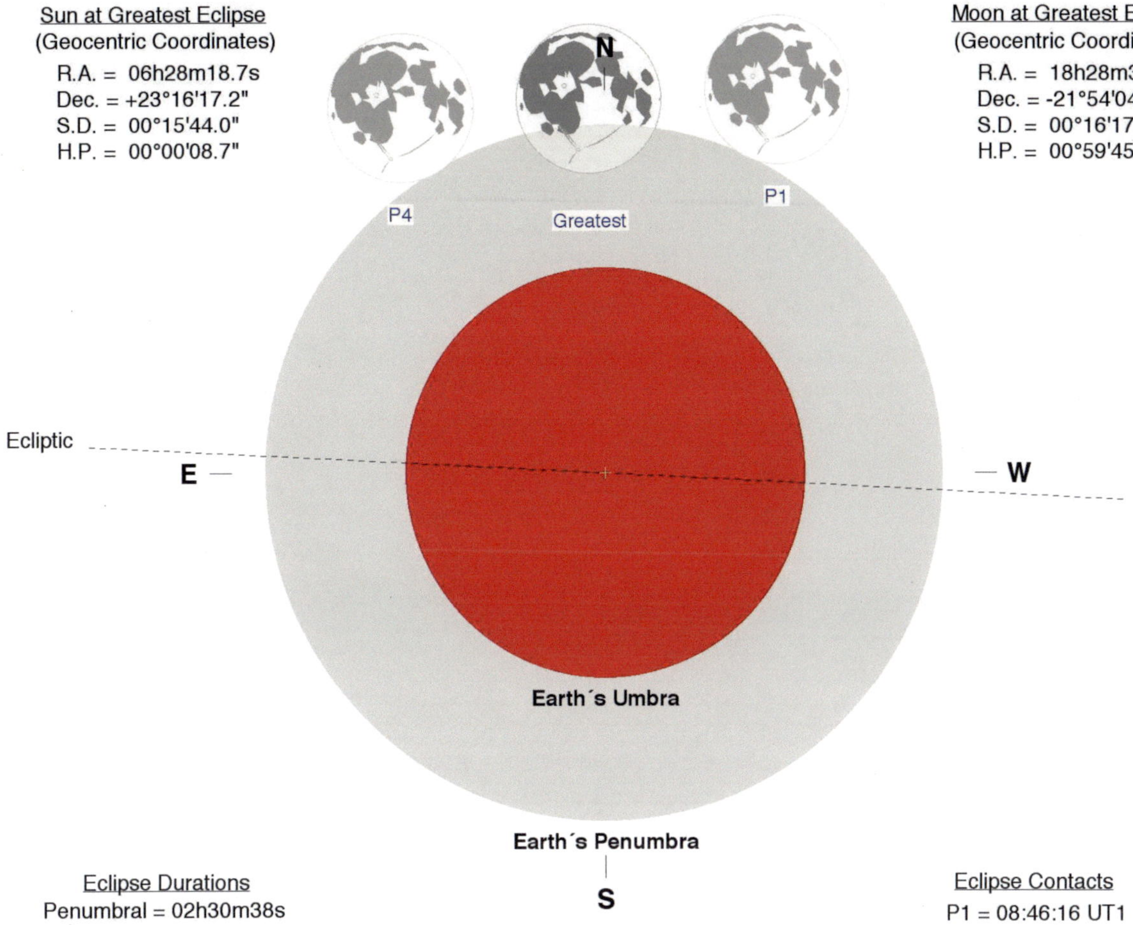

Eclipse Durations
Penumbral = 02h30m38s

Eclipse Contacts
P1 = 08:46:16 UT1
P4 = 11:16:54 UT1

Eph. = JPL DE430
Rule = Herald-Sinnott
ΔT = 89 s

0 15 30 45 60
Arc-Minutes

©2020 F. Espenak, www.EclipseWise.com

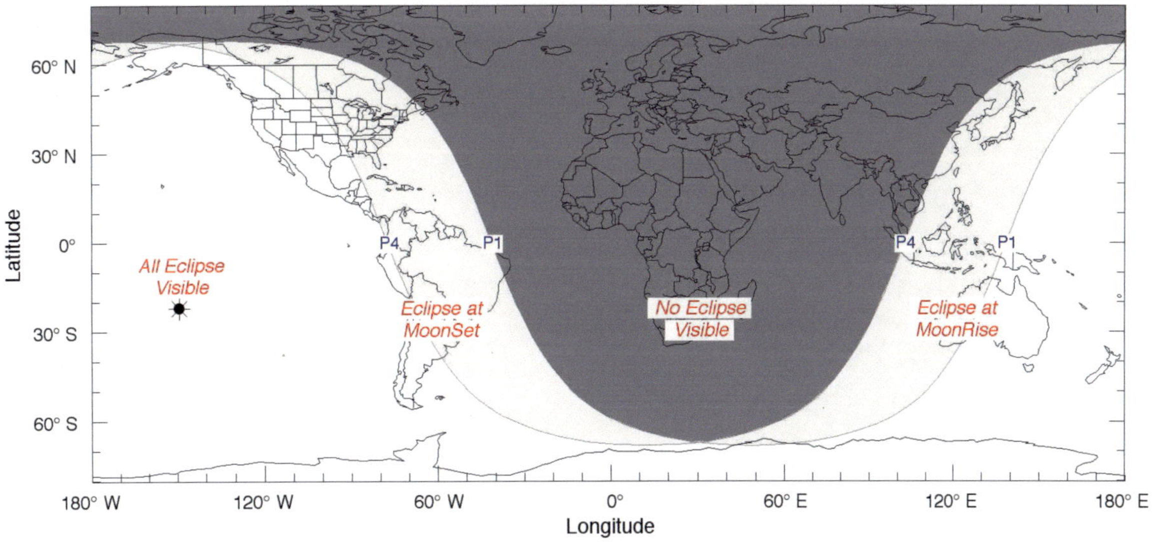

Penumbral Lunar Eclipse of 2056 Jul 26

Greatest Eclipse = 18:43:24.9 TD (= 18:41:55.4 UT1)

Penumbral Magnitude = 0.6448 Gamma = -1.2048 Saros Series = 149
Umbral Magnitude = -0.3477 Axis = 1.1693° Saros Member = 5 of 71

Sun at Greatest Eclipse
(Geocentric Coordinates)
R.A. = 08h27m22.5s
Dec. = +19°07'40.1"
S.D. = 00°15'44.9"
H.P. = 00°00'08.7"

Moon at Greatest Eclipse
(Geocentric Coordinates)
R.A. = 20h28m05.5s
Dec. = -20°17'05.8"
S.D. = 00°15'52.1"
H.P. = 00°58'14.4"

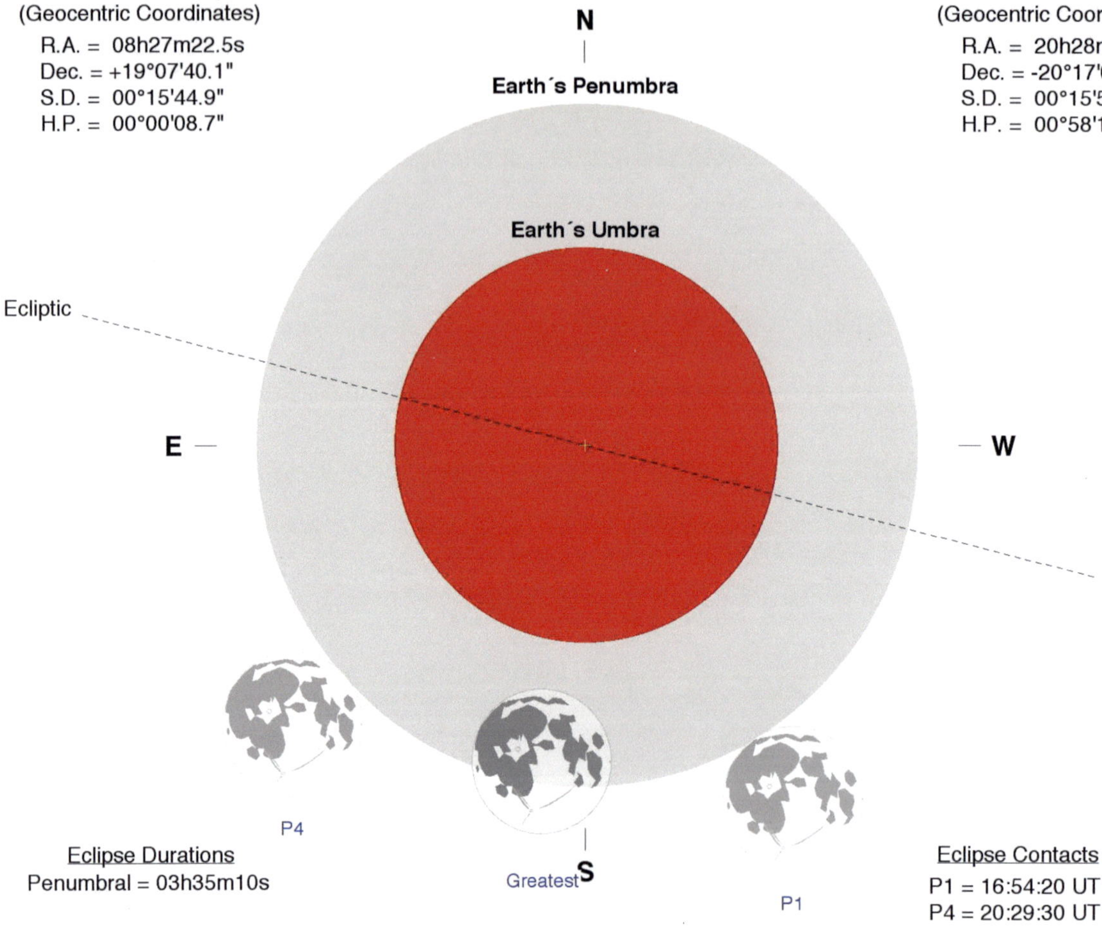

Eclipse Durations
Penumbral = 03h35m10s

Eclipse Contacts
P1 = 16:54:20 UT1
P4 = 20:29:30 UT1

Eph. = JPL DE430
Rule = Herald-Sinnott
ΔT = 89 s

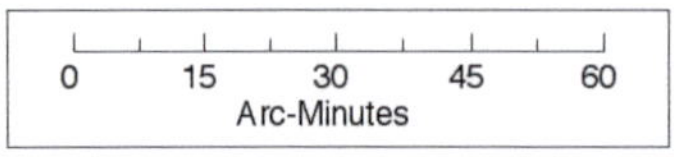

©2020 F. Espenak, www.EclipseWise.com

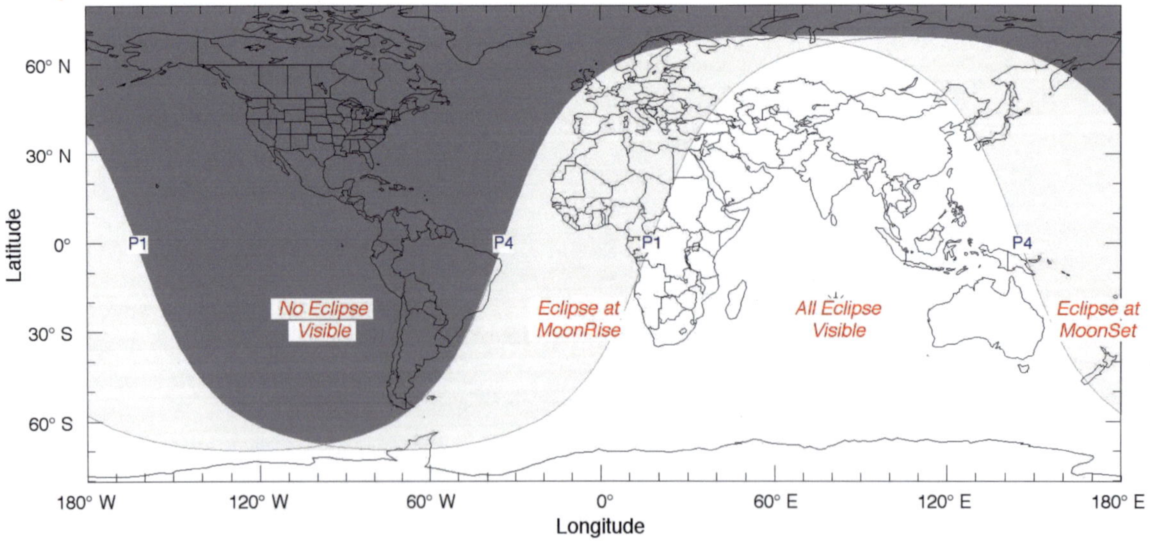

Penumbral Lunar Eclipse of 2056 Dec 22

Greatest Eclipse = 01:48:56.0 TD (= 01:47:26.2 UT1)

Penumbral Magnitude = 0.7872 Gamma = -1.1560 Saros Series = 116
Umbral Magnitude = -0.3093 Axis = 1.0483° Saros Member = 60 of 73

Sun at Greatest Eclipse
(Geocentric Coordinates)

R.A. = 18h04m03.3s
Dec. = -23°25'40.9"
S.D. = 00°16'15.5"
H.P. = 00°00'08.9"

Moon at Greatest Eclipse
(Geocentric Coordinates)

R.A. = 06h04m28.2s
Dec. = +22°23'02.6"
S.D. = 00°14'49.6"
H.P. = 00°54'24.9"

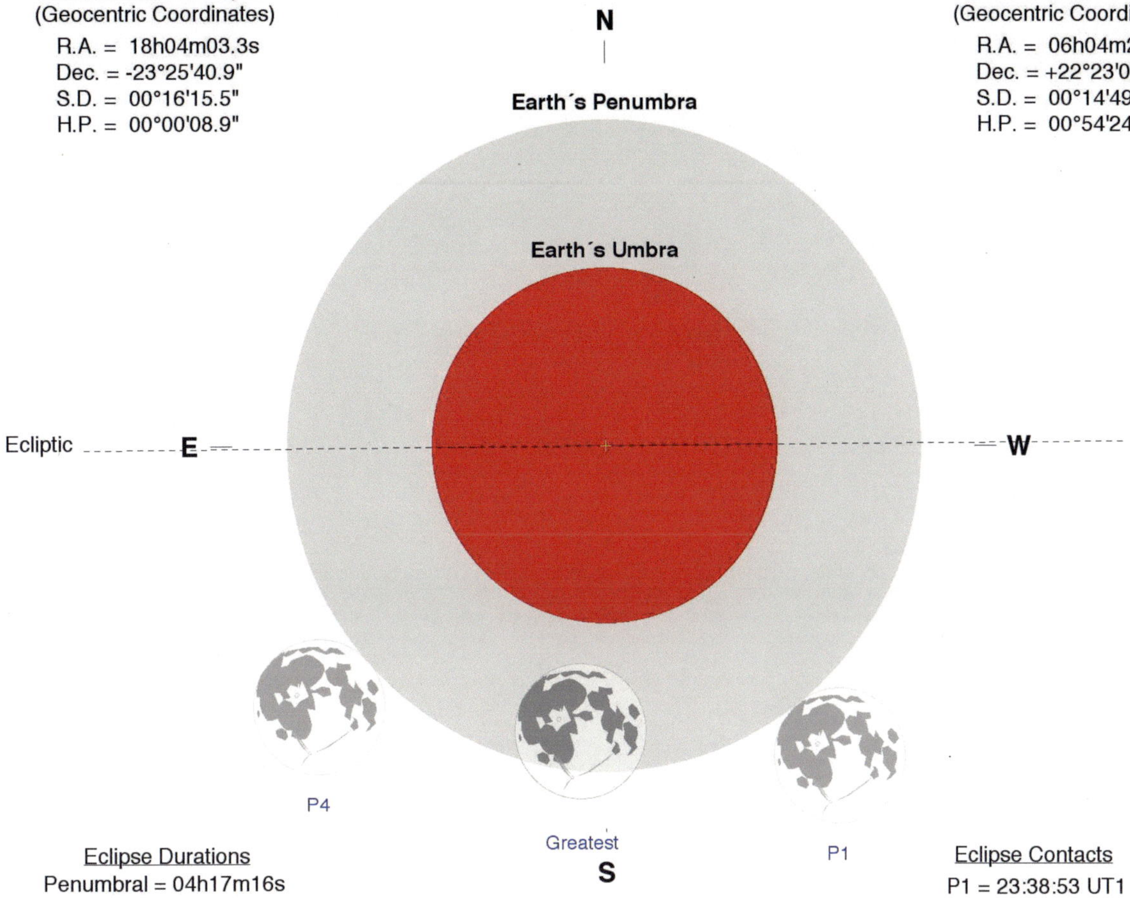

Eclipse Durations
Penumbral = 04h17m16s

Eclipse Contacts
P1 = 23:38:53 UT1
P4 = 03:56:09 UT1

Eph. = JPL DE430
Rule = Herald-Sinnott
ΔT = 90 s

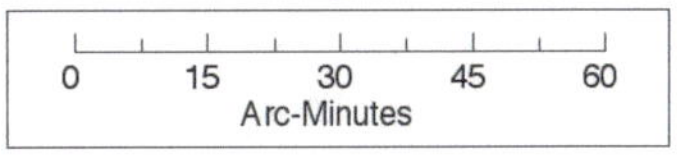

©2020 F. Espenak, www.EclipseWise.com

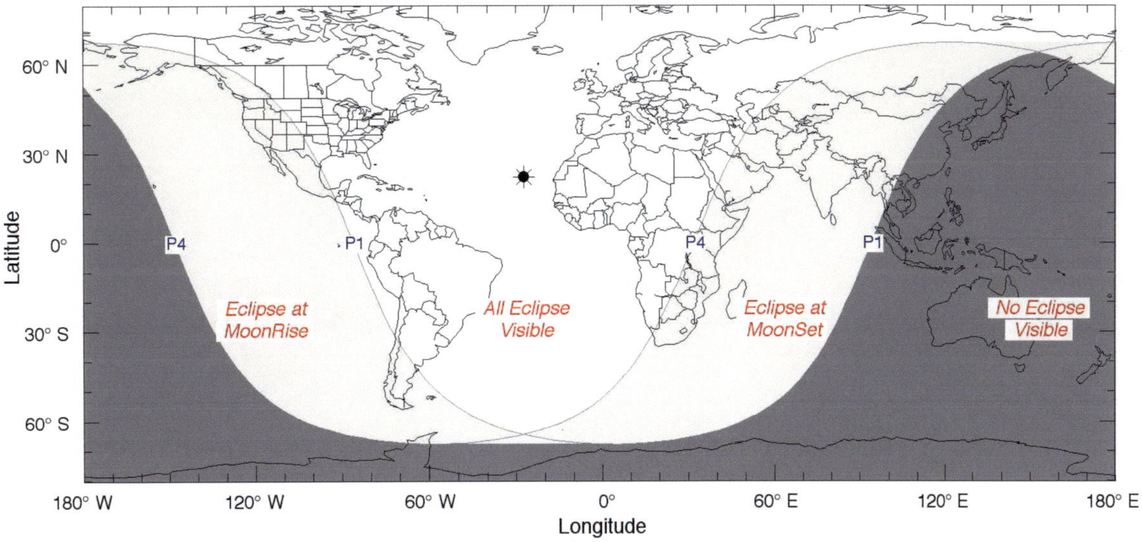

Partial Lunar Eclipse of 2057 Jun 17

Greatest Eclipse = 02:26:20.1 TD (= 02:24:50.0 UT1)

Penumbral Magnitude = 1.6982 Gamma = 0.6168 Saros Series = 121
Umbral Magnitude = 0.7570 Axis = 0.6310° Saros Member = 57 of 82

Sun at Greatest Eclipse
(Geocentric Coordinates)
R.A. = 05h44m25.1s
Dec. = +23°22'59.4"
S.D. = 00°15'44.7"
H.P. = 00°00'08.7"

Moon at Greatest Eclipse
(Geocentric Coordinates)
R.A. = 17h44m45.6s
Dec. = -22°45'25.3"
S.D. = 00°16'43.7"
H.P. = 01°01'23.6"

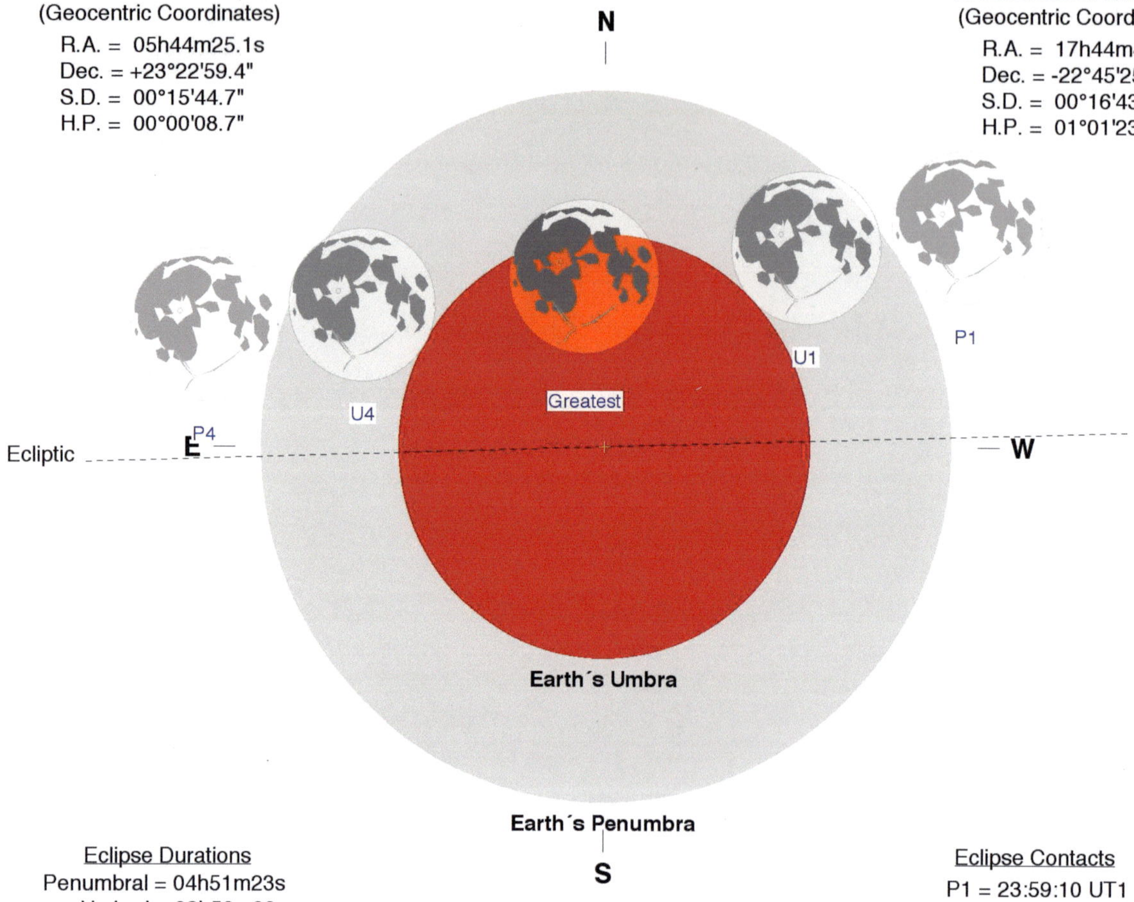

Eclipse Durations
Penumbral = 04h51m23s
Umbral = 02h50m00s

Eclipse Contacts
P1 = 23:59:10 UT1
U1 = 00:59:52 UT1
U4 = 03:49:53 UT1
P4 = 04:50:33 UT1

Eph. = JPL DE430
Rule = Herald-Sinnott
ΔT = 90 s

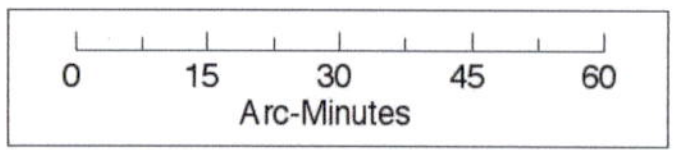

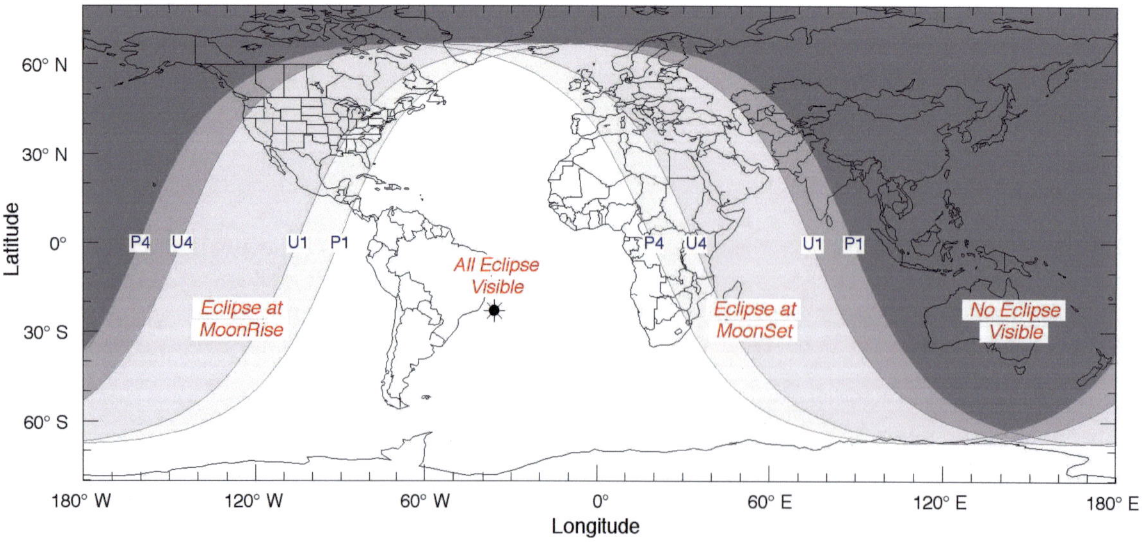

Partial Lunar Eclipse of 2057 Dec 11

Greatest Eclipse = 00:53:38.0 TD (= 00:52:07.6 UT1)

Penumbral Magnitude = 2.0194 Gamma = -0.4853 Saros Series = 126
Umbral Magnitude = 0.9197 Axis = 0.4384° Saros Member = 47 of 70

Sun at Greatest Eclipse
(Geocentric Coordinates)
R.A. = 17h14m07.3s
Dec. = -23°00'46.9"
S.D. = 00°16'14.5"
H.P. = 00°00'08.9"

Moon at Greatest Eclipse
(Geocentric Coordinates)
R.A. = 05h14m27.7s
Dec. = +22°34'54.1"
S.D. = 00°14'46.1"
H.P. = 00°54'12.2"

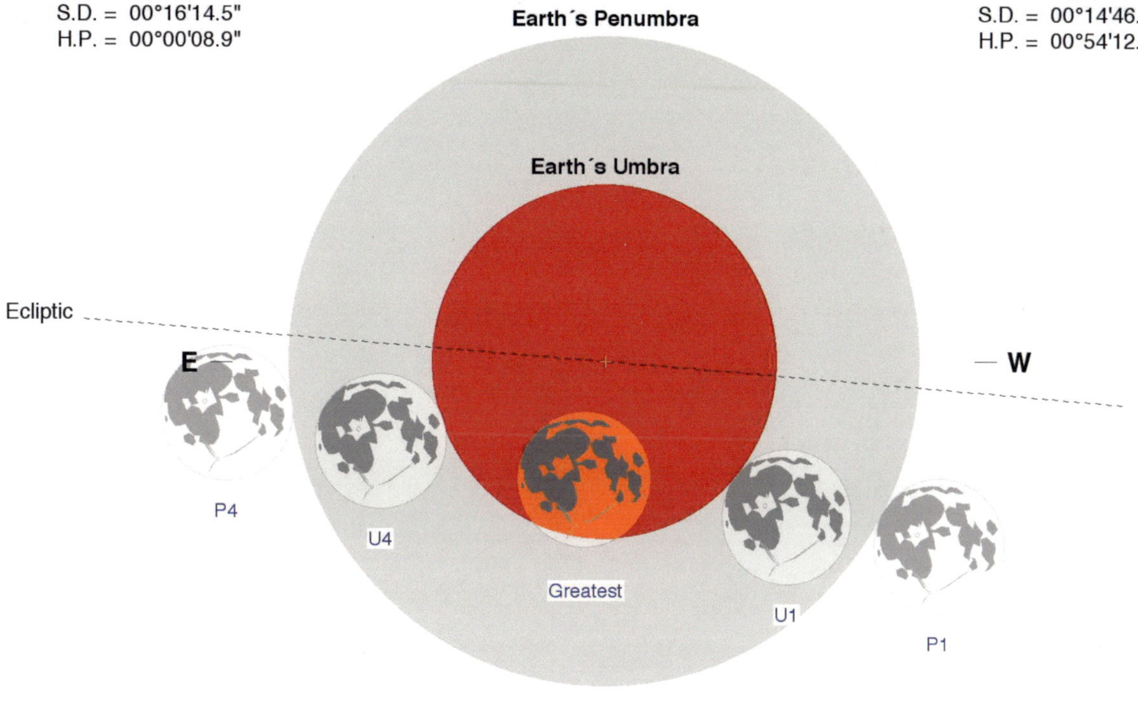

Eclipse Durations
Penumbral = 05h59m42s
Umbral = 03h24m50s

Eclipse Contacts
P1 = 21:52:20 UT1
U1 = 23:09:45 UT1
U4 = 02:34:35 UT1
P4 = 03:52:02 UT1

Eph. = JPL DE430
Rule = Herald-Sinnott
ΔT = 90 s

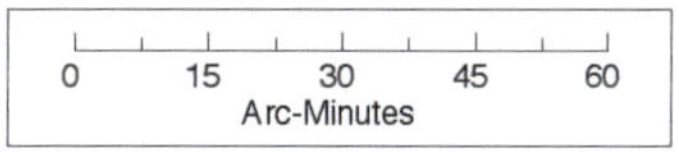

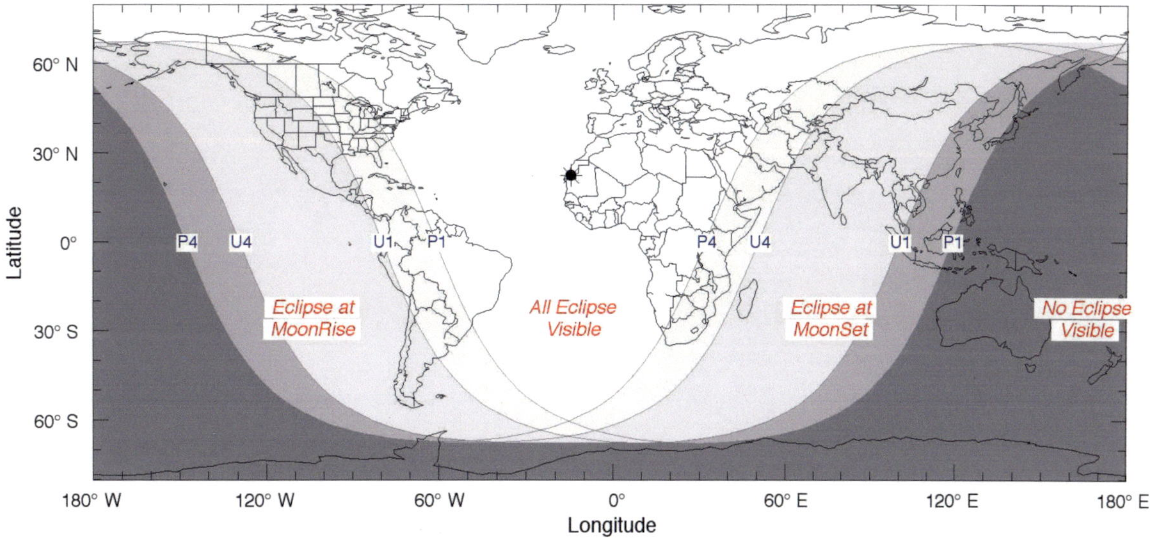

Total Lunar Eclipse of 2058 Jun 06

Greatest Eclipse = 19:15:48.0 TD (= 19:14:17.3 UT1)

Penumbral Magnitude = 2.6226   Gamma = -0.1181   Saros Series = 131
Umbral Magnitude = 1.6628   Axis = 0.1186°   Saros Member = 36 of 72

Sun at Greatest Eclipse
(Geocentric Coordinates)
R.A. = 05h00m41.7s
Dec. = +22°43'57.0"
S.D. = 00°15'45.8"
H.P. = 00°00'08.7"

Moon at Greatest Eclipse
(Geocentric Coordinates)
R.A. = 17h00m35.5s
Dec. = -22°50'55.4"
S.D. = 00°16'25.3"
H.P. = 01°00'16.2"

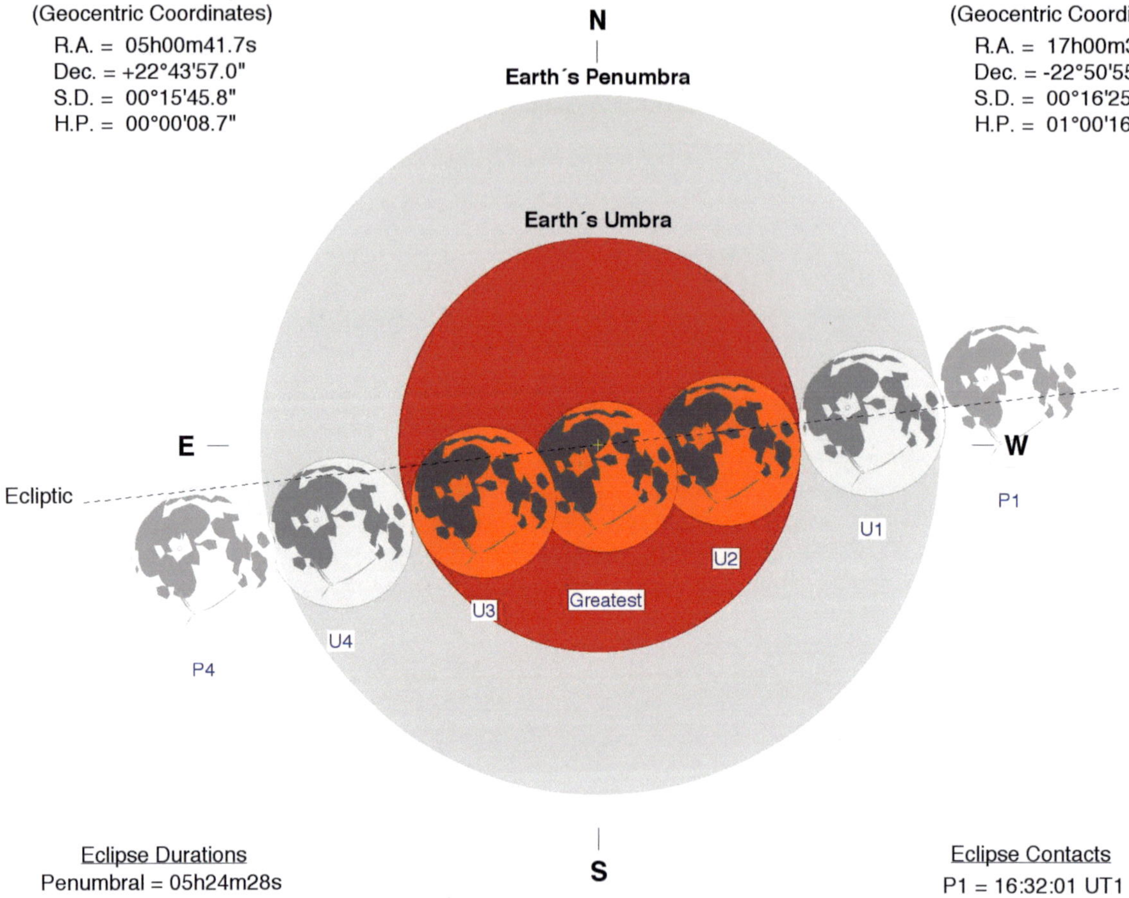

Eclipse Durations
Penumbral = 05h24m28s
Umbral = 03h34m09s
Total = 01h38m03s

Eph. = JPL DE430
Rule = Herald-Sinnott
ΔT = 91 s

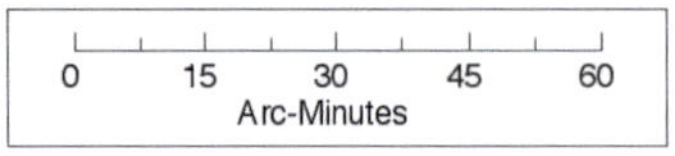

©2020 F. Espenak, www.EclipseWise.com

Eclipse Contacts
P1 = 16:32:01 UT1
U1 = 17:27:13 UT1
U2 = 18:25:14 UT1
U3 = 20:03:17 UT1
U4 = 21:01:22 UT1
P4 = 21:56:29 UT1

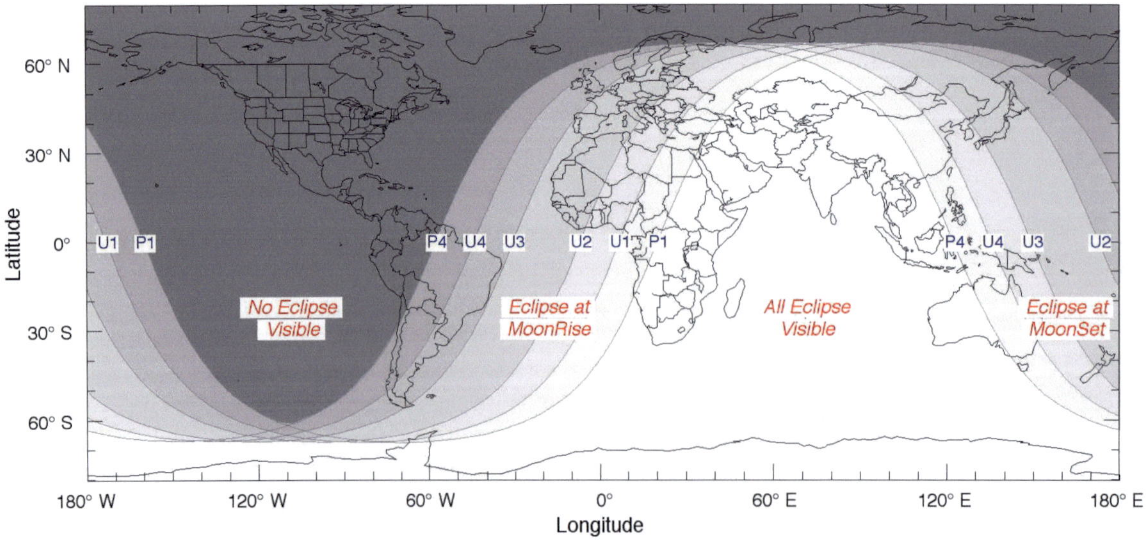

Total Lunar Eclipse of 2058 Nov 30

Greatest Eclipse = 03:16:17.9 TD (= 03:14:46.8 UT1)

Penumbral Magnitude = 2.4819 Gamma = 0.2208 Saros Series = 136
Umbral Magnitude = 1.4277 Axis = 0.2077° Saros Member = 22 of 72

Sun at Greatest Eclipse
(Geocentric Coordinates)
R.A. = 16h25m34.1s
Dec. = -21°39'35.3"
S.D. = 00°16'12.9"
H.P. = 00°00'08.9"

Moon at Greatest Eclipse
(Geocentric Coordinates)
R.A. = 04h25m20.3s
Dec. = +21°51'37.8"
S.D. = 00°15'22.8"
H.P. = 00°56'26.8"

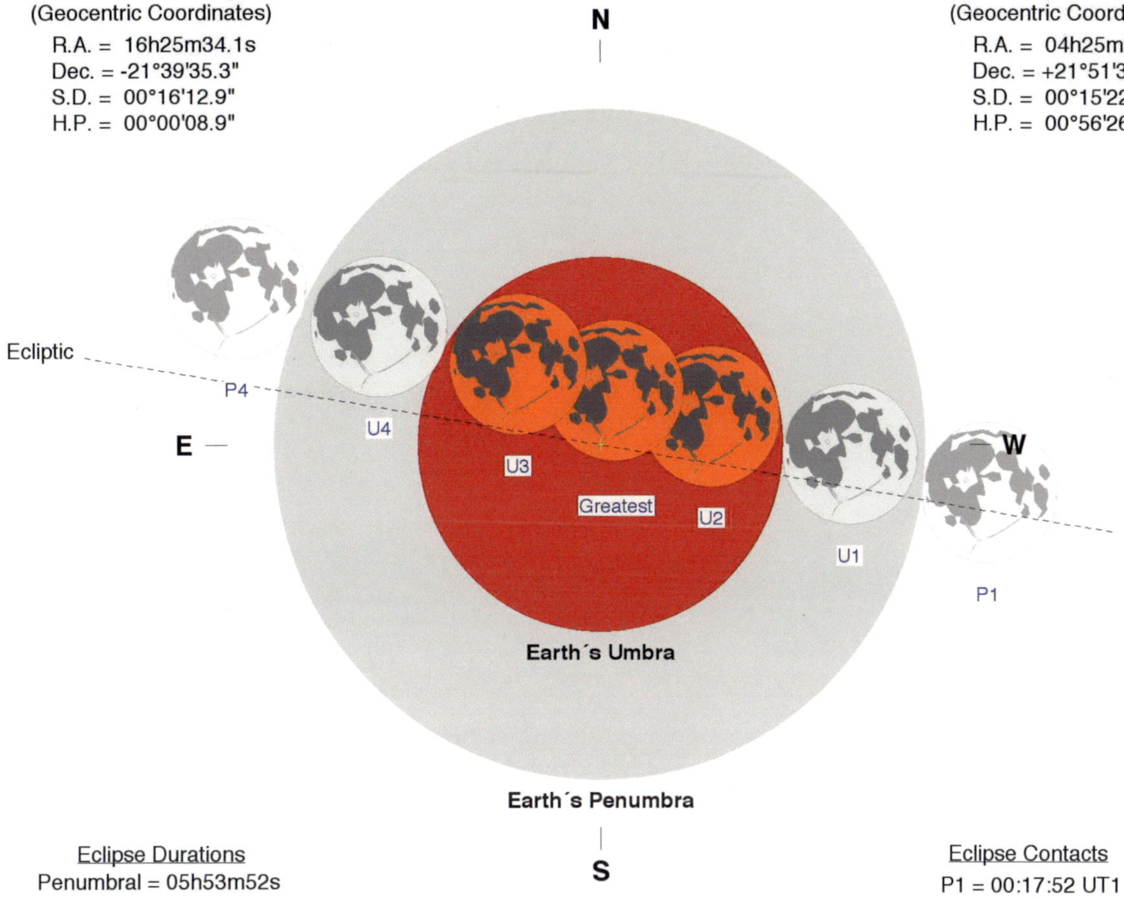

Eclipse Durations
Penumbral = 05h53m52s
Umbral = 03h41m28s
Total = 01h30m25s

Eph. = JPL DE430
Rule = Herald-Sinnott
ΔT = 91 s

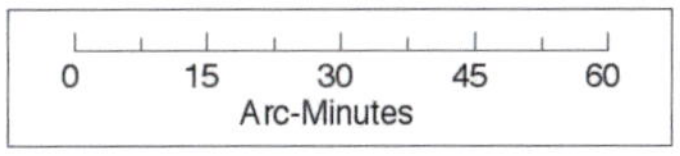

©2020 F. Espenak, www.EclipseWise.com

Eclipse Contacts
P1 = 00:17:52 UT1
U1 = 01:23:59 UT1
U2 = 02:29:28 UT1
U3 = 03:59:54 UT1
U4 = 05:05:27 UT1
P4 = 06:11:45 UT1

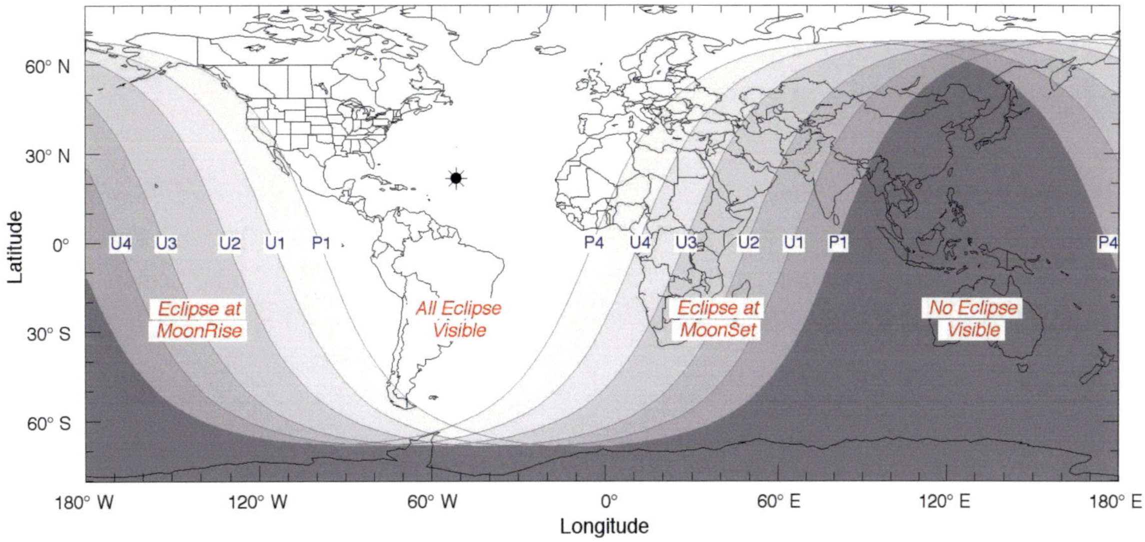

Partial Lunar Eclipse of 2059 May 27

Greatest Eclipse = 07:55:34.3 TD (= 07:54:02.8 UT1)

Penumbral Magnitude = 1.1963 Gamma = -0.9098 Saros Series = 141
Umbral Magnitude = 0.1846 Axis = 0.8684° Saros Member = 26 of 72

Sun at Greatest Eclipse
(Geocentric Coordinates)
R.A. = 04h16m47.9s
Dec. = +21°18'58.7"
S.D. = 00°15'47.3"
H.P. = 00°00'08.7"

Moon at Greatest Eclipse
(Geocentric Coordinates)
R.A. = 16h15m47.5s
Dec. = -22°09'09.5"
S.D. = 00°15'36.4"
H.P. = 00°57'16.6"

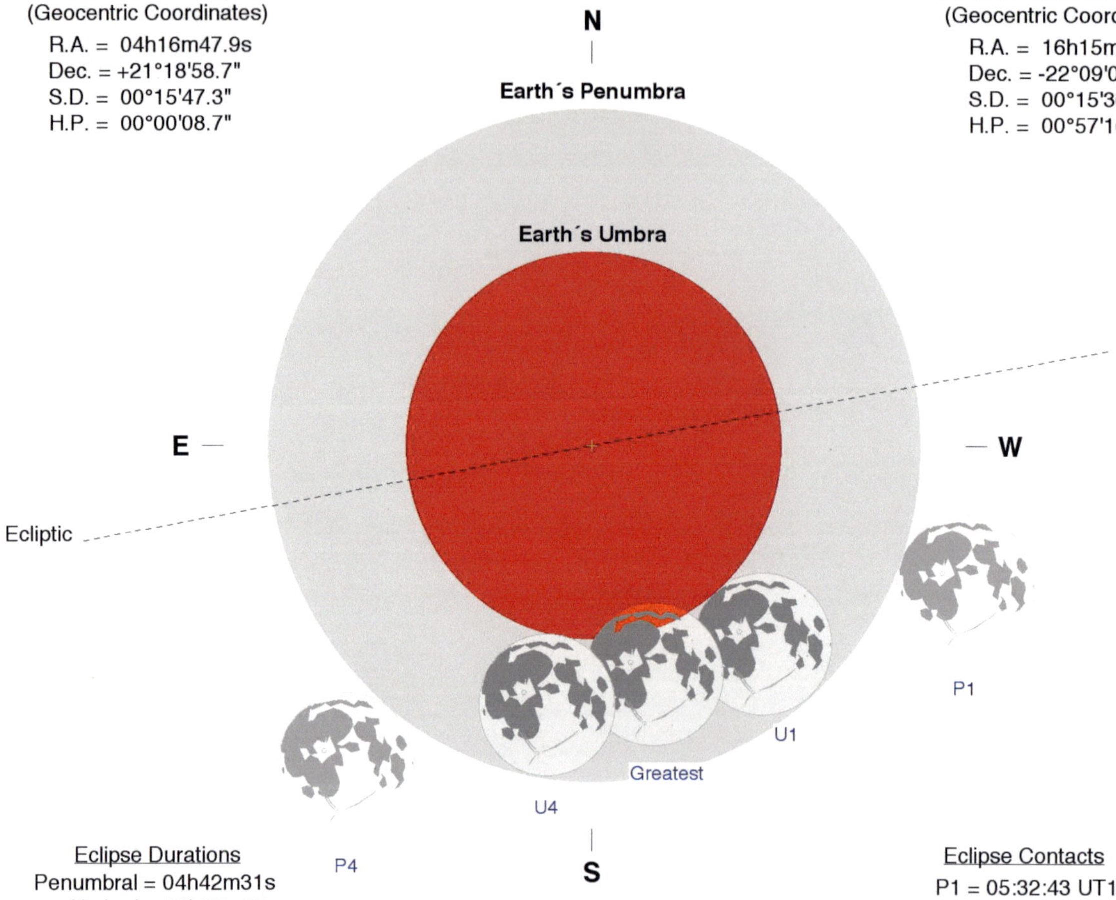

Eclipse Durations
Penumbral = 04h42m31s
Umbral = 01h37m55s

Eclipse Contacts
P1 = 05:32:43 UT1
U1 = 07:05:02 UT1
U4 = 08:42:57 UT1
P4 = 10:15:14 UT1

Eph. = JPL DE430
Rule = Herald-Sinnott
ΔT = 91 s

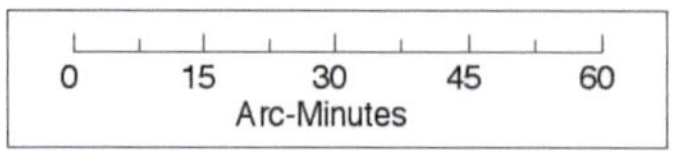

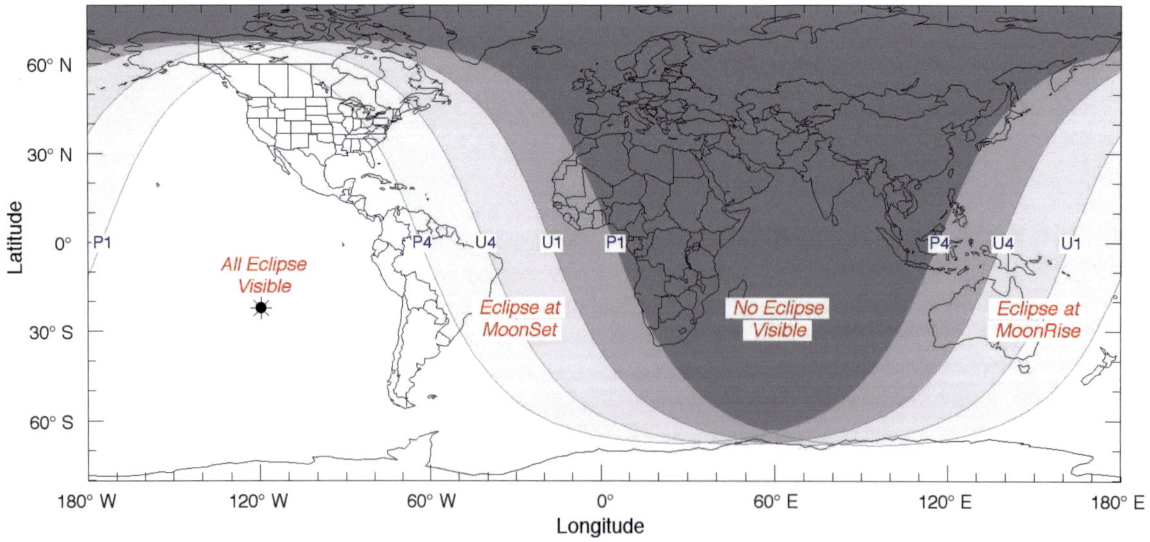

Partial Lunar Eclipse of 2059 Nov 19

Greatest Eclipse = 13:01:35.6 TD (= 13:00:03.8 UT1)

Penumbral Magnitude = 1.2055
Umbral Magnitude = 0.2097

Gamma = 0.9004
Axis = 0.8949°

Saros Series = 146
Saros Member = 13 of 72

Sun at Greatest Eclipse
(Geocentric Coordinates)
R.A. = 15h39m38.2s
Dec. = -19°31'36.5"
S.D. = 00°16'10.9"
H.P. = 00°00'08.9"

Moon at Greatest Eclipse
(Geocentric Coordinates)
R.A. = 03h38m24.3s
Dec. = +20°22'25.1"
S.D. = 00°16'15.0"
H.P. = 00°59'38.3"

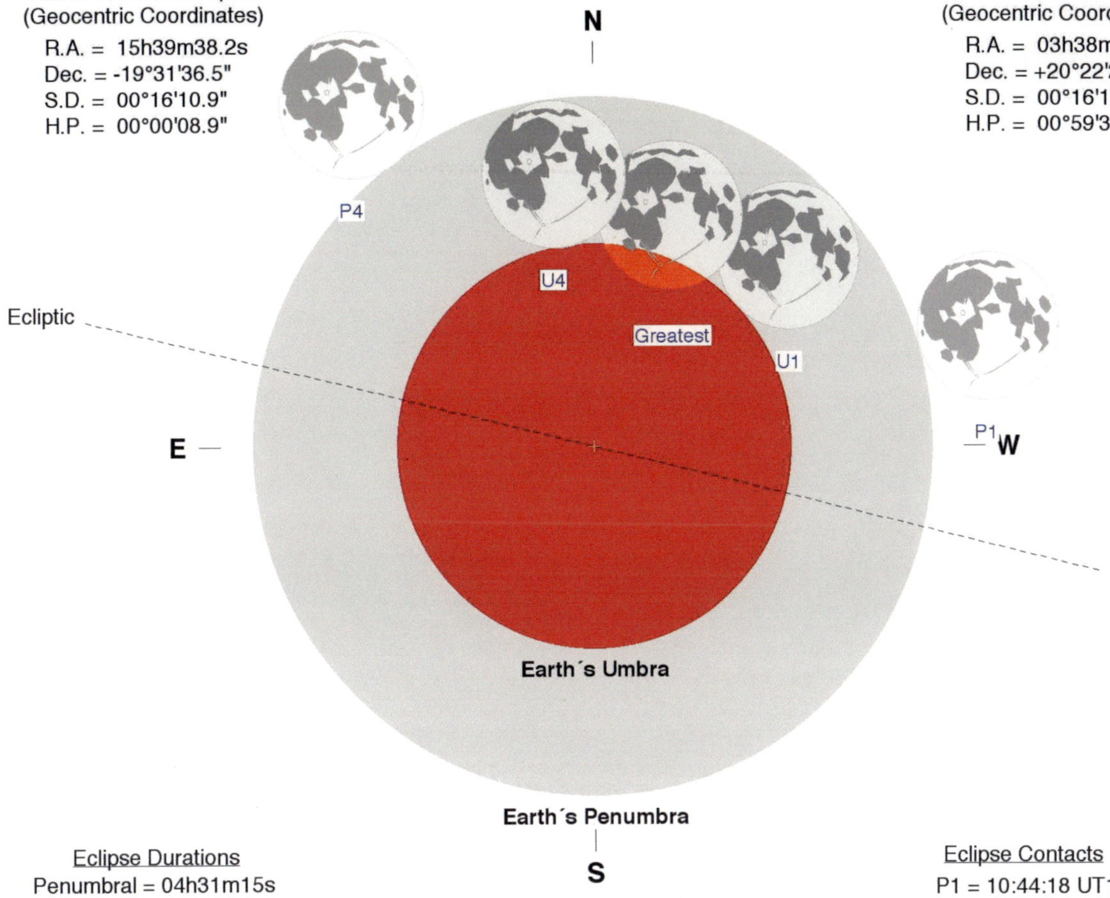

Eclipse Durations
Penumbral = 04h31m15s
Umbral = 01h39m56s

Eclipse Contacts
P1 = 10:44:18 UT1
U1 = 12:09:51 UT1
U4 = 13:49:47 UT1
P4 = 15:15:33 UT1

Eph. = JPL DE430
Rule = Herald-Sinnott
ΔT = 92 s

0 15 30 45 60
Arc-Minutes

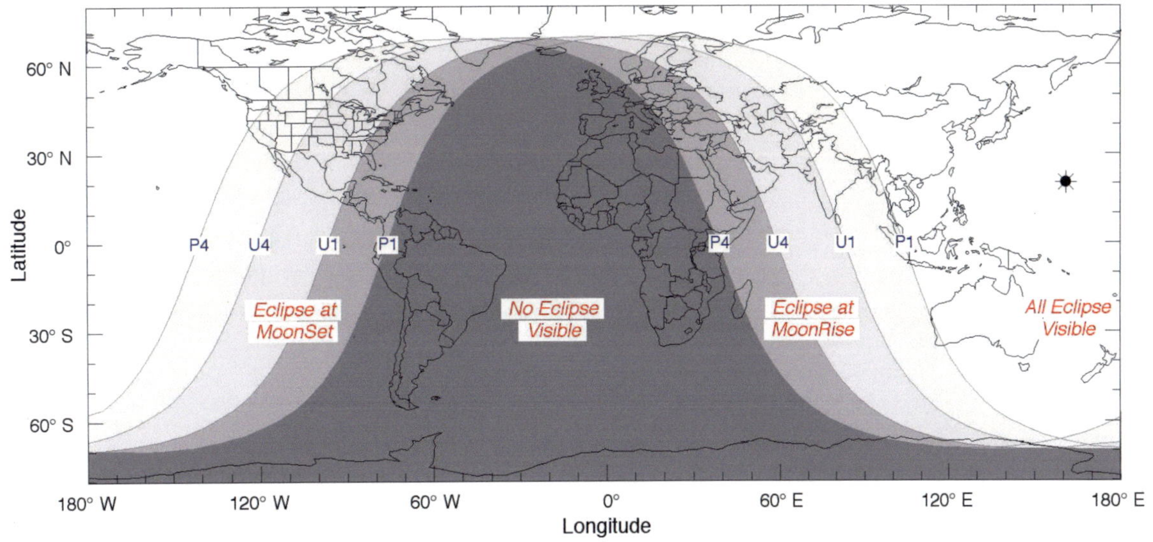

Penumbral Lunar Eclipse of 2060 Apr 15

Greatest Eclipse = 21:37:04.9 TD (= 21:35:32.8 UT1)

Penumbral Magnitude = 0.7694
Umbral Magnitude = -0.3136

Gamma = 1.1622
Axis = 1.0462°

Saros Series = 113
Saros Member = 66 of 71

Sun at Greatest Eclipse
(Geocentric Coordinates)
R.A. = 01h38m58.9s
Dec. = +10°17'00.0"
S.D. = 00°15'56.4"
H.P. = 00°00'08.8"

Moon at Greatest Eclipse
(Geocentric Coordinates)
R.A. = 13h40m53.7s
Dec. = -09°20'57.2"
S.D. = 00°14'43.1"
H.P. = 00°54'01.1"

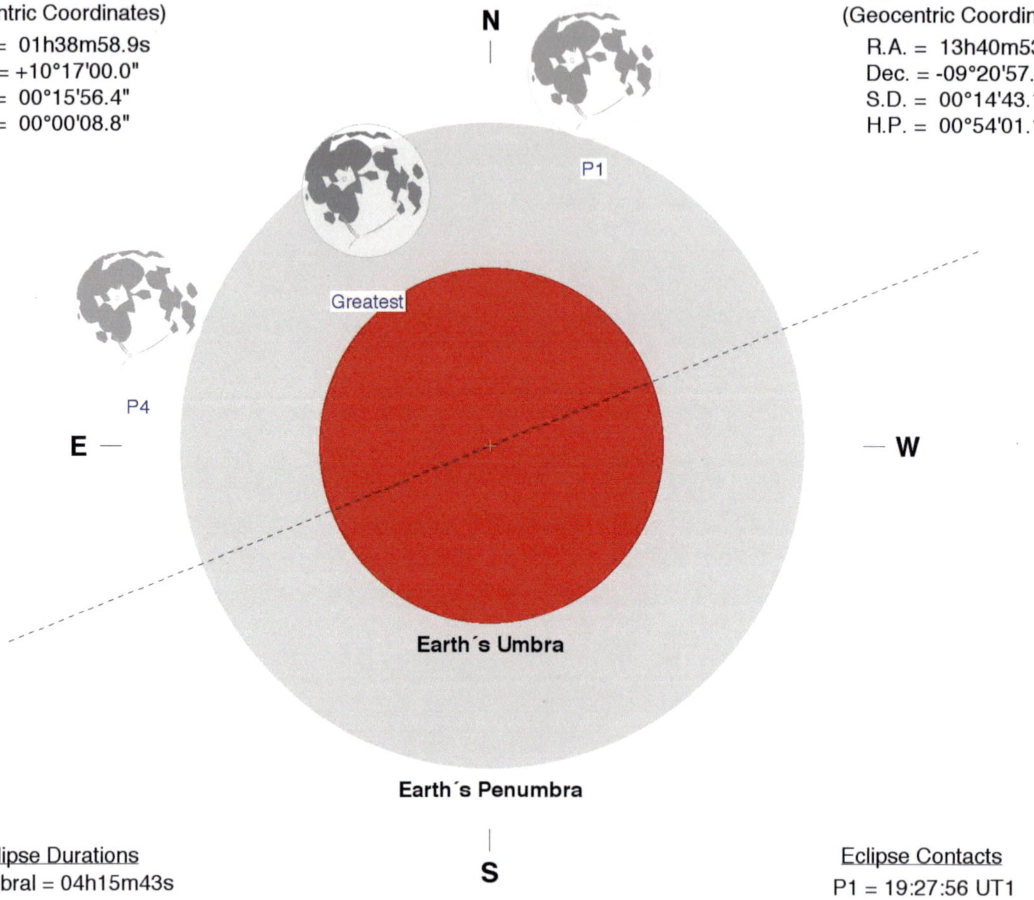

Eclipse Durations
Penumbral = 04h15m43s

Eclipse Contacts
P1 = 19:27:56 UT1
P4 = 23:43:39 UT1

Eph. = JPL DE430
Rule = Herald-Sinnott
ΔT = 92 s

0 15 30 45 60
Arc-Minutes

©2020 F. Espenak, www.EclipseWise.com

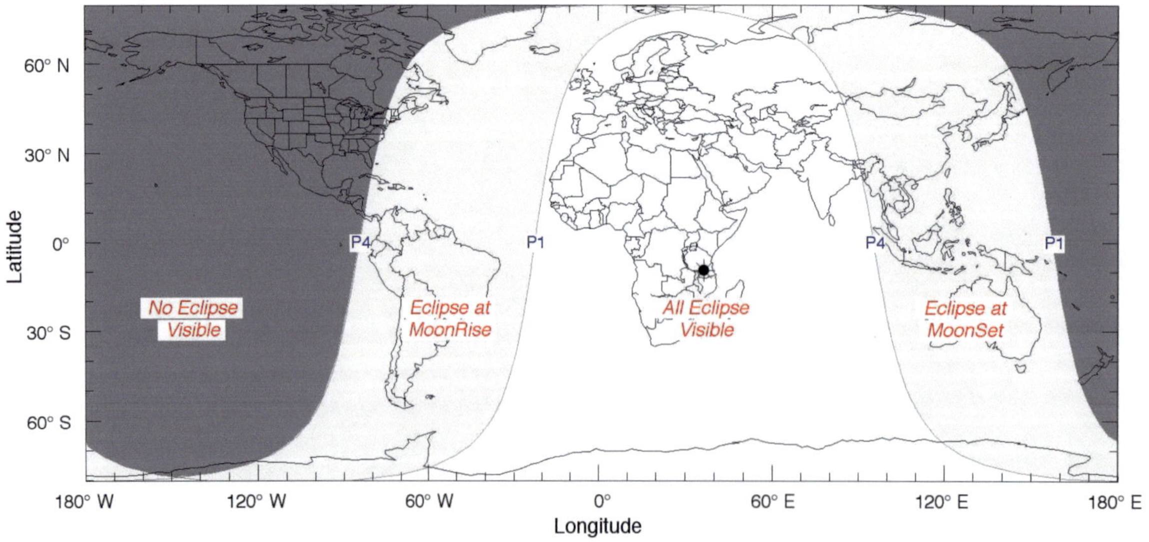

Penumbral Lunar Eclipse of 2060 Oct 09

Greatest Eclipse = 18:53:32.7 TD (= 18:52:00.2 UT1)

Penumbral Magnitude = 0.8816 Gamma = -1.0671 Saros Series = 118
Umbral Magnitude = -0.0779 Axis = 1.0892° Saros Member = 54 of 73

Sun at Greatest Eclipse
(Geocentric Coordinates)
R.A. = 13h03m35.8s
Dec. = -06°46'17.7"
S.D. = 00°16'00.8"
H.P. = 00°00'08.8"

Moon at Greatest Eclipse
(Geocentric Coordinates)
R.A. = 01h05m39.2s
Dec. = +05°48'35.1"
S.D. = 00°16'41.3"
H.P. = 01°01'15.0"

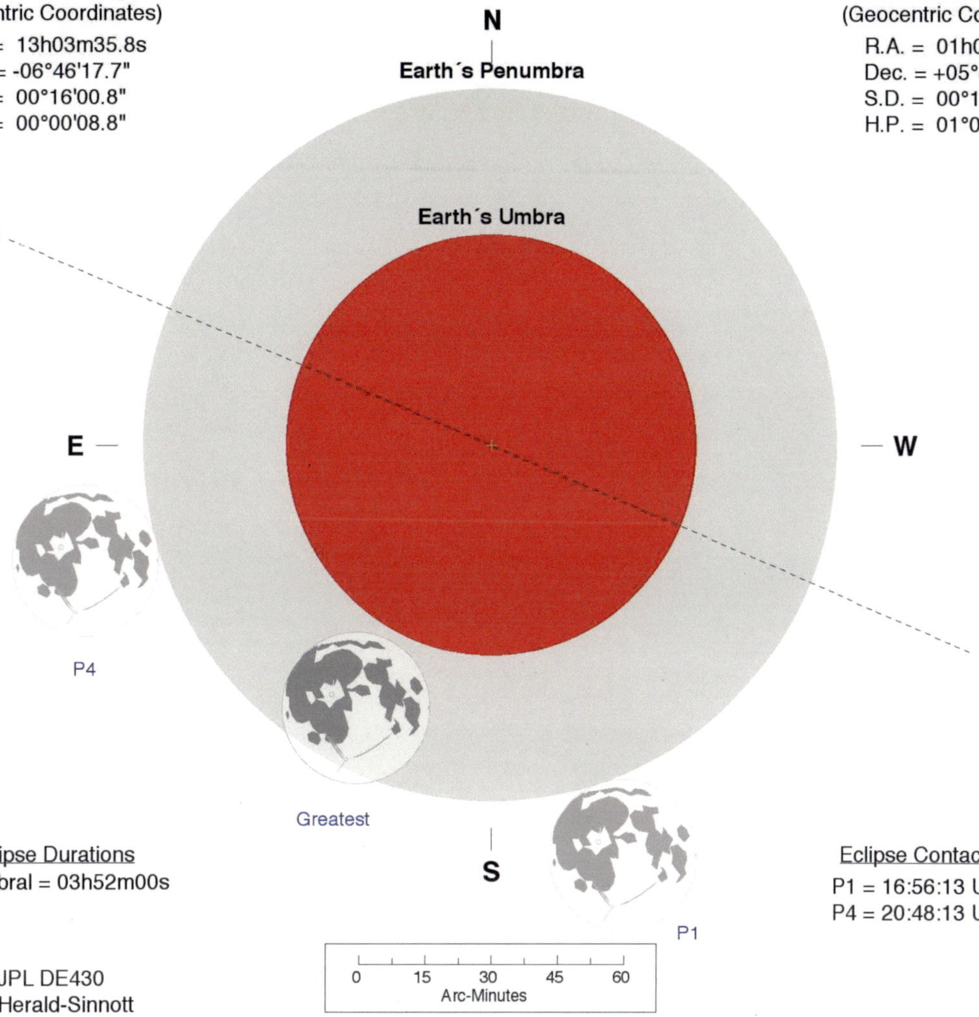

Eclipse Durations
Penumbral = 03h52m00s

Eclipse Contacts
P1 = 16:56:13 UT1
P4 = 20:48:13 UT1

Eph. = JPL DE430
Rule = Herald-Sinnott
ΔT = 92 s

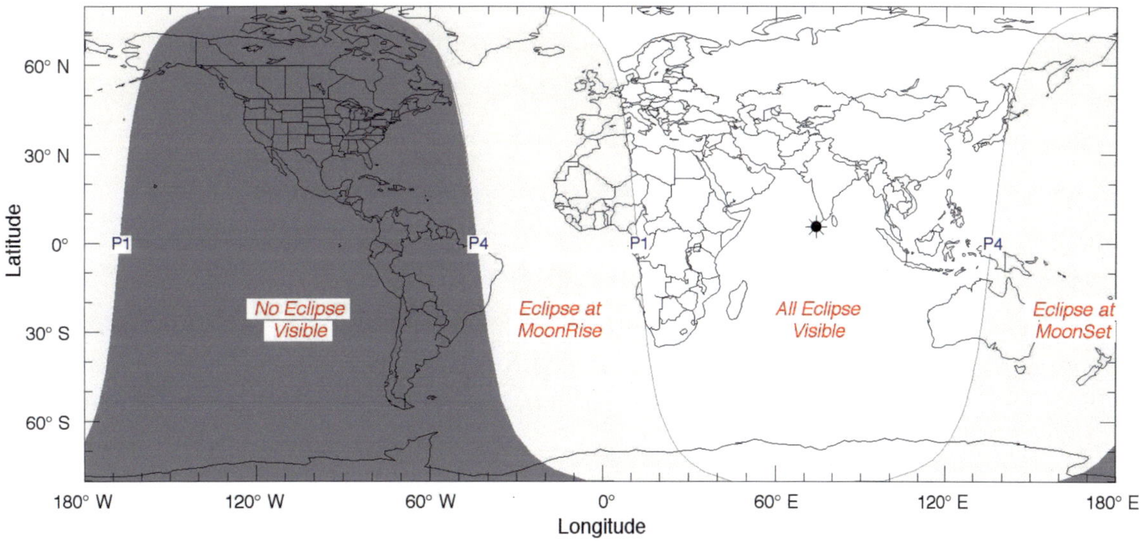

Penumbral Lunar Eclipse of 2060 Nov 08

Greatest Eclipse = 04:04:12.8 TD (= 04:02:40.3 UT1)

Penumbral Magnitude = 0.0286 Gamma = 1.5332 Saros Series = 156
Umbral Magnitude = -0.9356 Axis = 1.5699° Saros Member = 1 of 81

Sun at Greatest Eclipse
(Geocentric Coordinates)
R.A. = 14h56m11.8s
Dec. = -16°46'13.7"
S.D. = 00°16'08.5"
H.P. = 00°00'08.9"

Moon at Greatest Eclipse
(Geocentric Coordinates)
R.A. = 02h53m43.2s
Dec. = +18°13'31.2"
S.D. = 00°16'44.5"
H.P. = 01°01'26.6"

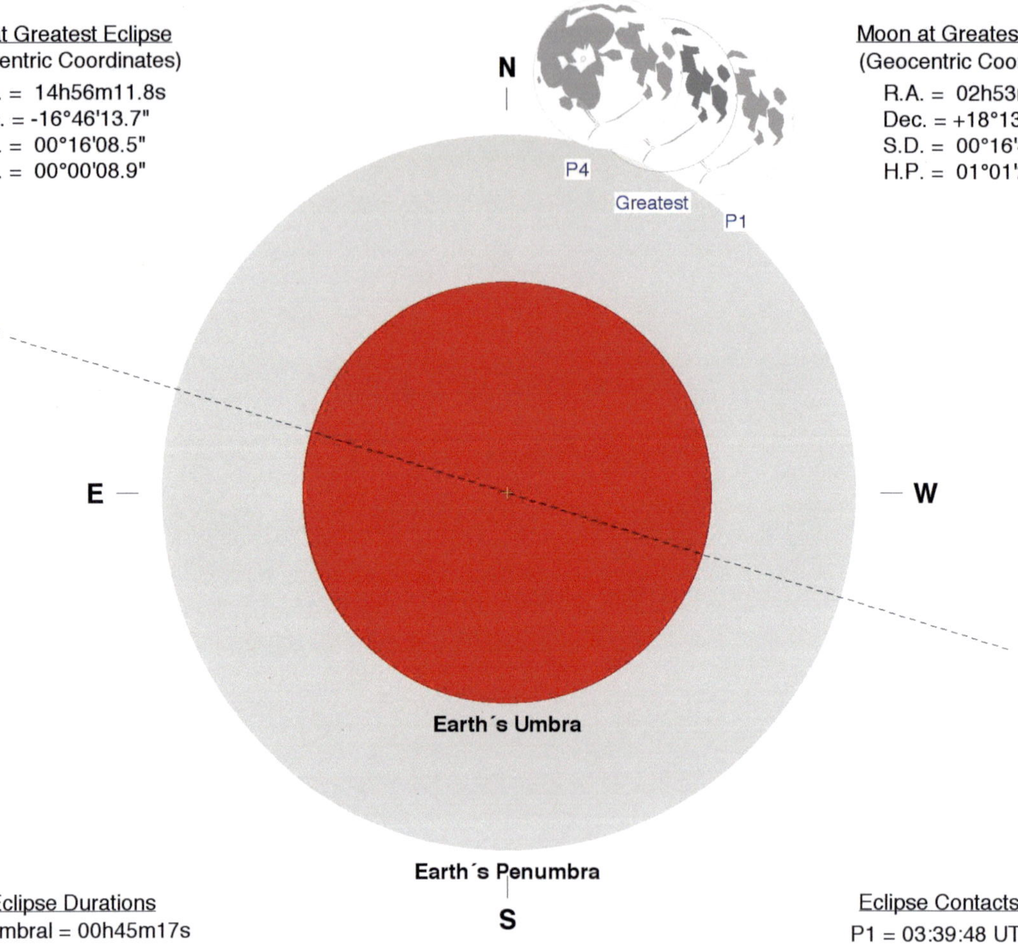

Eclipse Durations
Penumbral = 00h45m17s

Eclipse Contacts
P1 = 03:39:48 UT1
P4 = 04:25:04 UT1

Eph. = JPL DE430
Rule = Herald-Sinnott
ΔT = 93 s

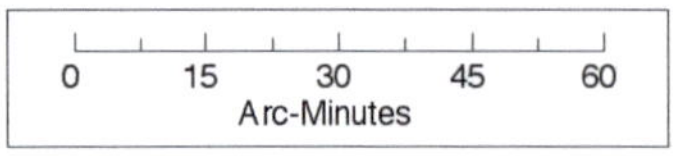

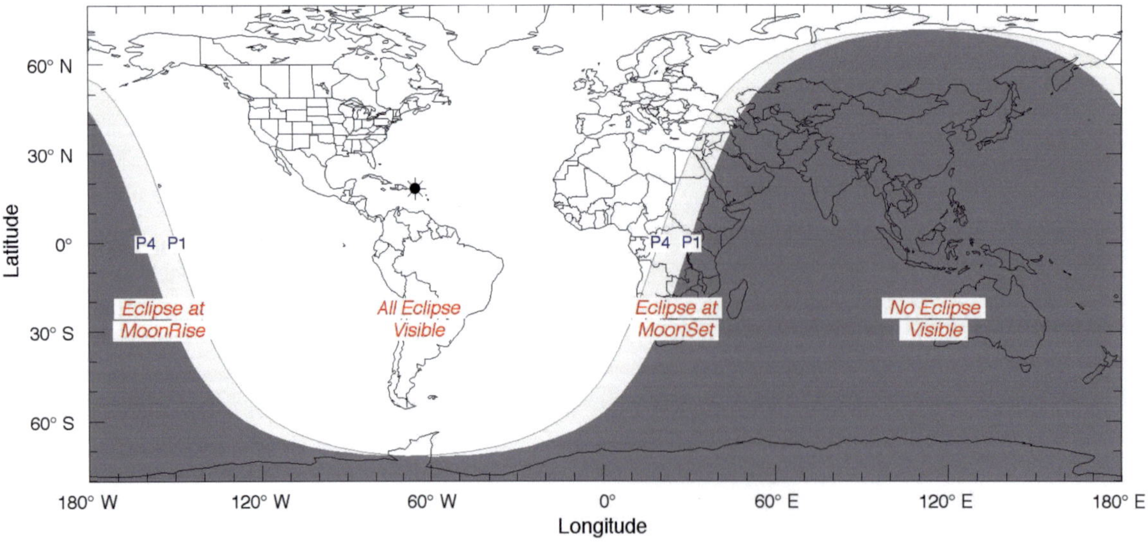

Total Lunar Eclipse of 2061 Apr 04

Greatest Eclipse = 21:54:04.8 TD (= 21:52:32.0 UT1)

Penumbral Magnitude = 2.1064
Umbral Magnitude = 1.0360

Gamma = 0.4300
Axis = 0.3929°

Saros Series = 123
Saros Member = 55 of 72

Sun at Greatest Eclipse
(Geocentric Coordinates)
R.A. = 00h57m38.4s
Dec. = +06°09'23.4"
S.D. = 00°15'59.4"
H.P. = 00°00'08.8"

Moon at Greatest Eclipse
(Geocentric Coordinates)
R.A. = 12h58m23.4s
Dec. = -05°48'38.8"
S.D. = 00°14'56.4"
H.P. = 00°54'49.7"

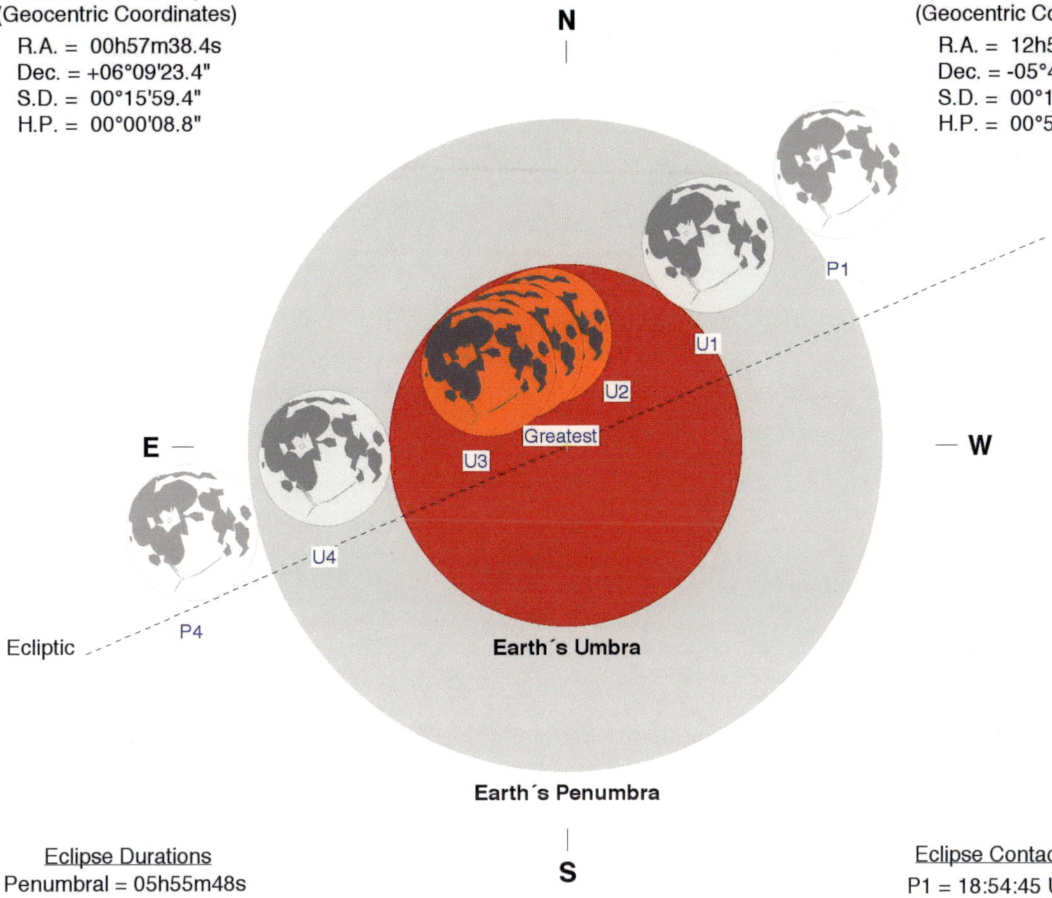

Eclipse Durations
Penumbral = 05h55m48s
Umbral = 03h30m19s
Total = 00h30m53s

Eph. = JPL DE430
Rule = Herald-Sinnott
ΔT = 93 s

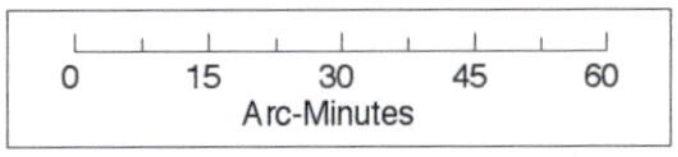

©2020 F. Espenak, www.EclipseWise.com

Eclipse Contacts
P1 = 18:54:45 UT1
U1 = 20:07:30 UT1
U2 = 21:37:23 UT1
U3 = 22:08:16 UT1
U4 = 23:37:48 UT1
P4 = 00:50:33 UT1

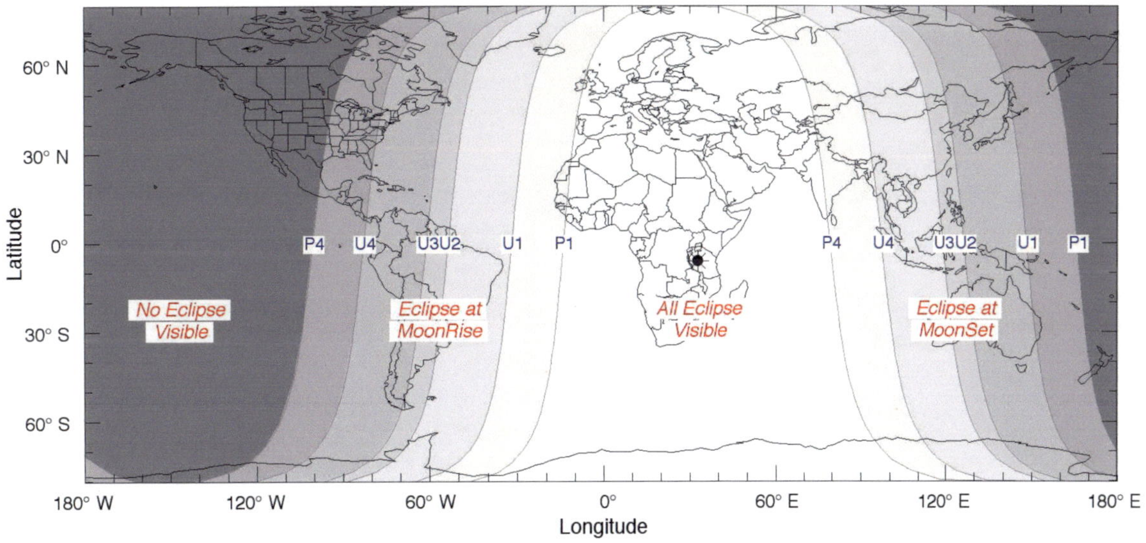

Total Lunar Eclipse of 2061 Sep 29

Greatest Eclipse = 09:38:13.1 TD (= 09:36:40.0 UT1)

Penumbral Magnitude = 2.1576
Umbral Magnitude = 1.1640

Gamma = -0.3810
Axis = 0.3745°

Saros Series = 128
Saros Member = 43 of 71

Sun at Greatest Eclipse
(Geocentric Coordinates)
R.A. = 12h24m54.1s
Dec. = -02°41'28.5"
S.D. = 00°15'57.9"
H.P. = 00°00'08.8"

Moon at Greatest Eclipse
(Geocentric Coordinates)
R.A. = 00h25m37.7s
Dec. = +02°21'48.8"
S.D. = 00°16'04.1"
H.P. = 00°58'58.5"

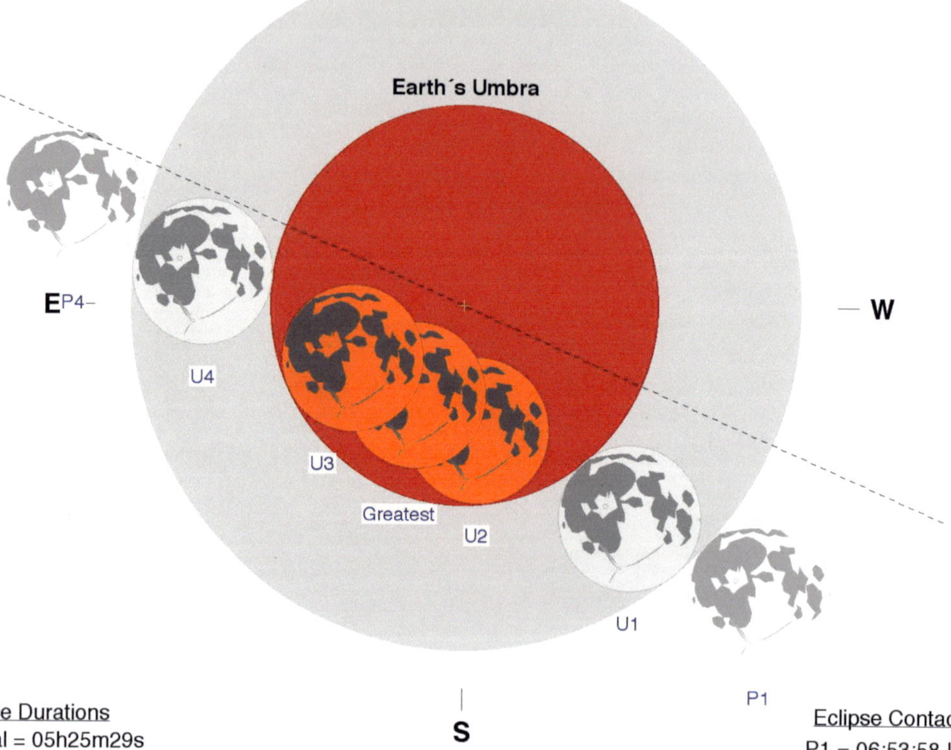

Eclipse Durations
Penumbral = 05h25m29s
Umbral = 03h23m03s
Total = 00h59m38s

Eph. = JPL DE430
Rule = Herald-Sinnott
ΔT = 93 s

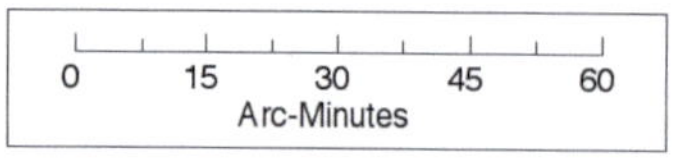

Eclipse Contacts
P1 = 06:53:58 UT1
U1 = 07:55:17 UT1
U2 = 09:07:07 UT1
U3 = 10:06:45 UT1
U4 = 11:18:20 UT1
P4 = 12:19:27 UT1

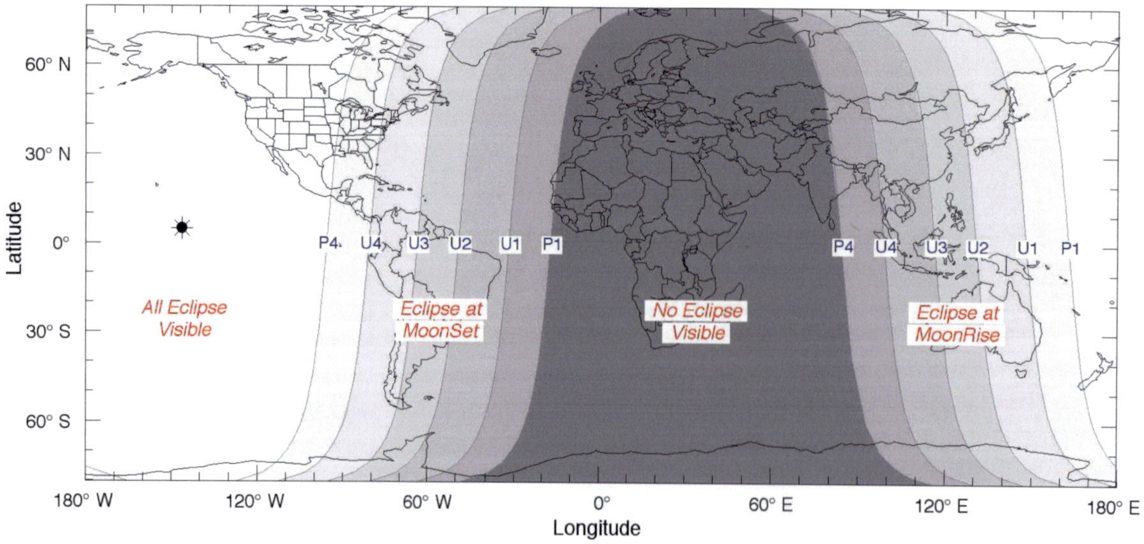

Total Lunar Eclipse of 2062 Mar 25

Greatest Eclipse = 03:33:50.4 TD (= 03:32:16.9 UT1)

Penumbral Magnitude = 2.2925
Umbral Magnitude = 1.2715

Gamma = -0.3150
Axis = 0.3027°

Saros Series = 133
Saros Member = 29 of 71

Sun at Greatest Eclipse
(Geocentric Coordinates)
R.A. = 00h17m31.2s
Dec. = +01°53'45.5"
S.D. = 00°16'02.5"
H.P. = 00°00'08.8"

Moon at Greatest Eclipse
(Geocentric Coordinates)
R.A. = 12h16m55.9s
Dec. = -02°09'38.2"
S.D. = 00°15'42.7"
H.P. = 00°57'39.7"

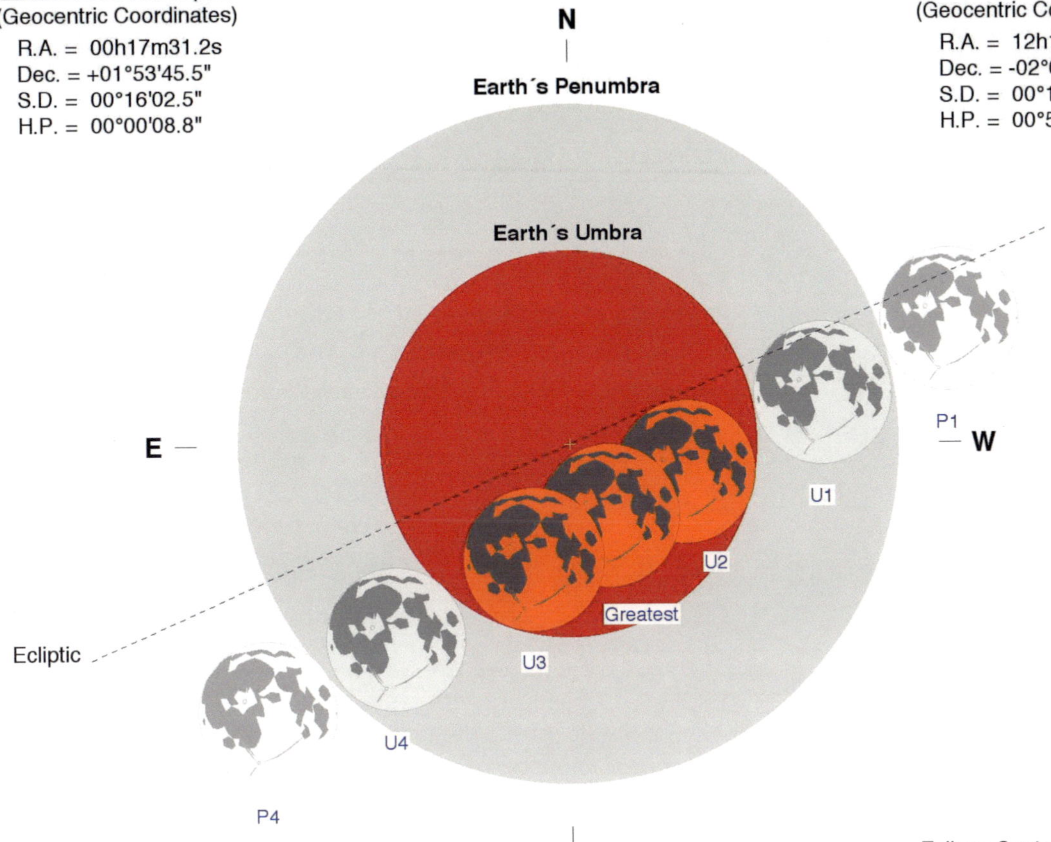

Eclipse Durations
Penumbral = 05h39m03s
Umbral = 03h31m58s
Total = 01h15m16s

Eph. = JPL DE430
Rule = Herald-Sinnott
ΔT = 94 s

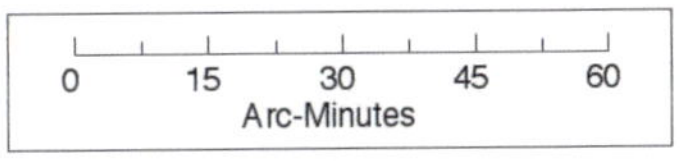

©2020 F. Espenak, www.EclipseWise.com

Eclipse Contacts
P1 = 00:42:44 UT1
U1 = 01:46:10 UT1
U2 = 02:54:25 UT1
U3 = 04:09:41 UT1
U4 = 05:18:08 UT1
P4 = 06:21:47 UT1

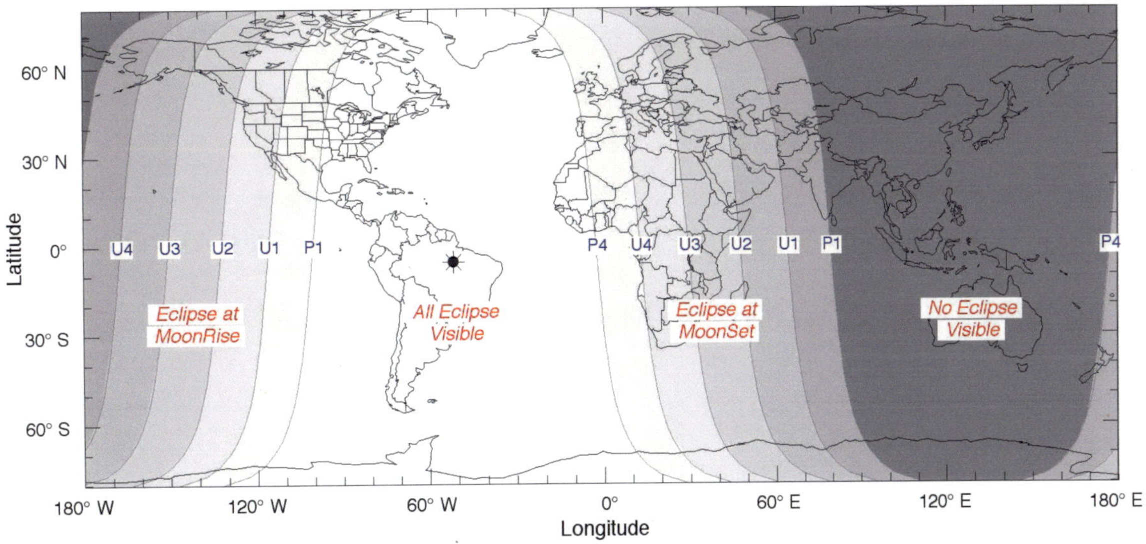

Total Lunar Eclipse of 2062 Sep 18

Greatest Eclipse = 18:34:02.1 TD (= 18:32:28.2 UT1)

Penumbral Magnitude = 2.1979	Gamma = 0.3736	Saros Series = 138
Umbral Magnitude = 1.1515	Axis = 0.3476°	Saros Member = 31 of 82

Sun at Greatest Eclipse
(Geocentric Coordinates)

R.A. = 11h45m50.3s
Dec. = +01°31'59.3"
S.D. = 00°15'55.0"
H.P. = 00°00'08.8"

Moon at Greatest Eclipse
(Geocentric Coordinates)

R.A. = 23h45m09.8s
Dec. = -01°13'45.9"
S.D. = 00°15'12.7"
H.P. = 00°55'49.7"

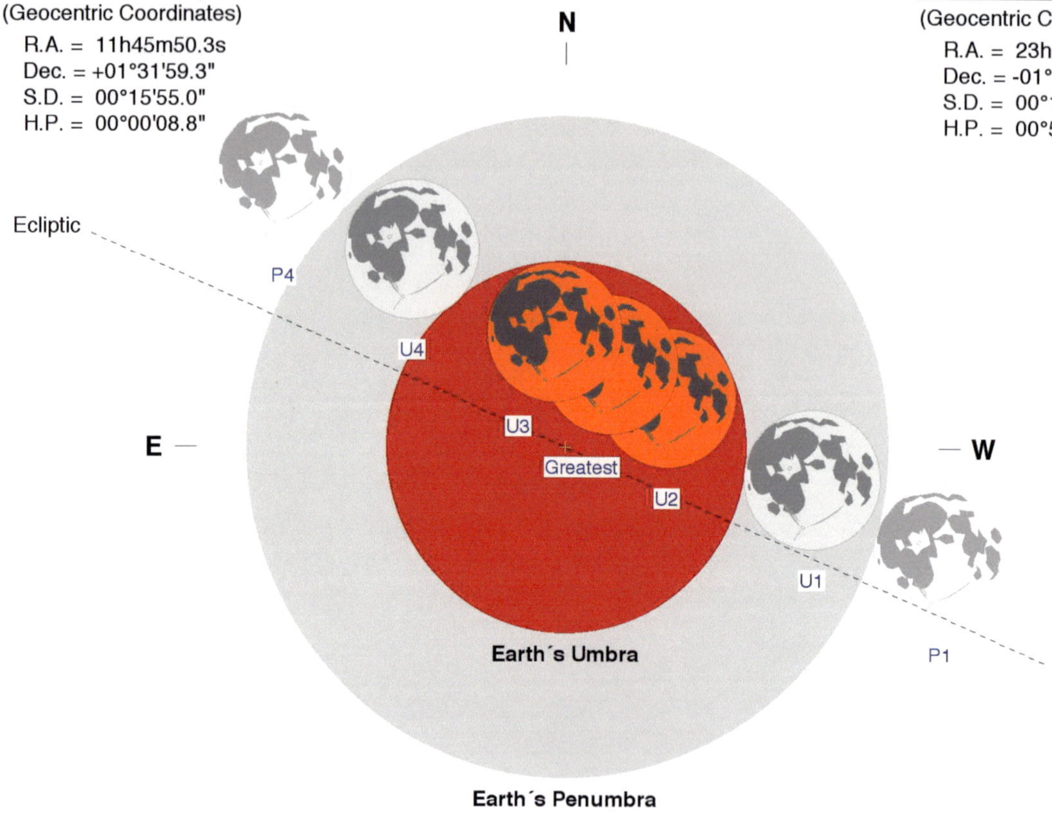

Eclipse Durations
Penumbral = 05h50m00s
Umbral = 03h33m06s
Total = 01h00m11s

Eph. = JPL DE430
Rule = Herald-Sinnott
ΔT = 94 s

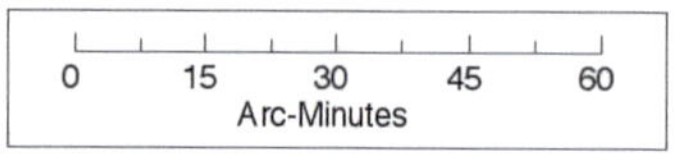

©2020 F. Espenak, www.EclipseWise.com

Eclipse Contacts
P1 = 15:37:21 UT1
U1 = 16:45:49 UT1
U2 = 18:02:08 UT1
U3 = 19:02:19 UT1
U4 = 20:18:56 UT1
P4 = 21:27:21 UT1

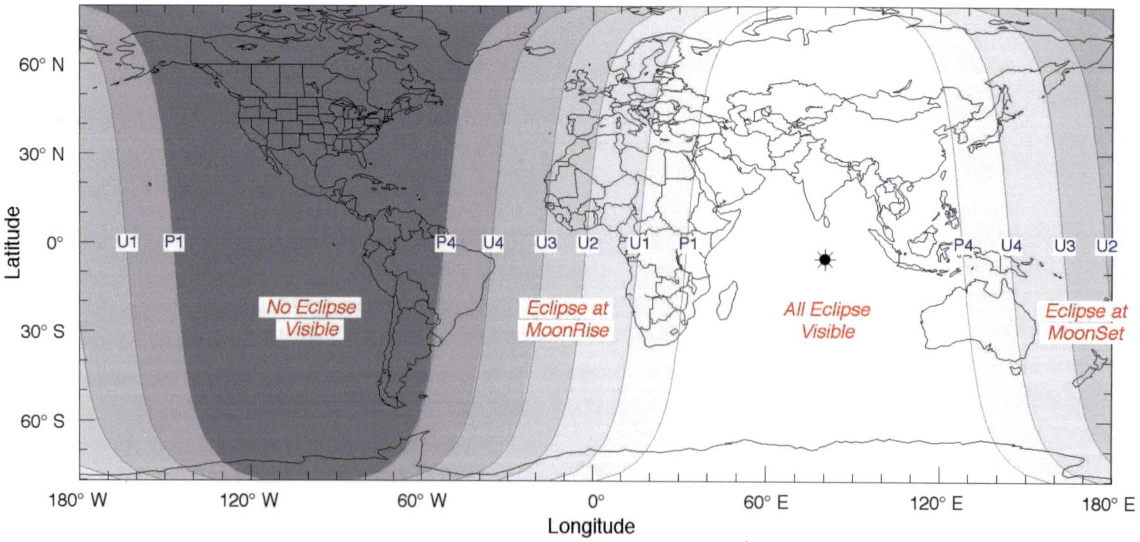

Partial Lunar Eclipse of 2063 Mar 14

Greatest Eclipse = 16:05:48.6 TD (= 16:04:14.4 UT1)

Penumbral Magnitude = 1.0108 Gamma = -1.0008 Saros Series = 143
Umbral Magnitude = 0.0363 Axis = 1.0105° Saros Member = 20 of 72

Sun at Greatest Eclipse
(Geocentric Coordinates)
R.A. = 23h38m23.2s
Dec. = -02°20'14.4"
S.D. = 00°16'05.4"
H.P. = 00°00'08.8"

Moon at Greatest Eclipse
(Geocentric Coordinates)
R.A. = 11h36m26.0s
Dec. = +01°27'08.7"
S.D. = 00°16'30.6"
H.P. = 01°00'35.4"

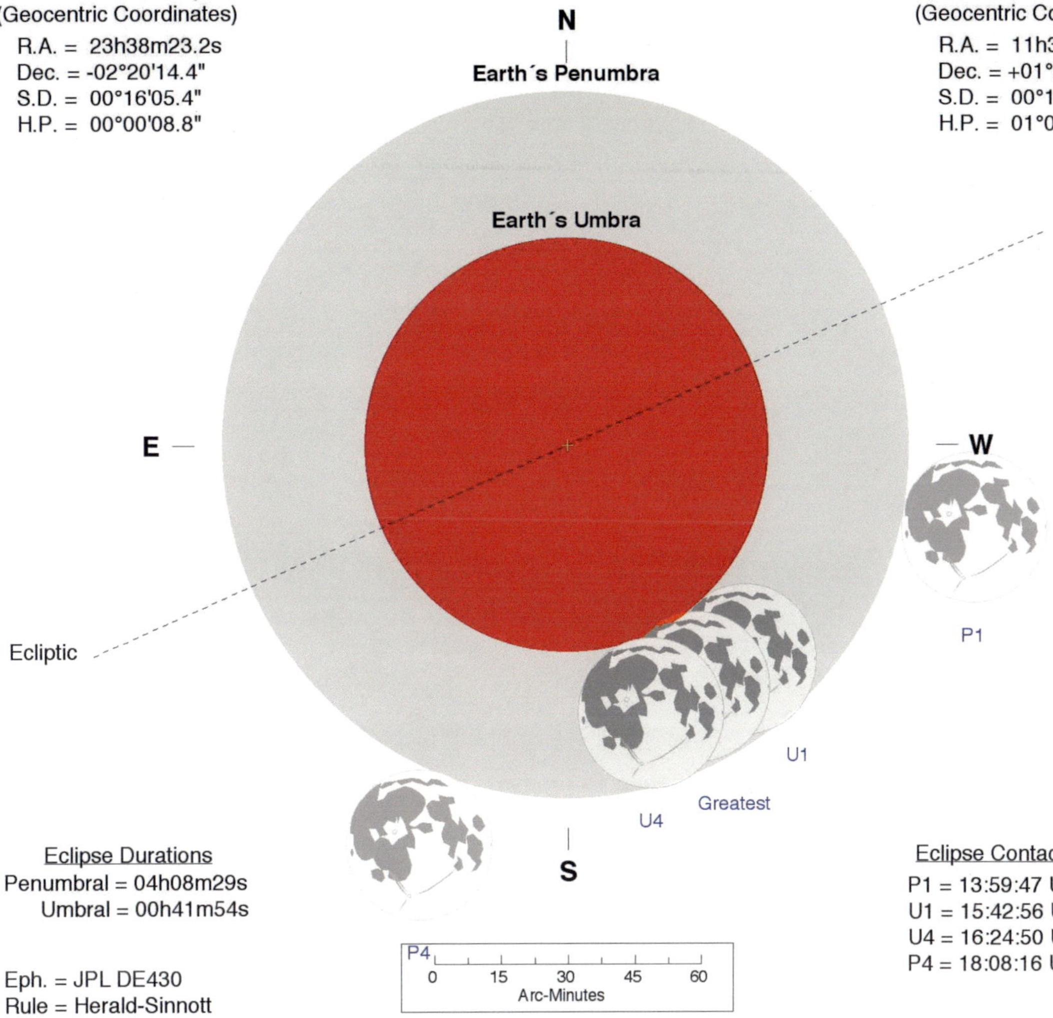

Eclipse Durations
Penumbral = 04h08m29s
Umbral = 00h41m54s

Eclipse Contacts
P1 = 13:59:47 UT1
U1 = 15:42:56 UT1
U4 = 16:24:50 UT1
P4 = 18:08:16 UT1

Eph. = JPL DE430
Rule = Herald-Sinnott
ΔT = 94 s

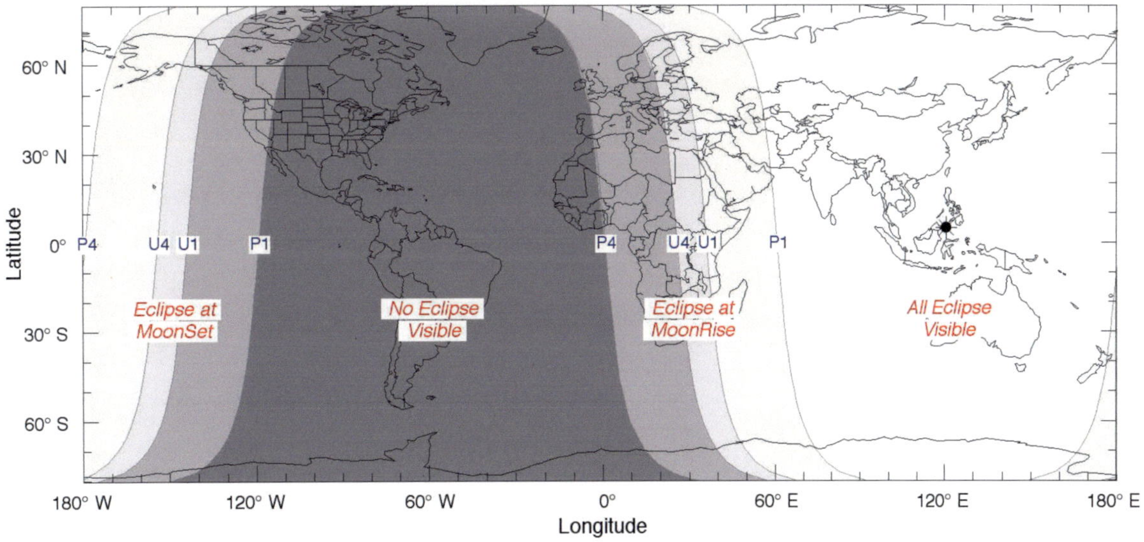

Penumbral Lunar Eclipse of 2063 Sep 07

Greatest Eclipse = 20:41:10.9 TD (= 20:39:36.3 UT1)

Penumbral Magnitude = 0.8121 Gamma = 1.1375 Saros Series = 148
Umbral Magnitude = -0.2657 Axis = 1.0243° Saros Member = 6 of 70

Sun at Greatest Eclipse
(Geocentric Coordinates)
R.A. = 11h05m48.9s
Dec. = +05°47'49.4"
S.D. = 00°15'52.2"
H.P. = 00°00'08.7"

Moon at Greatest Eclipse
(Geocentric Coordinates)
R.A. = 23h03m51.6s
Dec. = -04°53'44.4"
S.D. = 00°14'43.4"
H.P. = 00°54'02.2"

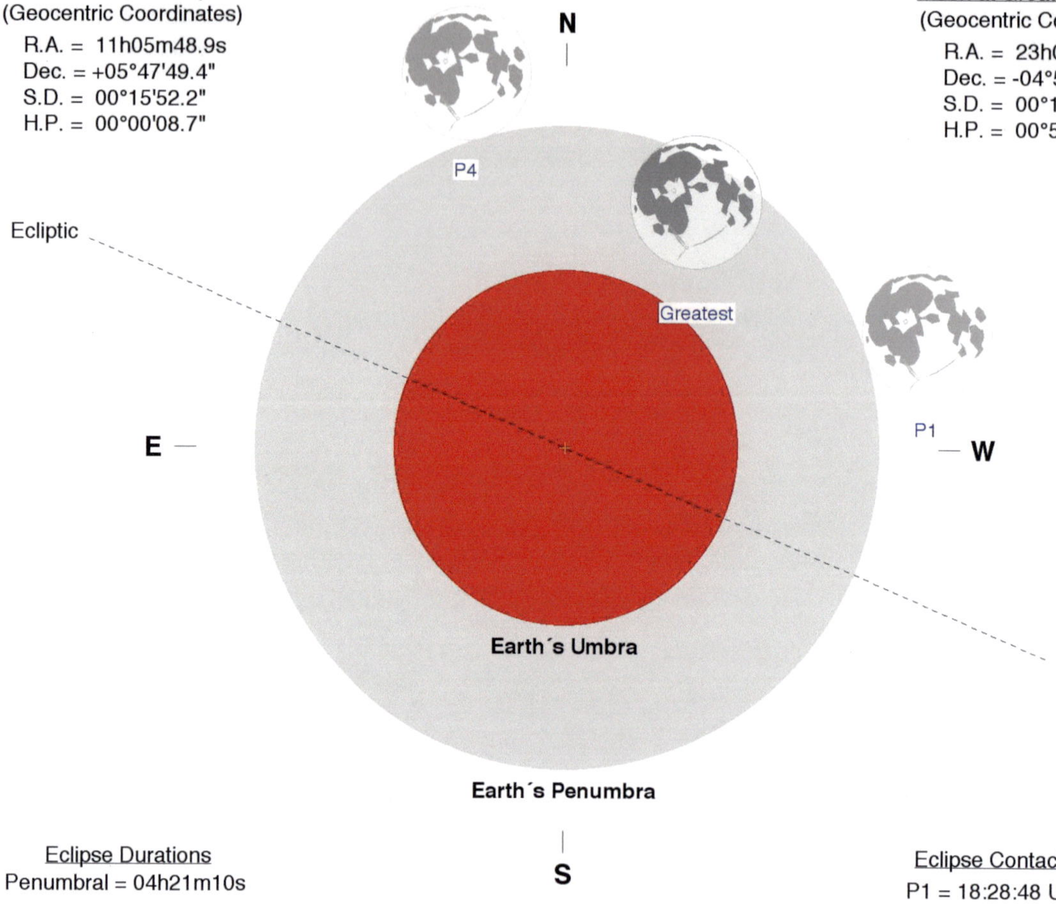

Eclipse Durations
Penumbral = 04h21m10s

Eclipse Contacts
P1 = 18:28:48 UT1
P4 = 22:49:58 UT1

Eph. = JPL DE430
Rule = Herald-Sinnott
ΔT = 95 s

0 15 30 45 60
Arc-Minutes

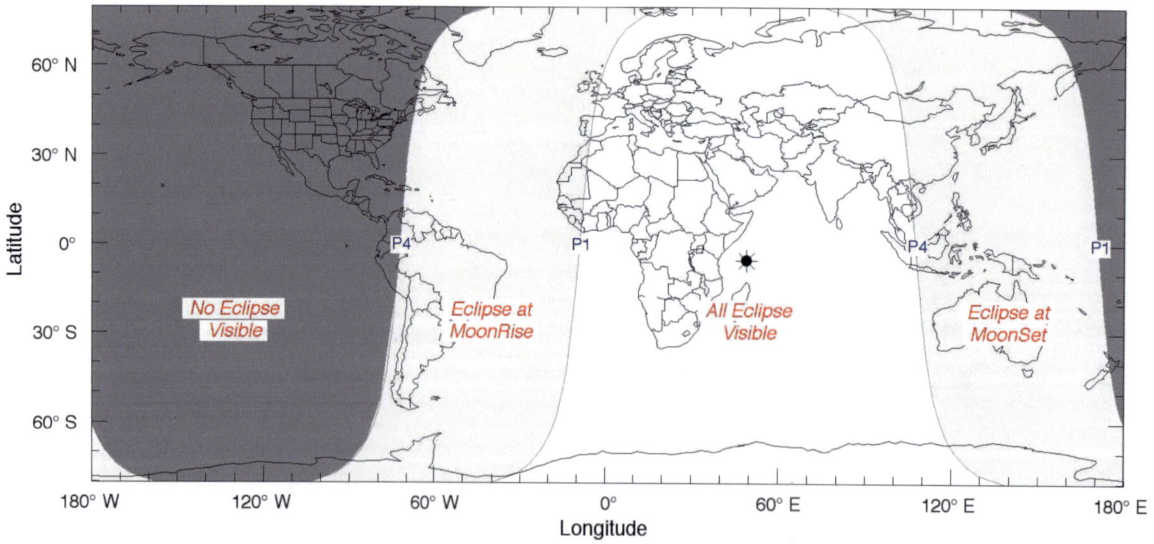

Partial Lunar Eclipse of 2064 Feb 02

Greatest Eclipse = 21:48:57.3 TD (= 21:47:22.4 UT1)

Penumbral Magnitude = 1.0215 Gamma = 0.9969 Saros Series = 115
Umbral Magnitude = 0.0395 Axis = 1.0078° Saros Member = 60 of 72

Sun at Greatest Eclipse
(Geocentric Coordinates)
R.A. = 21h05m13.1s
Dec. = -16°40'07.5"
S.D. = 00°16'13.8"
H.P. = 00°00'08.9"

Moon at Greatest Eclipse
(Geocentric Coordinates)
R.A. = 09h06m49.3s
Dec. = +17°36'03.4"
S.D. = 00°16'31.7"
H.P. = 01°00'39.5"

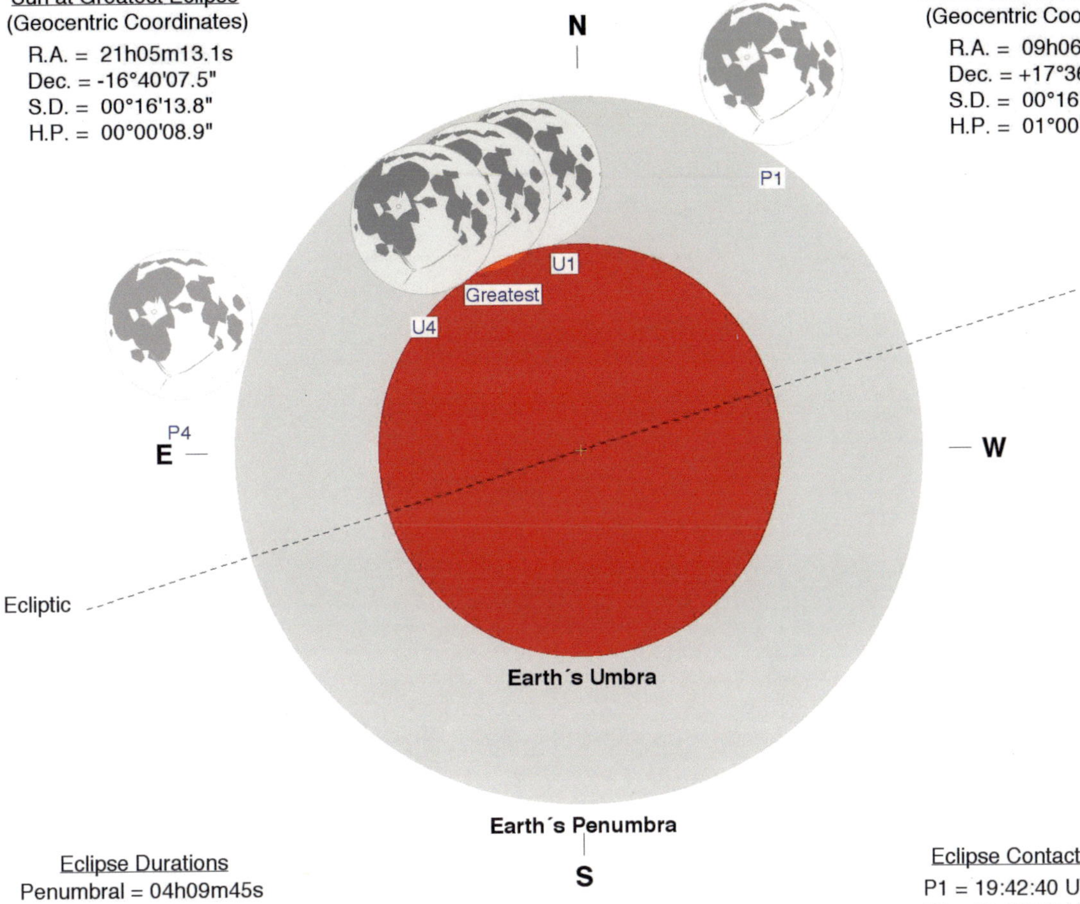

Eclipse Durations
Penumbral = 04h09m45s
Umbral = 00h43m41s

Eclipse Contacts
P1 = 19:42:40 UT1
U1 = 21:25:49 UT1
U4 = 22:09:30 UT1
P4 = 23:52:25 UT1

Eph. = JPL DE430
Rule = Herald-Sinnott
ΔT = 95 s

0 15 30 45 60
Arc-Minutes

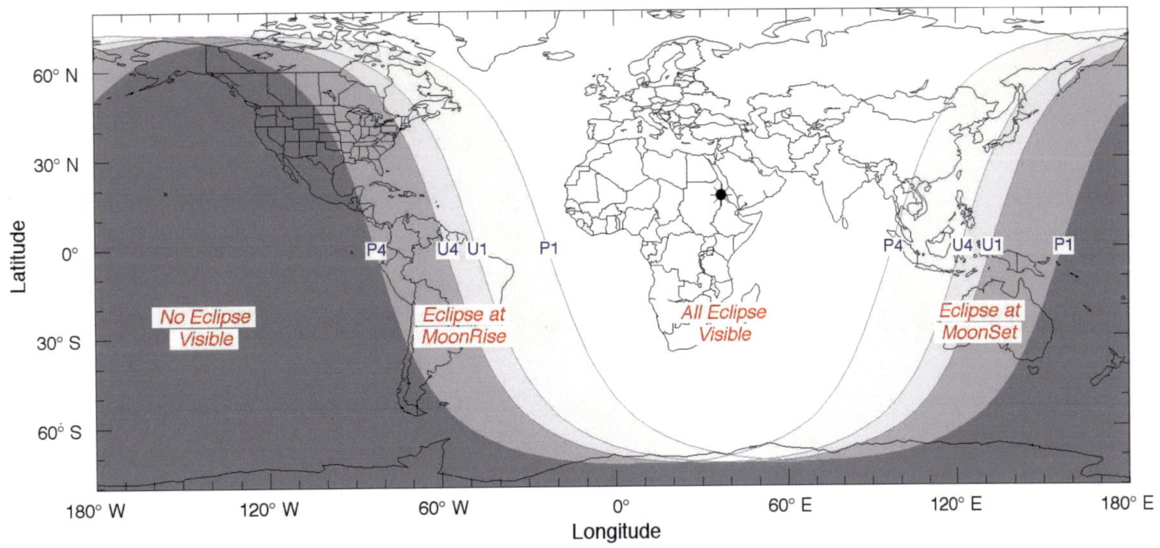

Partial Lunar Eclipse of 2064 Jul 28

Greatest Eclipse = 07:52:48.2 TD (= 07:51:12.9 UT1)

Penumbral Magnitude = 1.1378   Gamma = -0.9473   Saros Series = 120
Umbral Magnitude = 0.1055   Axis = 0.8840°   Saros Member = 60 of 83

Sun at Greatest Eclipse
(Geocentric Coordinates)
R.A. = 08h33m43.4s
Dec. = +18°45'12.2"
S.D. = 00°15'45.1"
H.P. = 00°00'08.7"

Moon at Greatest Eclipse
(Geocentric Coordinates)
R.A. = 20h35m00.2s
Dec. = -19°35'03.3"
S.D. = 00°15'15.5"
H.P. = 00°55'59.9"

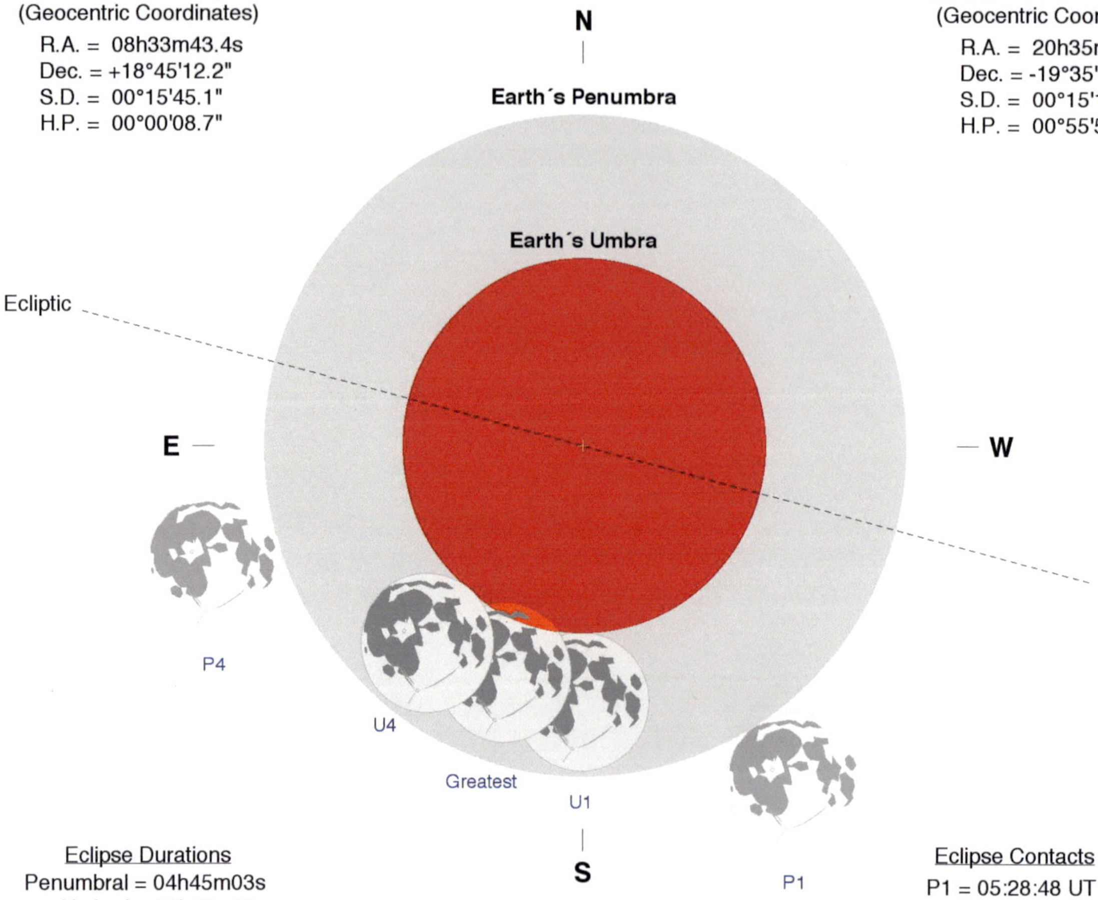

Eclipse Durations
Penumbral = 04h45m03s
Umbral = 01h16m33s

Eclipse Contacts
P1 = 05:28:48 UT1
U1 = 07:13:03 UT1
U4 = 08:29:36 UT1
P4 = 10:13:51 UT1

Eph. = JPL DE430
Rule = Herald-Sinnott
ΔT = 95 s

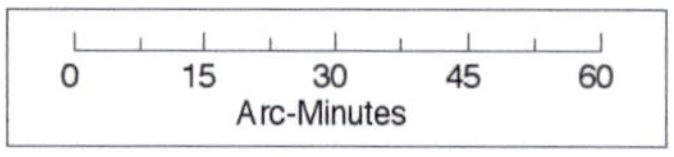

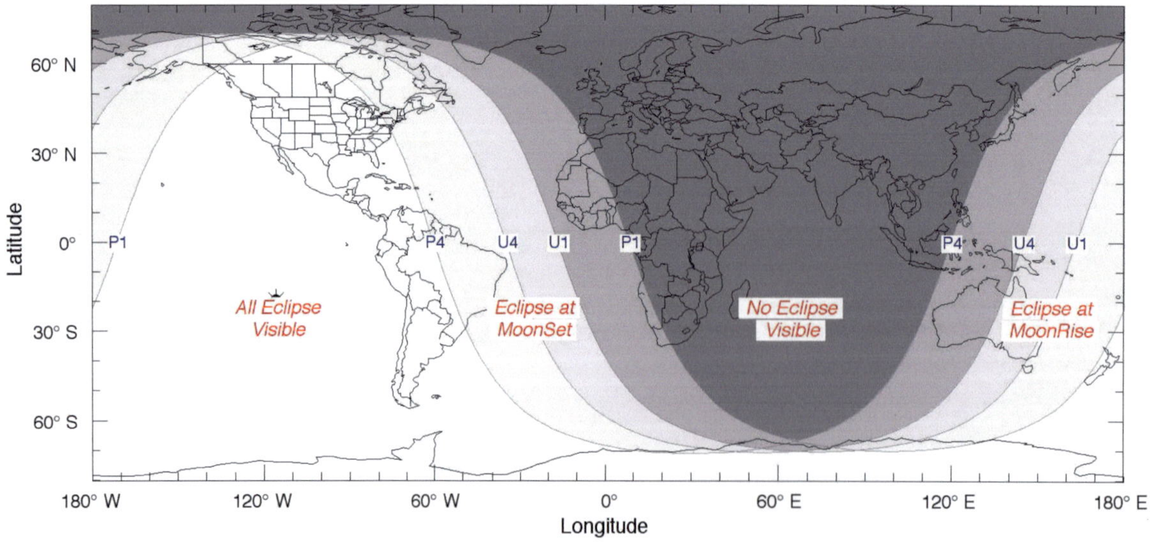

Total Lunar Eclipse of 2065 Jan 22

Greatest Eclipse = 09:58:58.5 TD (= 09:57:22.9 UT1)

Penumbral Magnitude = 2.2579　Gamma = 0.3371　Saros Series = 125
Umbral Magnitude = 1.2248　Axis = 0.3243°　Saros Member = 51 of 72

Sun at Greatest Eclipse
(Geocentric Coordinates)
R.A. = 20h20m52.4s
Dec. = -19°29'52.6"
S.D. = 00°16'15.1"
H.P. = 00°00'08.9"

Moon at Greatest Eclipse
(Geocentric Coordinates)
R.A. = 08h21m19.3s
Dec. = +19°48'16.7"
S.D. = 00°15'43.9"
H.P. = 00°57'44.2"

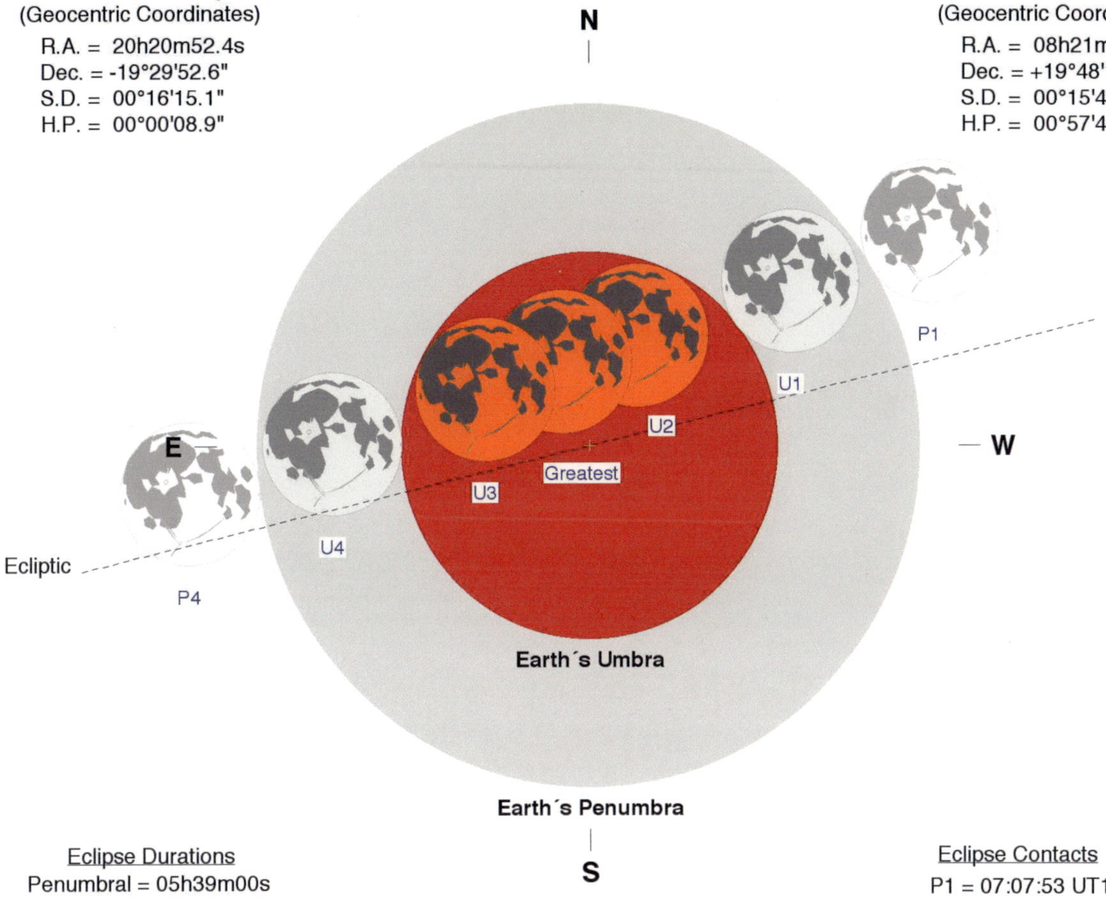

Eclipse Durations
Penumbral = 05h39m00s
Umbral = 03h29m43s
Total = 01h09m26s

Eph. = JPL DE430
Rule = Herald-Sinnott
ΔT = 96 s

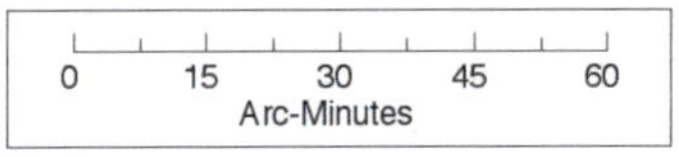

©2020 F. Espenak, www.EclipseWise.com

Eclipse Contacts
P1 = 07:07:53 UT1
U1 = 08:12:37 UT1
U2 = 09:22:50 UT1
U3 = 10:32:16 UT1
U4 = 11:42:20 UT1
P4 = 12:46:53 UT1

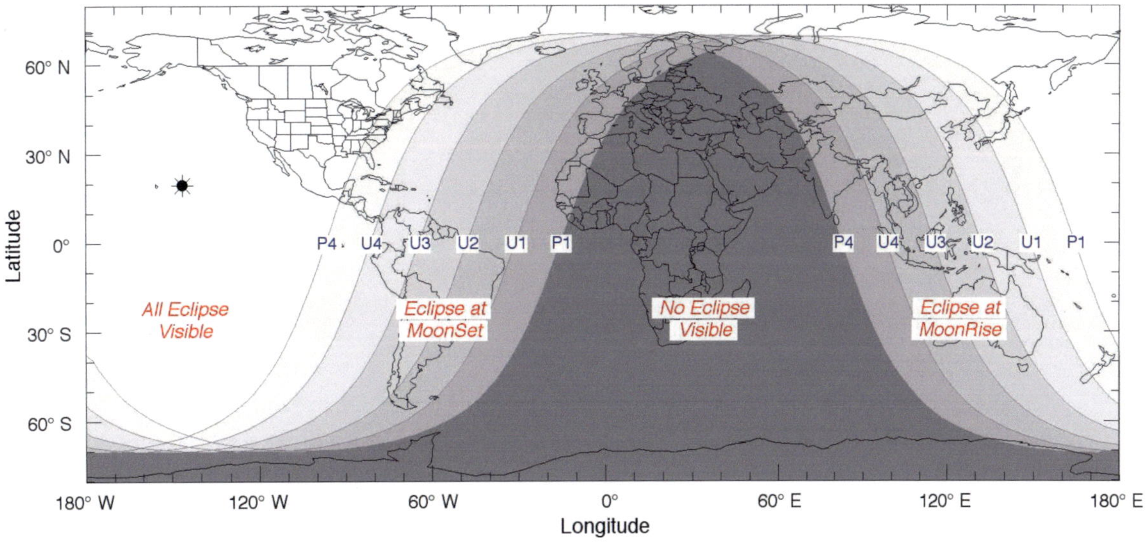

Total Lunar Eclipse of 2065 Jul 17

Greatest Eclipse = 17:48:40.5 TD (= 17:47:04.5 UT1)

Penumbral Magnitude = 2.5907 Gamma = -0.1402 Saros Series = 130
Umbral Magnitude = 1.6138 Axis = 0.1382° Saros Member = 37 of 71

Sun at Greatest Eclipse
(Geocentric Coordinates)
R.A. = 07h50m48.2s
Dec. = +20°59'34.9"
S.D. = 00°15'44.3"
H.P. = 00°00'08.7"

Moon at Greatest Eclipse
(Geocentric Coordinates)
R.A. = 19h50m58.2s
Dec. = -21°07'32.3"
S.D. = 00°16'06.6"
H.P. = 00°59'07.5"

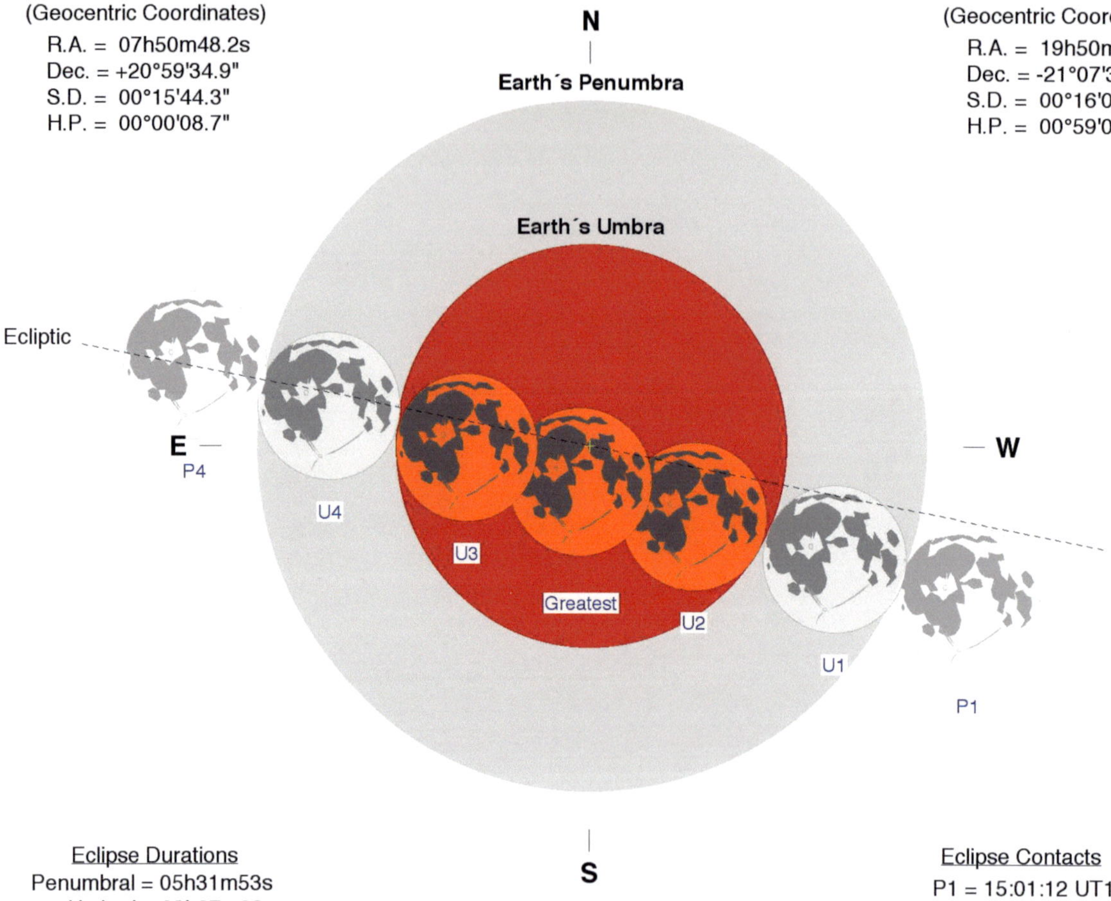

Eclipse Durations
Penumbral = 05h31m53s
Umbral = 03h37m02s
Total = 01h37m43s

Eph. = JPL DE430
Rule = Herald-Sinnott
ΔT = 96 s

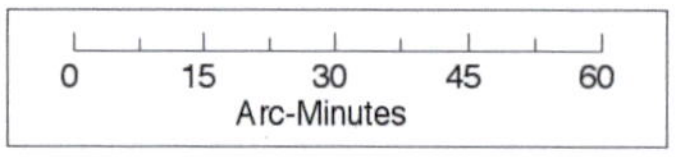

©2020 F. Espenak, www.EclipseWise.com

Eclipse Contacts
P1 = 15:01:12 UT1
U1 = 15:58:34 UT1
U2 = 16:58:15 UT1
U3 = 18:35:59 UT1
U4 = 19:35:36 UT1
P4 = 20:33:05 UT1

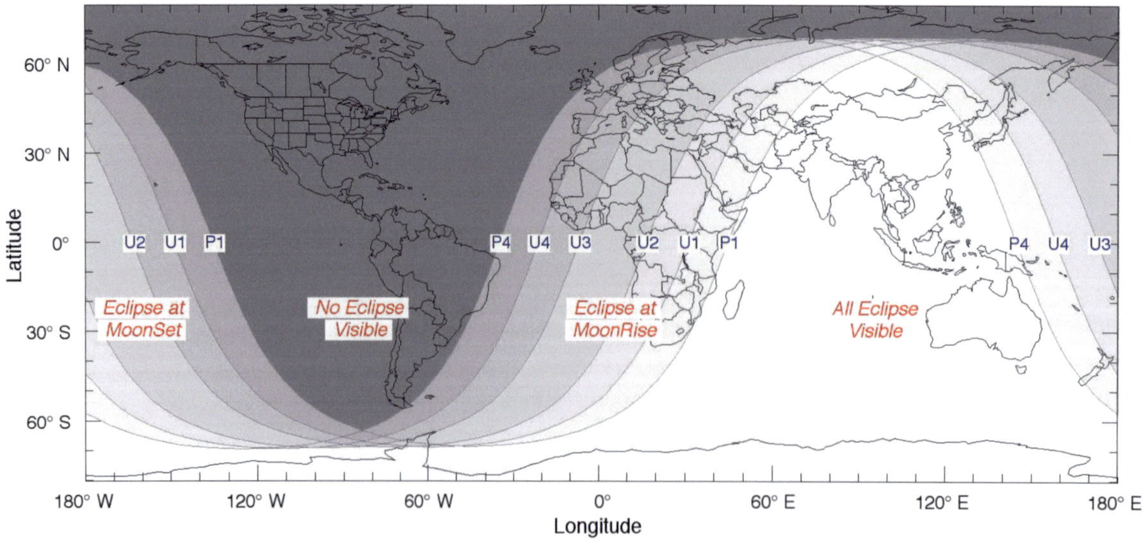

Total Lunar Eclipse of 2066 Jan 11

Greatest Eclipse = 15:04:47.3 TD (= 15:03:10.9 UT1)

Penumbral Magnitude = 2.2276   Gamma = -0.3687   Saros Series = 135
Umbral Magnitude = 1.1395   Axis = 0.3371°   Saros Member = 26 of 71

Sun at Greatest Eclipse
(Geocentric Coordinates)
R.A. = 19h33m47.1s
Dec. = -21°40'59.0"
S.D. = 00°16'15.8"
H.P. = 00°00'08.9"

Moon at Greatest Eclipse
(Geocentric Coordinates)
R.A. = 07h33m24.8s
Dec. = +21°21'26.1"
S.D. = 00°14'56.8"
H.P. = 00°54'51.2"

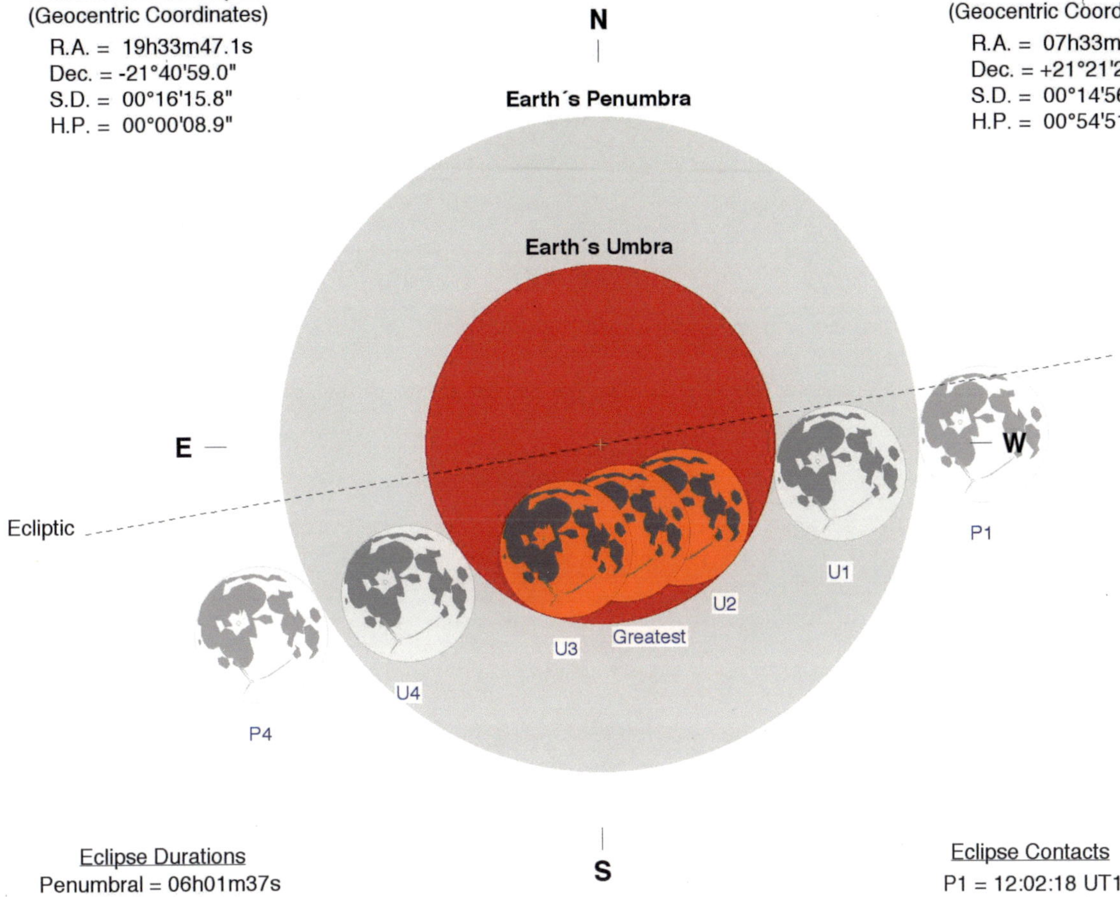

Eclipse Durations
Penumbral = 06h01m37s
Umbral = 03h35m59s
Total = 00h58m39s

Eph. = JPL DE430
Rule = Herald-Sinnott
ΔT = 96 s

0 15 30 45 60
Arc-Minutes

Eclipse Contacts
P1 = 12:02:18 UT1
U1 = 13:15:09 UT1
U2 = 14:33:44 UT1
U3 = 15:32:23 UT1
U4 = 16:51:08 UT1
P4 = 18:03:55 UT1

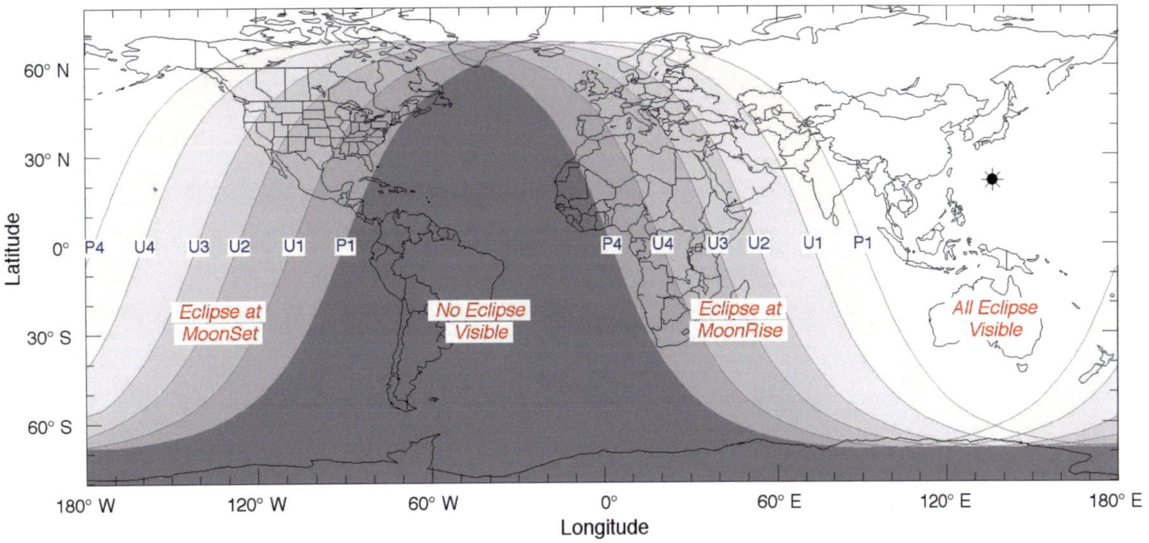

Partial Lunar Eclipse of 2066 Jul 07

Greatest Eclipse = 09:30:29.3 TD (= 09:28:52.6 UT1)

Penumbral Magnitude = 1.7196
Umbral Magnitude = 0.7770

Gamma = 0.6056
Axis = 0.6181°

Saros Series = 140
Saros Member = 27 of 77

Sun at Greatest Eclipse
(Geocentric Coordinates)
R.A. = 07h07m48.8s
Dec. = +22°30'58.3"
S.D. = 00°15'43.9"
H.P. = 00°00'08.7"

Moon at Greatest Eclipse
(Geocentric Coordinates)
R.A. = 19h07m14.9s
Dec. = -21°54'43.6"
S.D. = 00°16'41.4"
H.P. = 01°01'15.2"

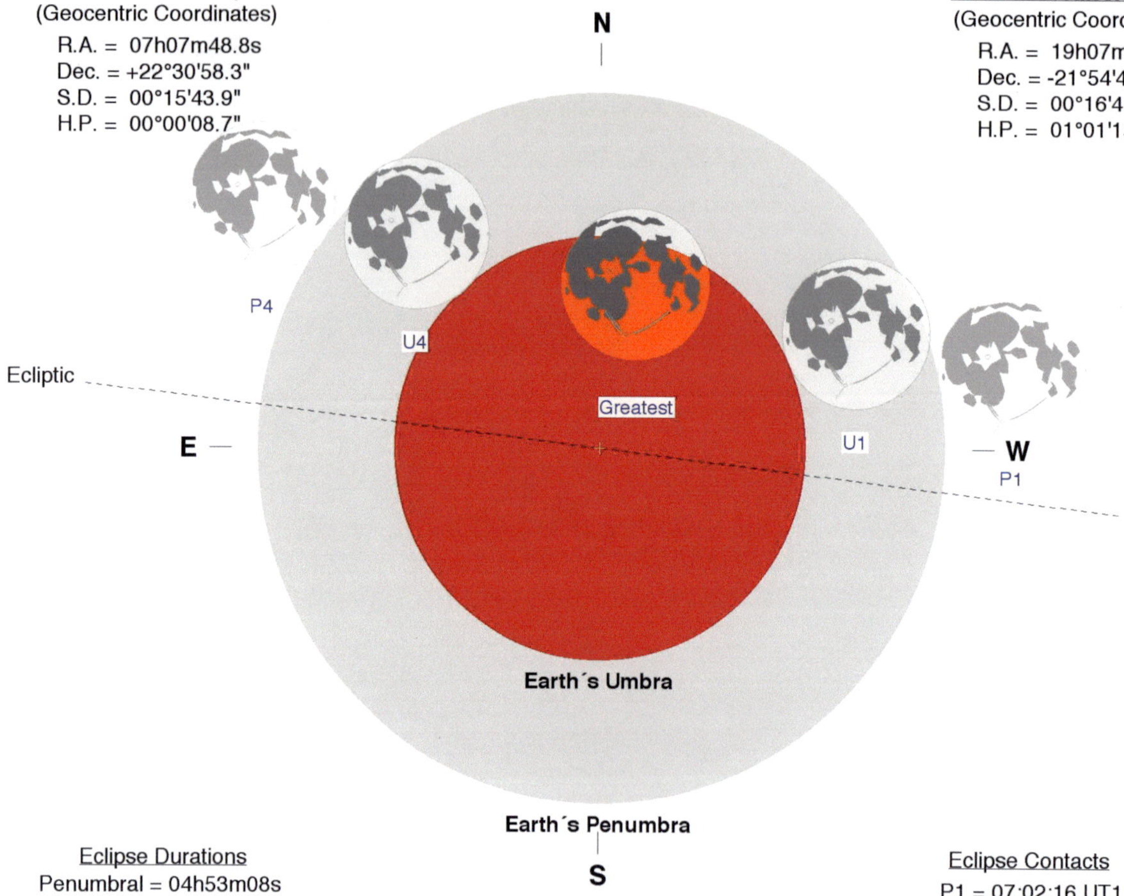

Eclipse Durations
Penumbral = 04h53m08s
Umbral = 02h52m01s

Eclipse Contacts
P1 = 07:02:16 UT1
U1 = 08:02:47 UT1
U4 = 10:54:48 UT1
P4 = 11:55:24 UT1

Eph. = JPL DE430
Rule = Herald-Sinnott
ΔT = 97 s

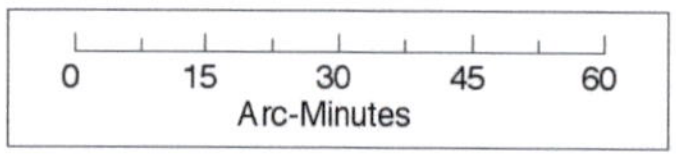

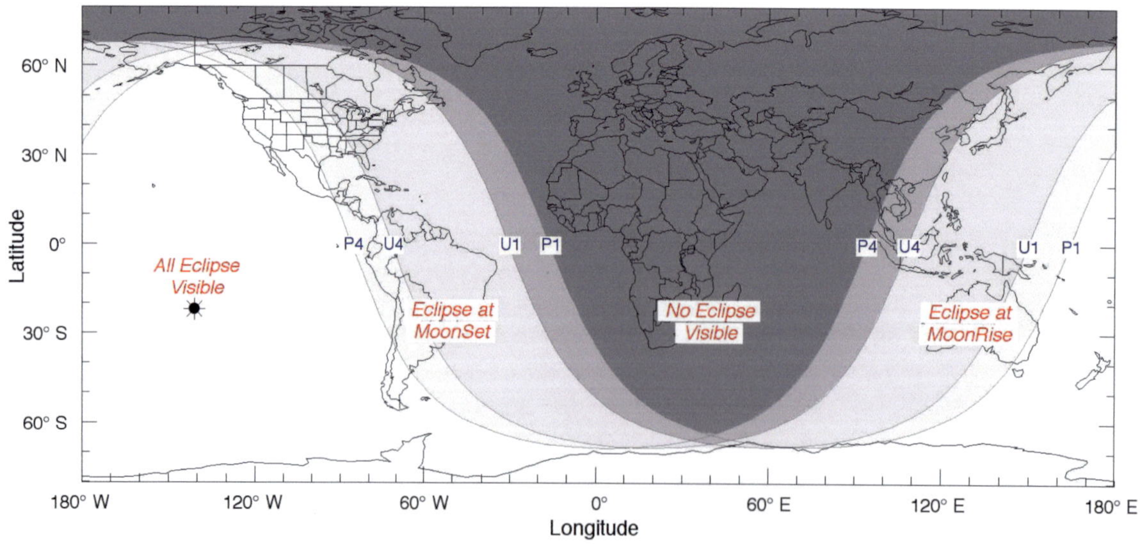

Penumbral Lunar Eclipse of 2066 Dec 31

Greatest Eclipse = 14:30:09.5 TD (= 14:28:32.4 UT1)

Penumbral Magnitude = 0.9789 Gamma = -1.0540 Saros Series = 145
Umbral Magnitude = -0.1264 Axis = 0.9485° Saros Member = 14 of 71

Sun at Greatest Eclipse
(Geocentric Coordinates)
R.A. = 18h44m27.3s
Dec. = -23°02'13.7"
S.D. = 00°16'15.9"
H.P. = 00°00'08.9"

Moon at Greatest Eclipse
(Geocentric Coordinates)
R.A. = 06h43m44.2s
Dec. = +22°06'11.2"
S.D. = 00°14'42.9"
H.P. = 00°54'00.3"

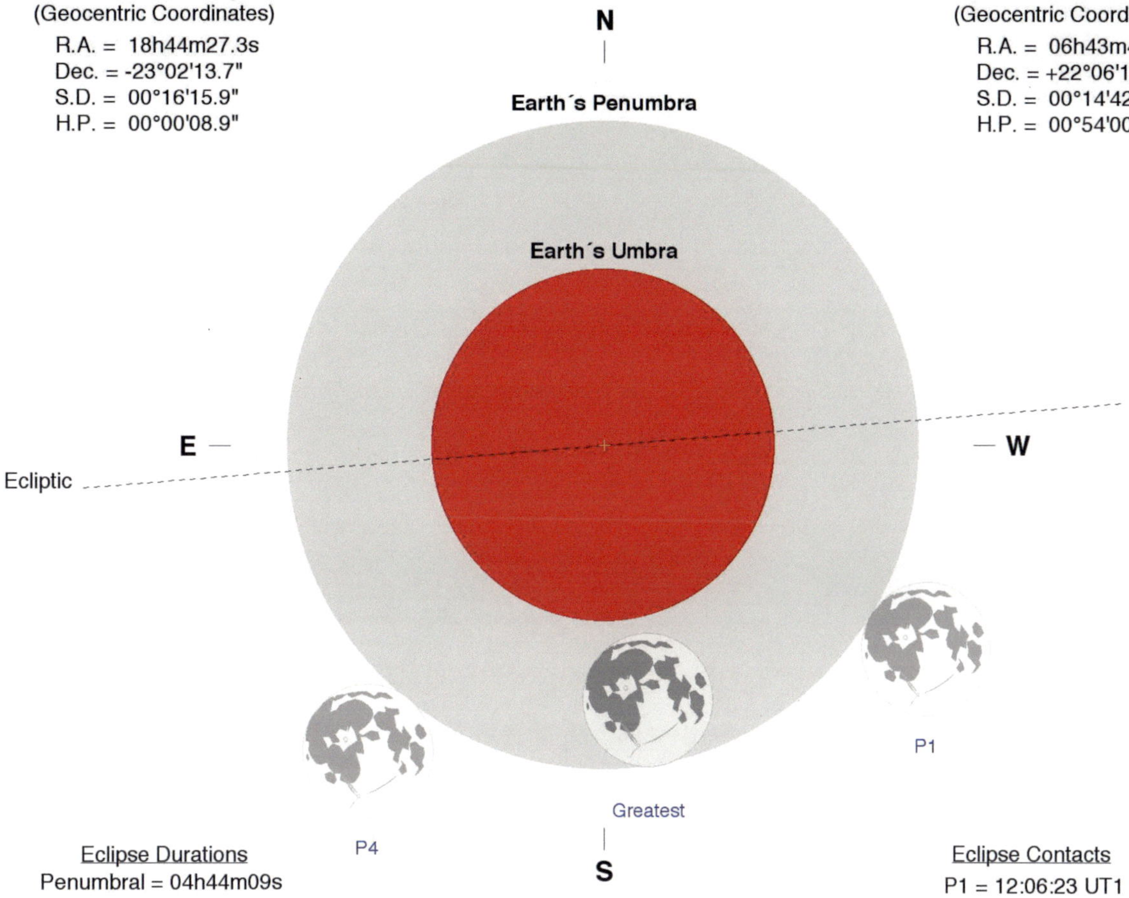

Eclipse Durations
Penumbral = 04h44m09s

Eclipse Contacts
P1 = 12:06:23 UT1
P4 = 16:50:32 UT1

Eph. = JPL DE430
Rule = Herald-Sinnott
ΔT = 97 s

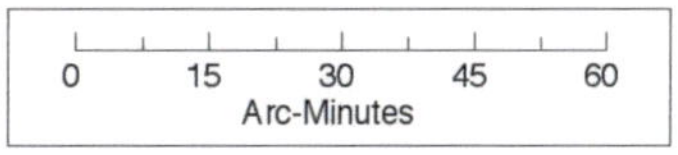

©2020 F. Espenak, www.EclipseWise.com

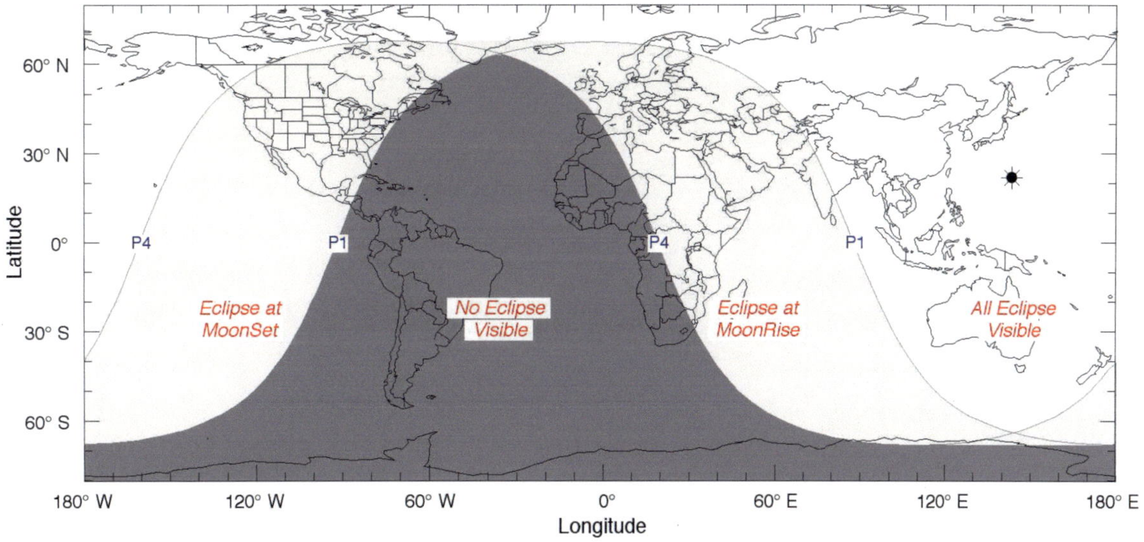

Penumbral Lunar Eclipse of 2067 May 28

Greatest Eclipse = 18:56:07.3 TD (= 18:54:29.9 UT1)

Penumbral Magnitude = 0.6416
Umbral Magnitude = -0.3316

Gamma = -1.2013
Axis = 1.1917°

Saros Series = 112
Saros Member = 68 of 72

Sun at Greatest Eclipse
(Geocentric Coordinates)
R.A. = 04h23m02.8s
Dec. = +21°33'42.4"
S.D. = 00°15'47.0"
H.P. = 00°00'08.7"

Moon at Greatest Eclipse
(Geocentric Coordinates)
R.A. = 16h22m41.8s
Dec. = -22°45'02.7"
S.D. = 00°16'13.2"
H.P. = 00°59'31.5"

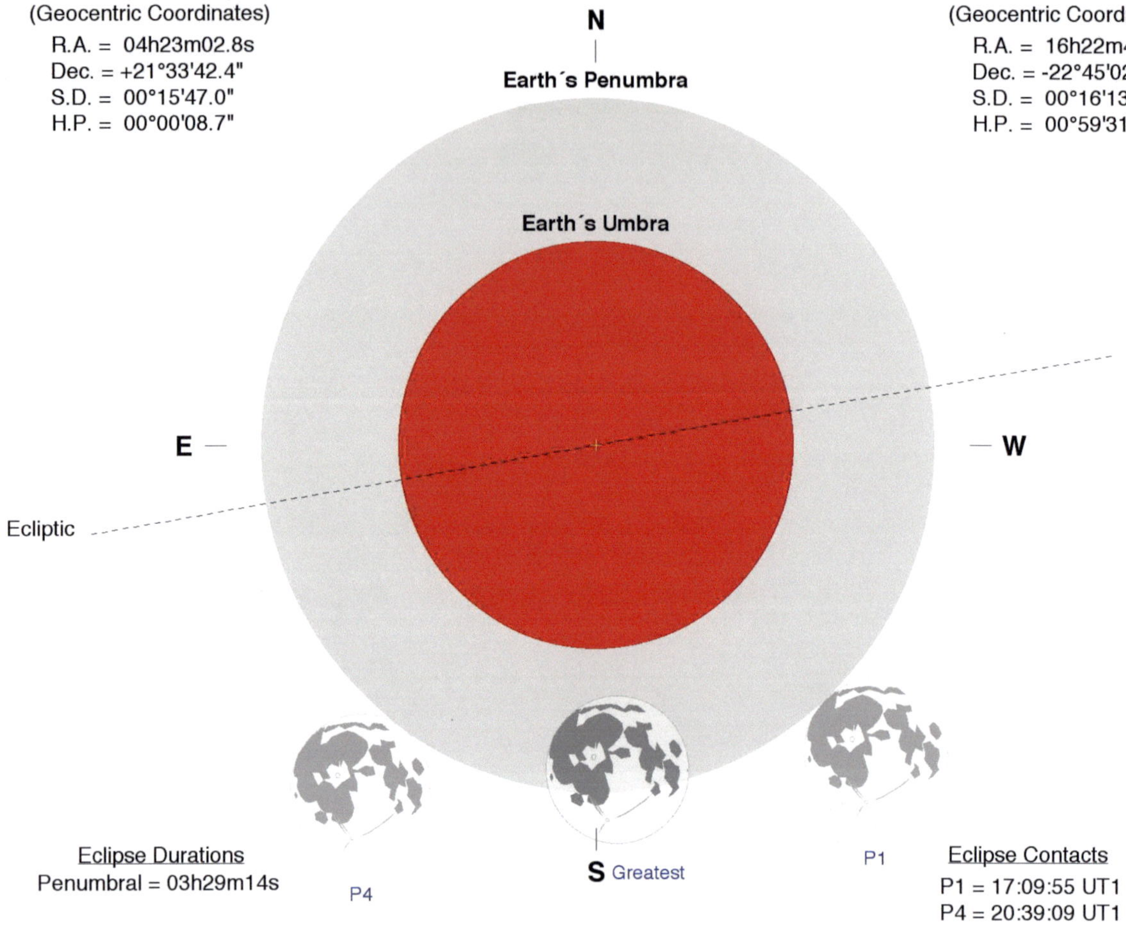

Eclipse Durations
Penumbral = 03h29m14s

Eclipse Contacts
P1 = 17:09:55 UT1
P4 = 20:39:09 UT1

Eph. = JPL DE430
Rule = Herald-Sinnott
ΔT = 97 s

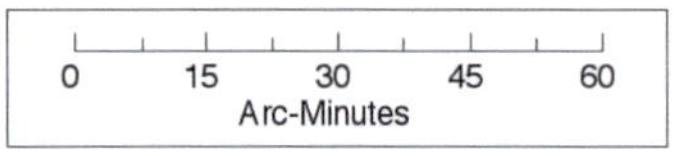

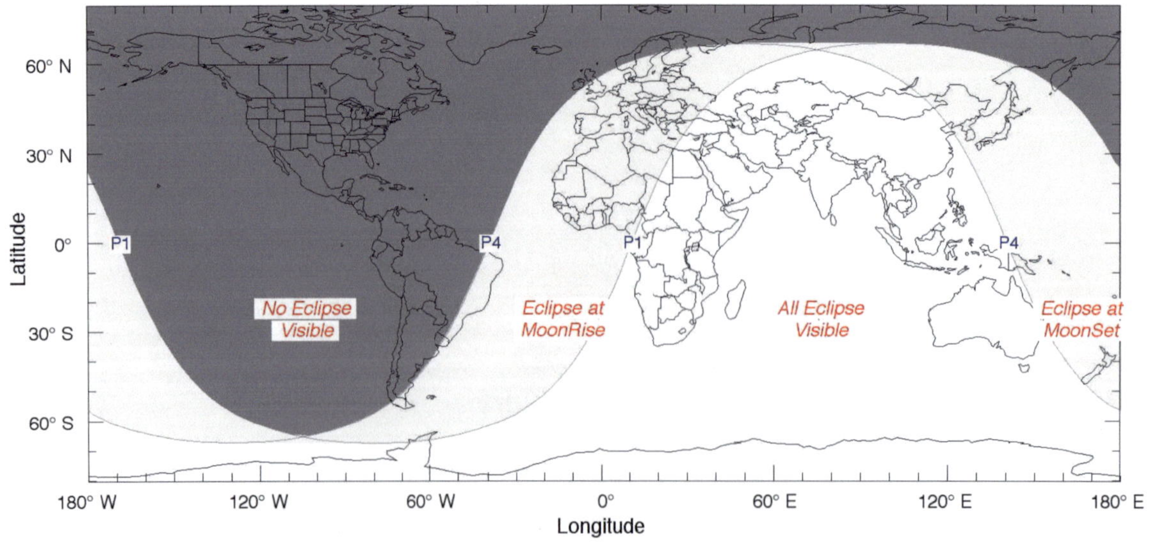

Penumbral Lunar Eclipse of 2067 Jun 27

Greatest Eclipse = 02:41:05.5 TD (= 02:39:28.0 UT1)

Penumbral Magnitude = 0.3770
Umbral Magnitude = -0.5736

Gamma = 1.3394
Axis = 1.3559°

Saros Series = 150
Saros Member = 4 of 71

Sun at Greatest Eclipse
(Geocentric Coordinates)
R.A. = 06h24m21.9s
Dec. = +23°18'42.3"
S.D. = 00°15'44.1"
H.P. = 00°00'08.7"

Moon at Greatest Eclipse
(Geocentric Coordinates)
R.A. = 18h23m33.9s
Dec. = -21°58'06.1"
S.D. = 00°16'33.1"
H.P. = 01°00'44.6"

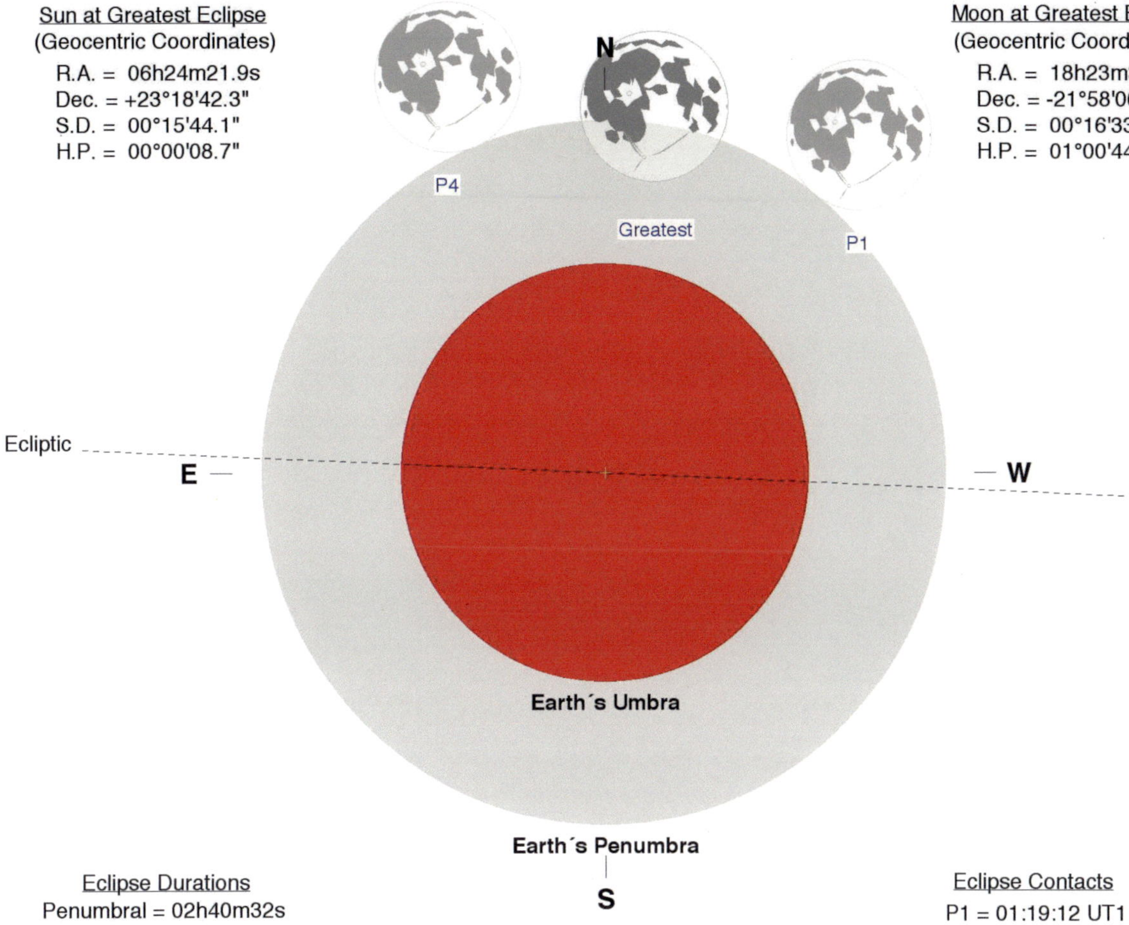

Eclipse Durations
Penumbral = 02h40m32s

Eclipse Contacts
P1 = 01:19:12 UT1
P4 = 03:59:44 UT1

Eph. = JPL DE430
Rule = Herald-Sinnott
ΔT = 97 s

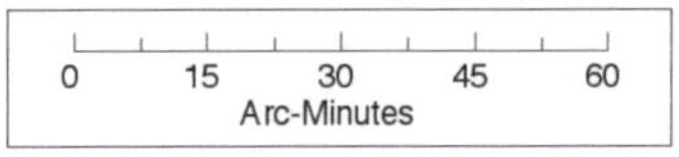

©2020 F. Espenak, www.EclipseWise.com

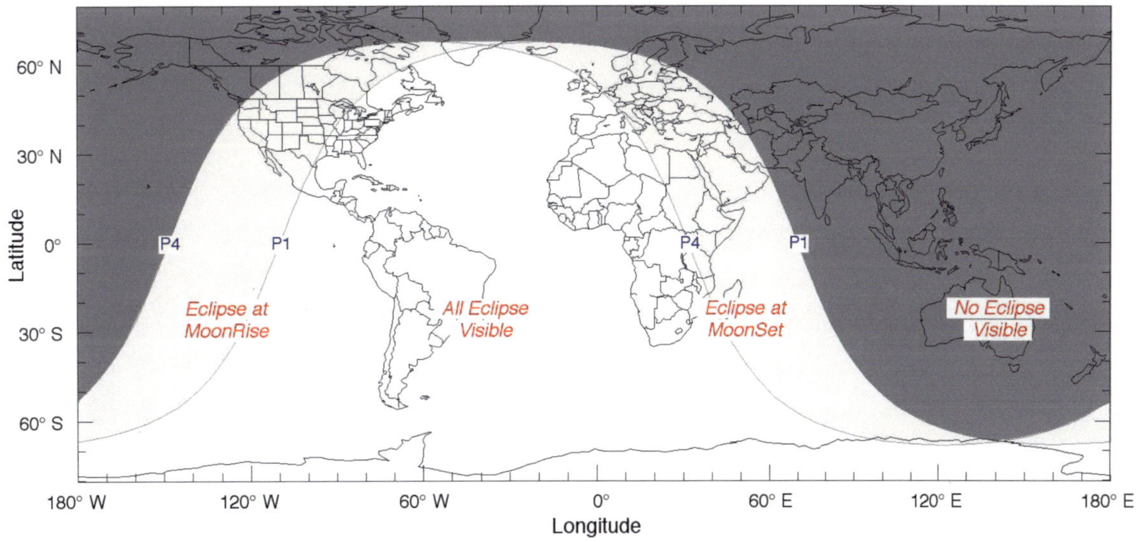

Penumbral Lunar Eclipse of 2067 Nov 21

Greatest Eclipse = 00:04:41.6 TD (= 00:03:03.8 UT1)

Penumbral Magnitude = 0.6557 Gamma = 1.2107 Saros Series = 117
Umbral Magnitude = -0.3798 Axis = 1.1575° Saros Member = 55 of 71

Sun at Greatest Eclipse
(Geocentric Coordinates)
R.A. = 15h46m01.0s
Dec. = -19°52'08.2"
S.D. = 00°16'11.2"
H.P. = 00°00'08.9"

Moon at Greatest Eclipse
(Geocentric Coordinates)
R.A. = 03h45m24.1s
Dec. = +21°01'03.2"
S.D. = 00°15'37.9"
H.P. = 00°57'22.2"

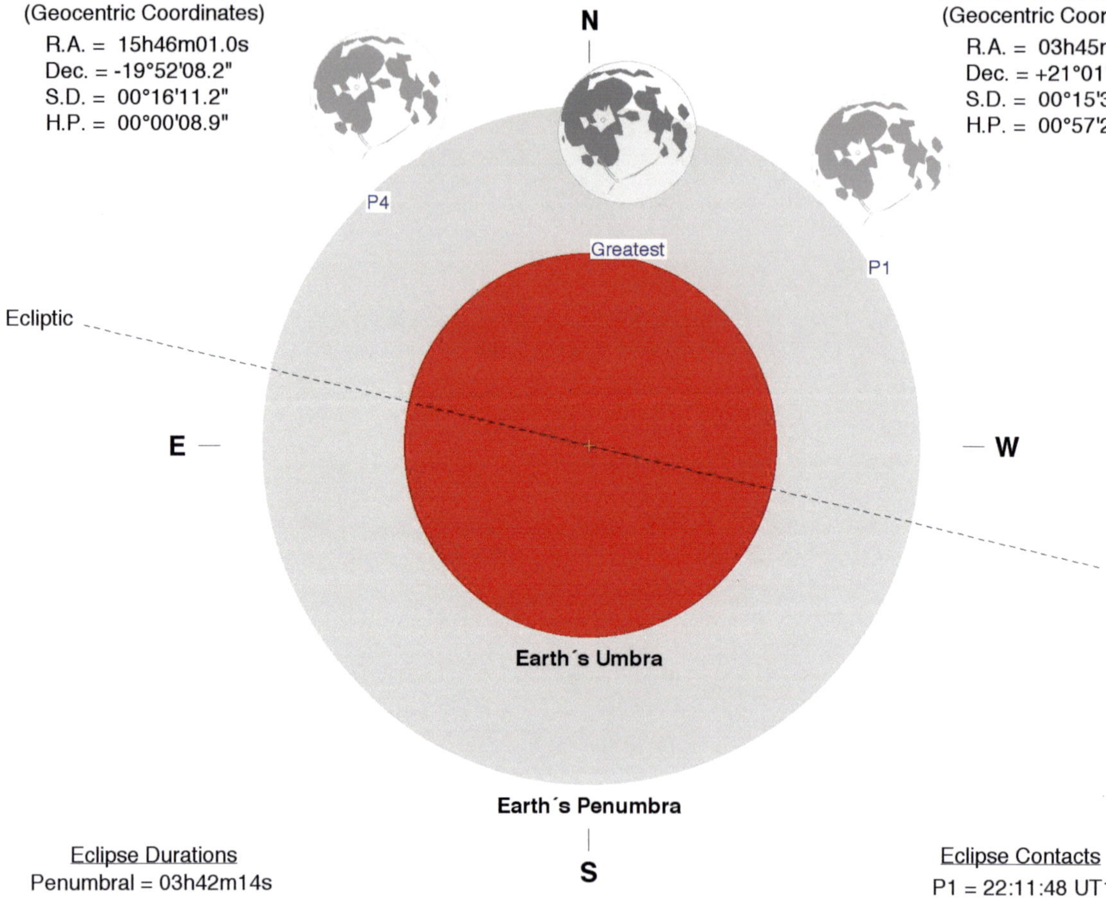

Eclipse Durations
Penumbral = 03h42m14s

Eclipse Contacts
P1 = 22:11:48 UT1
P4 = 01:54:02 UT1

Eph. = JPL DE430
Rule = Herald-Sinnott
ΔT = 98 s

0 15 30 45 60
Arc-Minutes

©2020 F. Espenak, www.EclipseWise.com

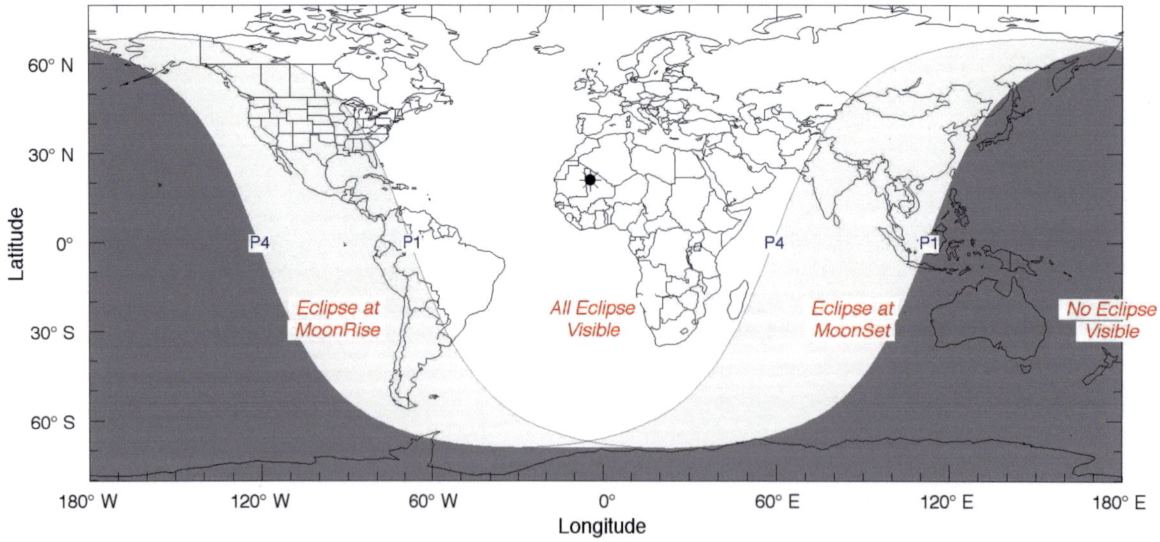

Partial Lunar Eclipse of 2068 May 17

Greatest Eclipse = 05:42:17.0 TD (= 05:40:38.8 UT1)

Penumbral Magnitude = 1.9839
Umbral Magnitude = 0.9545

Gamma = -0.4852
Axis = 0.4559°

Saros Series = 122
Saros Member = 59 of 74

Sun at Greatest Eclipse
(Geocentric Coordinates)
R.A. = 03h39m33.2s
Dec. = +19°31'07.8"
S.D. = 00°15'49.0"
H.P. = 00°00'08.7"

Moon at Greatest Eclipse
(Geocentric Coordinates)
R.A. = 15h39m17.9s
Dec. = -19°58'14.9"
S.D. = 00°15'21.9"
H.P. = 00°56'23.4"

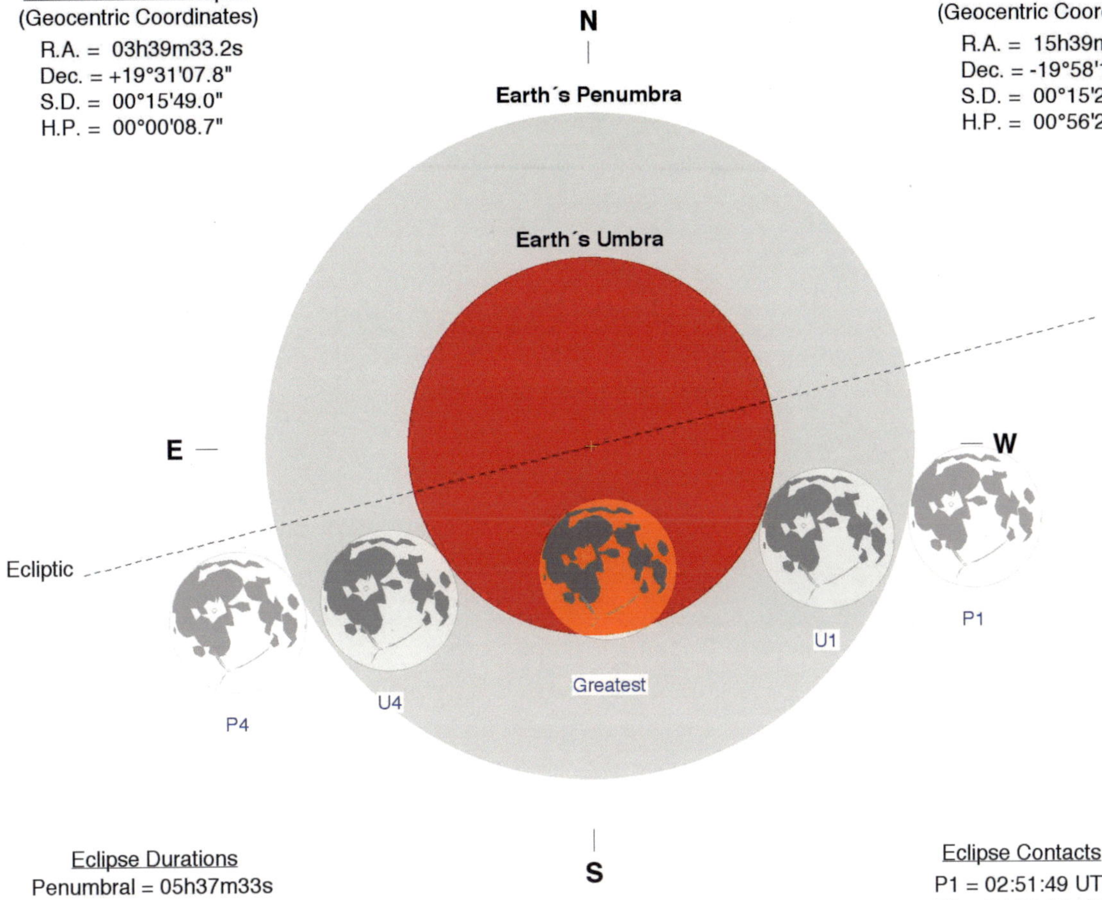

Eclipse Durations
Penumbral = 05h37m33s
Umbral = 03h19m46s

Eclipse Contacts
P1 = 02:51:49 UT1
U1 = 04:00:46 UT1
U4 = 07:20:31 UT1
P4 = 08:29:21 UT1

Eph. = JPL DE430
Rule = Herald-Sinnott
ΔT = 98 s

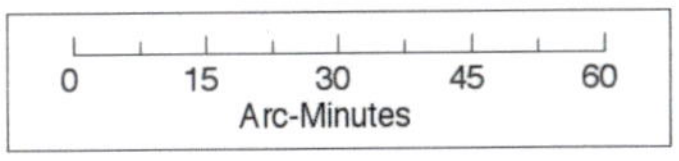

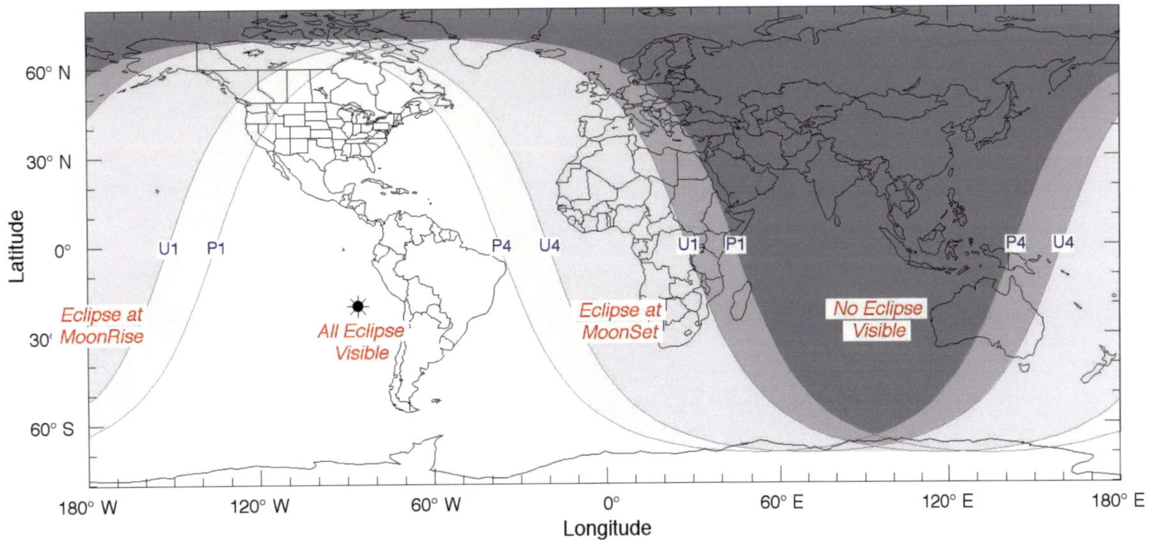

Total Lunar Eclipse of 2068 Nov 09

Greatest Eclipse = 11:46:59.7 TD (= 11:45:21.2 UT1)

Penumbral Magnitude = 1.9974
Umbral Magnitude = 1.0161

Gamma = 0.4645
Axis = 0.4675°

Saros Series = 127
Saros Member = 45 of 72

Sun at Greatest Eclipse
(Geocentric Coordinates)
R.A. = 15h01m47.2s
Dec. = -17°09'37.3"
S.D. = 00°16'08.8"
H.P. = 00°00'08.9"

Moon at Greatest Eclipse
(Geocentric Coordinates)
R.A. = 03h01m25.7s
Dec. = +17°37'12.0"
S.D. = 00°16'27.4"
H.P. = 01°00'23.7"

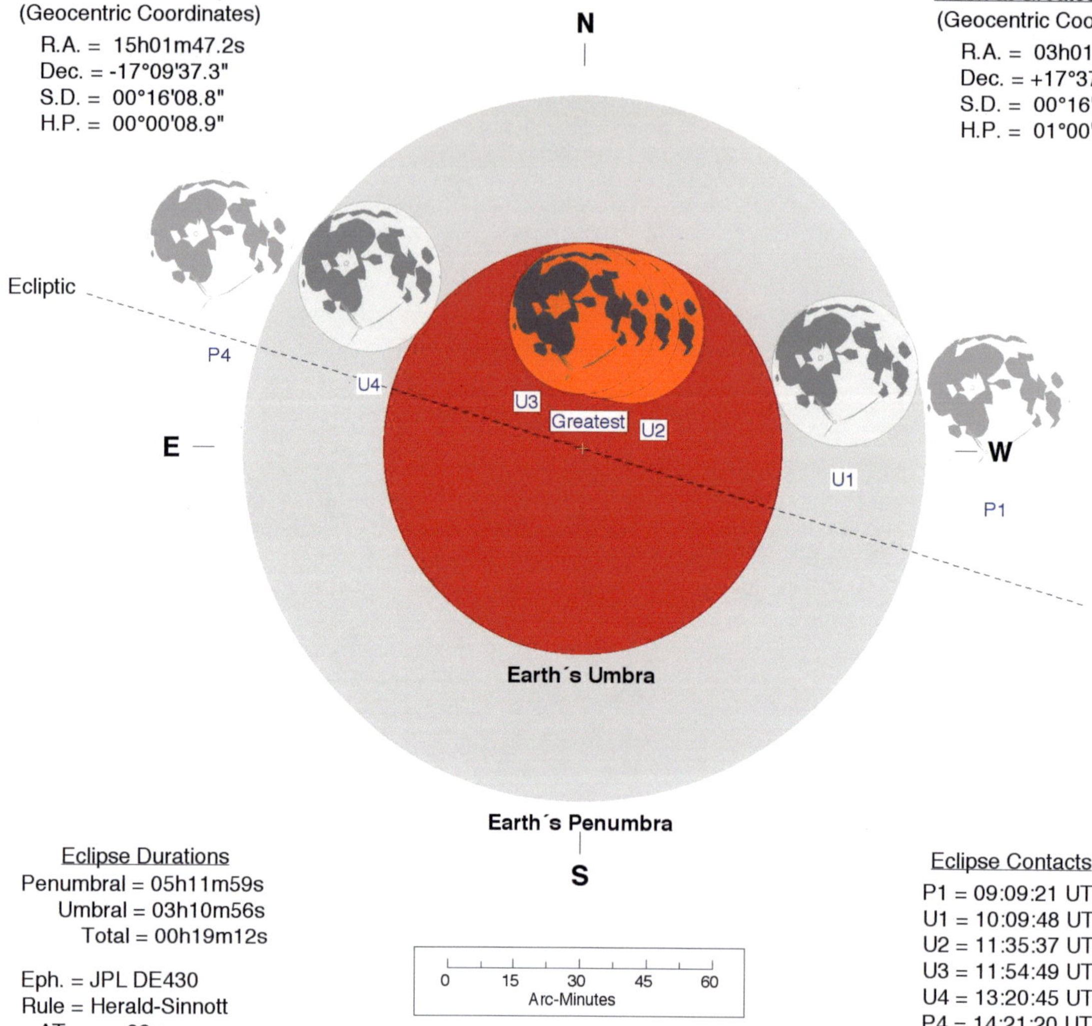

Eclipse Durations
Penumbral = 05h11m59s
Umbral = 03h10m56s
Total = 00h19m12s

Eph. = JPL DE430
Rule = Herald-Sinnott
ΔT = 99 s

Eclipse Contacts
P1 = 09:09:21 UT1
U1 = 10:09:48 UT1
U2 = 11:35:37 UT1
U3 = 11:54:49 UT1
U4 = 13:20:45 UT1
P4 = 14:21:20 UT1

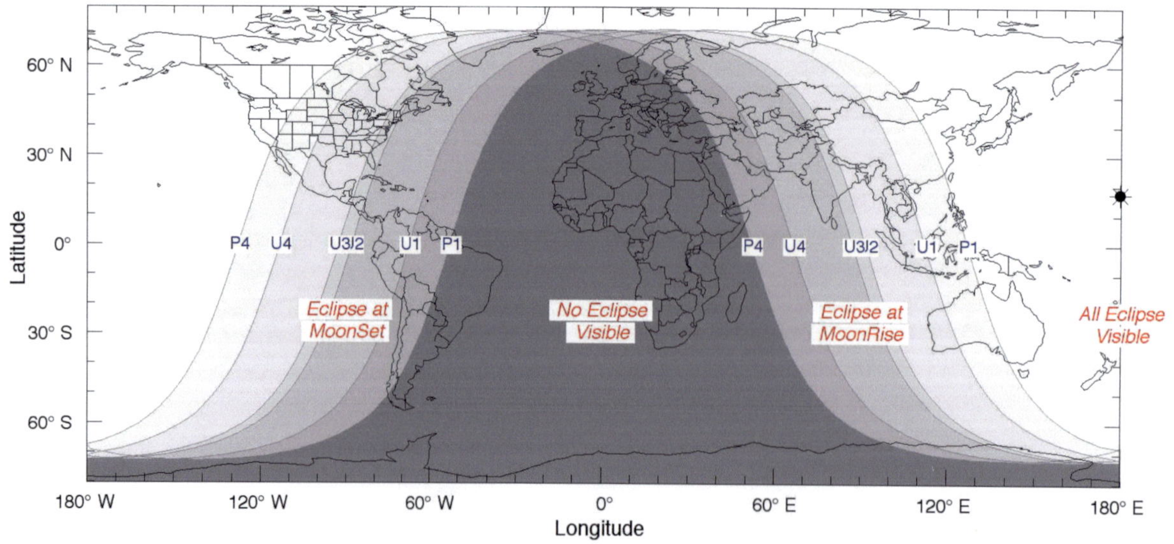

Total Lunar Eclipse of 2069 May 06

Greatest Eclipse = 09:09:56.6 TD (= 09:08:17.6 UT1)

Penumbral Magnitude = 2.3977 Gamma = 0.2717 Saros Series = 132
Umbral Magnitude = 1.3242 Axis = 0.2454° Saros Member = 33 of 71

Sun at Greatest Eclipse
(Geocentric Coordinates)
R.A. = 02h55m56.2s
Dec. = +16°44'53.2"
S.D. = 00°15'51.4"
H.P. = 00°00'08.7"

Moon at Greatest Eclipse
(Geocentric Coordinates)
R.A. = 14h56m07.8s
Dec. = -16°30'25.4"
S.D. = 00°14'46.2"
H.P. = 00°54'12.3"

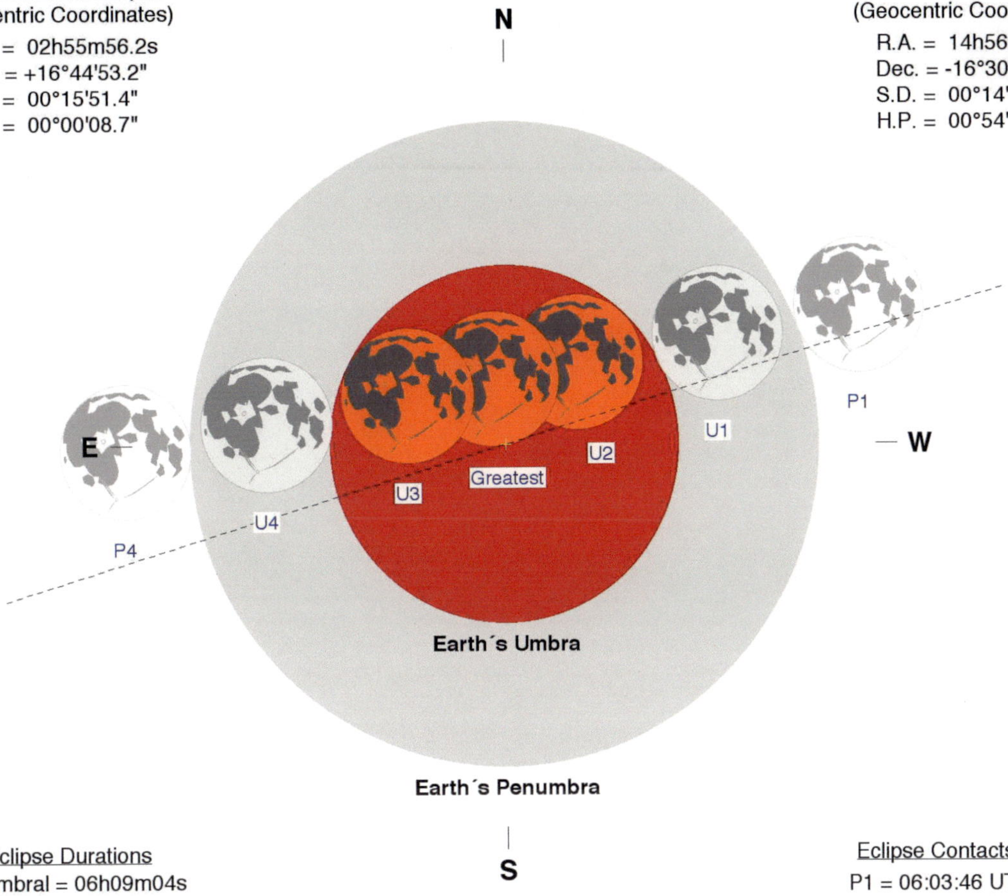

Eclipse Durations
Penumbral = 06h09m04s
Umbral = 03h47m01s
Total = 01h25m02s

Eph. = JPL DE430
Rule = Herald-Sinnott
ΔT = 99 s

0 15 30 45 60
Arc-Minutes

©2020 F. Espenak, www.EclipseWise.com

Eclipse Contacts
P1 = 06:03:46 UT1
U1 = 07:14:50 UT1
U2 = 08:25:52 UT1
U3 = 09:50:54 UT1
U4 = 11:01:51 UT1
P4 = 12:12:49 UT1

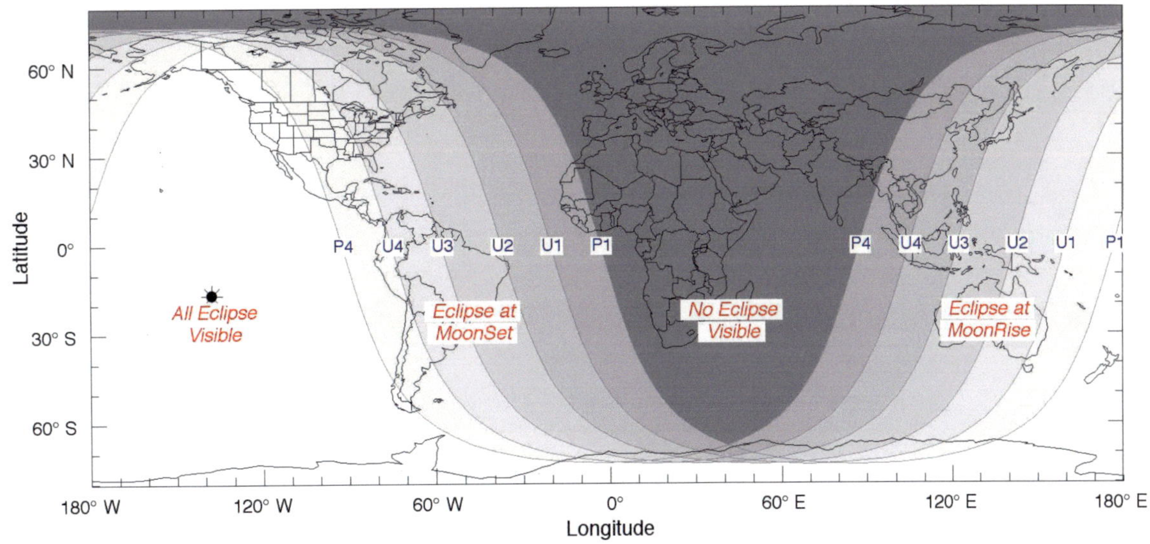

Total Lunar Eclipse of 2069 Oct 30

Greatest Eclipse = 03:35:05.7 TD (= 03:33:26.4 UT1)

Penumbral Magnitude = 2.4247
Umbral Magnitude = 1.4628

Gamma = -0.2263
Axis = 0.2317°

Saros Series = 137
Saros Member = 29 of 78

Sun at Greatest Eclipse
(Geocentric Coordinates)
R.A. = 14h19m49.6s
Dec. = -13°56'35.4"
S.D. = 00°16'06.2"
H.P. = 00°00'08.9"

Moon at Greatest Eclipse
(Geocentric Coordinates)
R.A. = 02h20m02.8s
Dec. = +13°43'03.8"
S.D. = 00°16'44.5"
H.P. = 01°01'26.5"

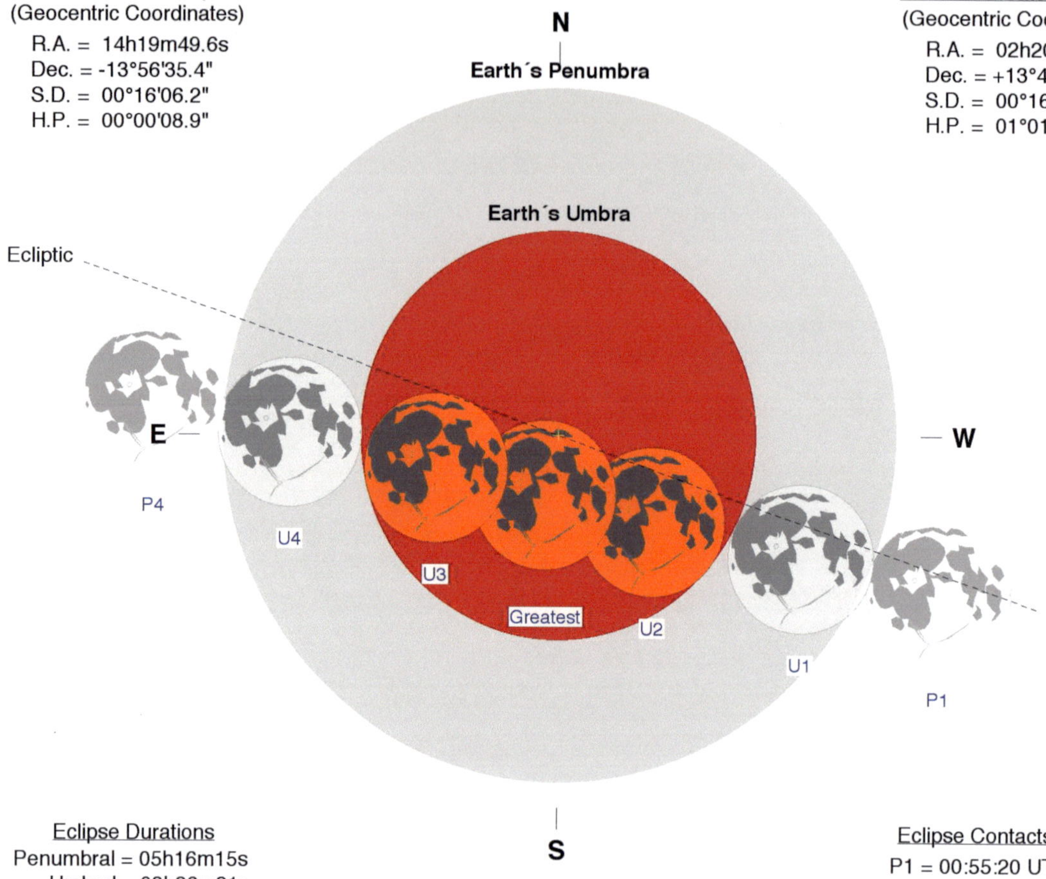

Eclipse Durations
Penumbral = 05h16m15s
Umbral = 03h26m21s
Total = 01h27m26s

Eph. = JPL DE430
Rule = Herald-Sinnott
ΔT = 99 s

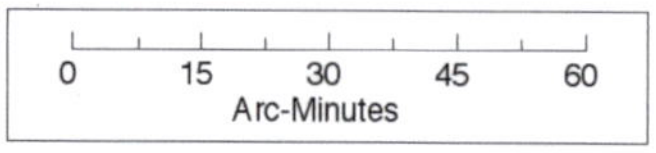

Eclipse Contacts
P1 = 00:55:20 UT1
U1 = 01:50:18 UT1
U2 = 02:49:48 UT1
U3 = 04:17:14 UT1
U4 = 05:16:39 UT1
P4 = 06:11:35 UT1

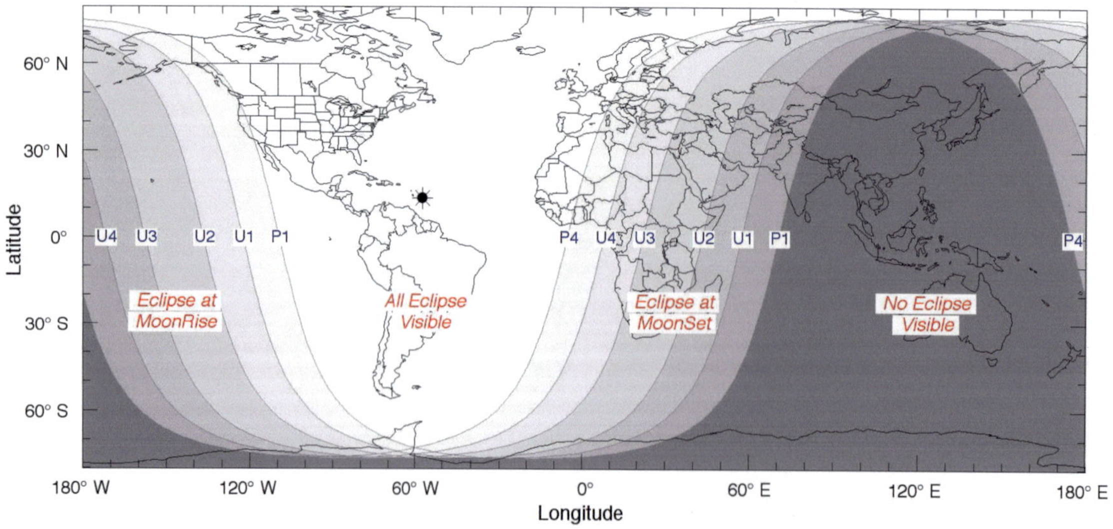

Penumbral Lunar Eclipse of 2070 Apr 25

Greatest Eclipse = 09:21:24.5 TD (= 09:19:44.8 UT1)

Penumbral Magnitude = 1.0527
Umbral Magnitude = -0.0197

Gamma = 1.0044
Axis = 0.9109°

Saros Series = 142
Saros Member = 21 of 73

Sun at Greatest Eclipse
(Geocentric Coordinates)
R.A. = 02h12m57.7s
Dec. = +13°21'41.4"
S.D. = 00°15'54.1"
H.P. = 00°00'08.7"

Moon at Greatest Eclipse
(Geocentric Coordinates)
R.A. = 14h13m51.0s
Dec. = -12°28'35.7"
S.D. = 00°14'49.7"
H.P. = 00°54'25.3"

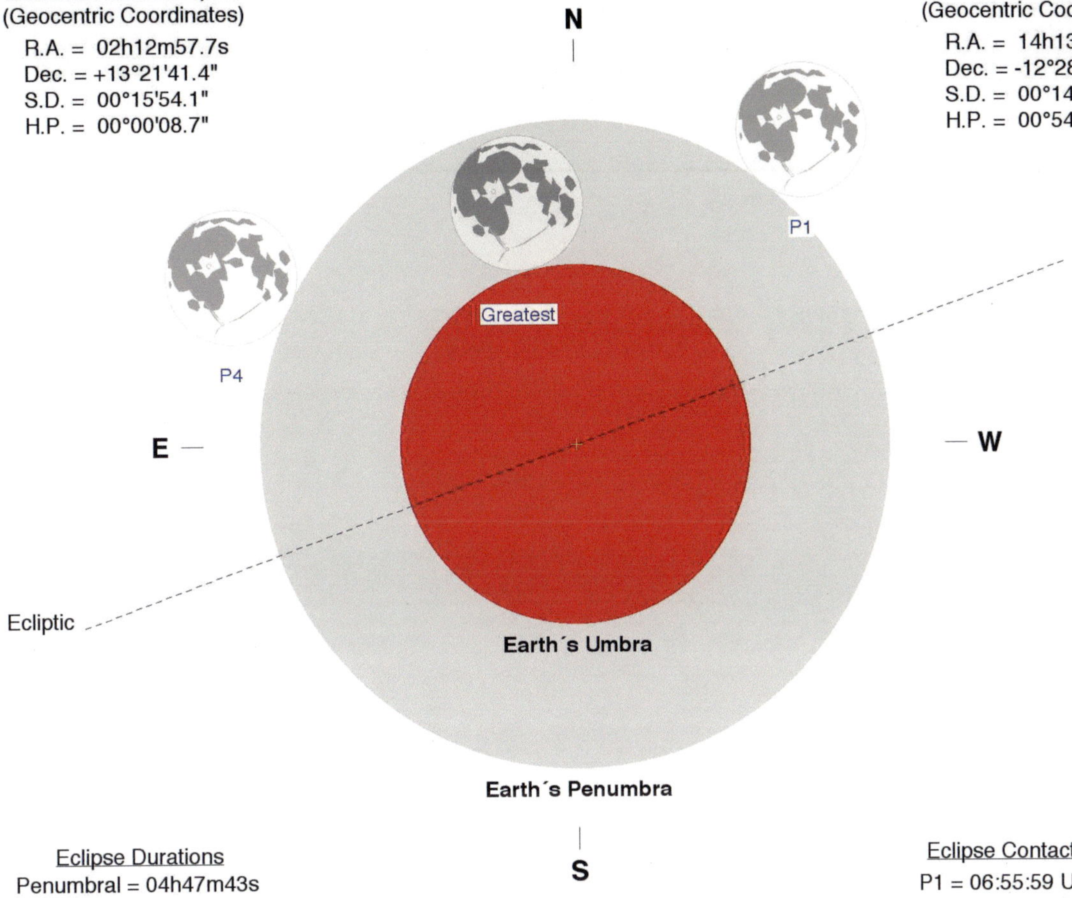

Eclipse Durations
Penumbral = 04h47m43s

Eclipse Contacts
P1 = 06:55:59 UT1
P4 = 11:43:42 UT1

Eph. = JPL DE430
Rule = Herald-Sinnott
ΔT = 100 s

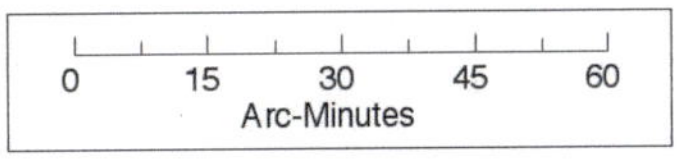

©2020 F. Espenak, www.EclipseWise.com

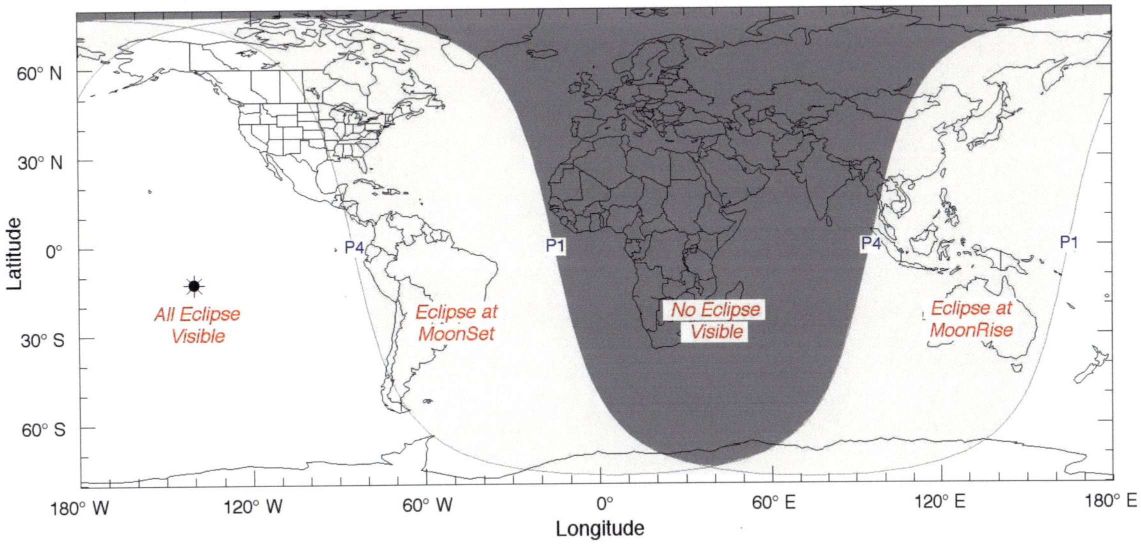

Partial Lunar Eclipse of 2070 Oct 19

Greatest Eclipse = 18:51:11.9 TD (= 18:49:31.8 UT1)

Penumbral Magnitude = 1.1270
Umbral Magnitude = 0.1395

Gamma = -0.9406
Axis = 0.9354°

Saros Series = 147
Saros Member = 11 of 70

Sun at Greatest Eclipse
(Geocentric Coordinates)
R.A. = 13h39m13.7s
Dec. = -10°18'14.5"
S.D. = 00°16'03.4"
H.P. = 00°00'08.8"

Moon at Greatest Eclipse
(Geocentric Coordinates)
R.A. = 01h40m14.8s
Dec. = +09°24'10.4"
S.D. = 00°16'15.6"
H.P. = 00°59'40.4"

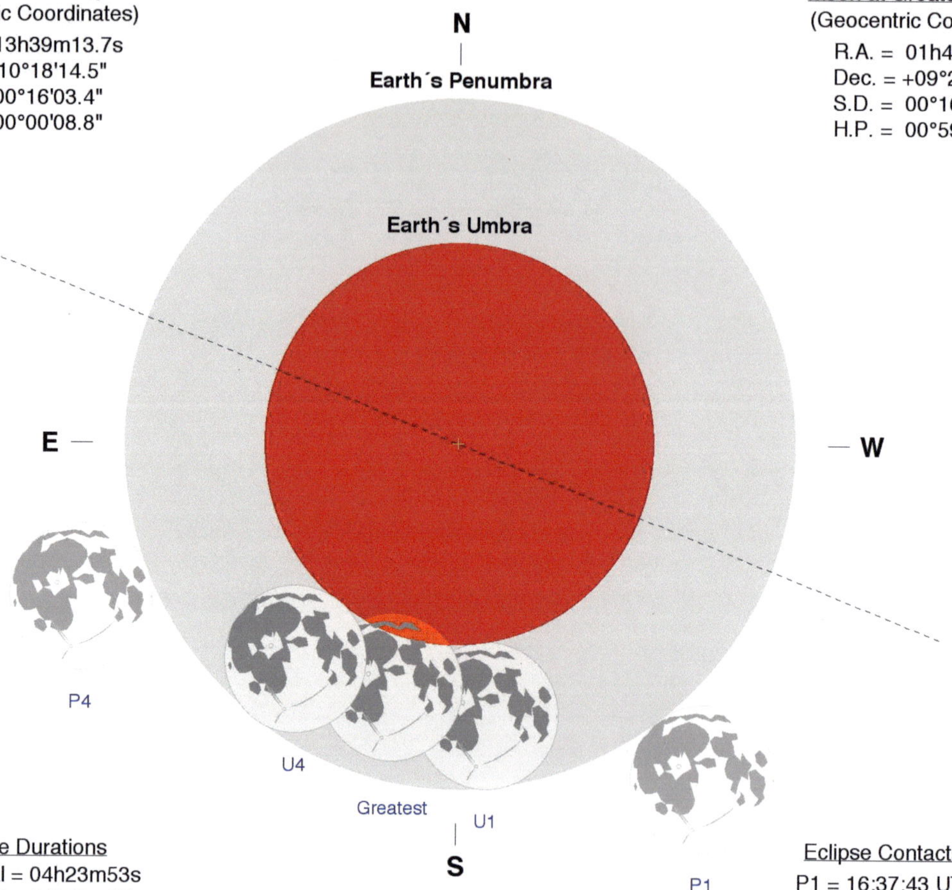

Eclipse Durations
Penumbral = 04h23m53s
Umbral = 01h22m17s

Eclipse Contacts
P1 = 16:37:43 UT1
U1 = 18:08:38 UT1
U4 = 19:30:55 UT1
P4 = 21:01:36 UT1

Eph. = JPL DE430
Rule = Herald-Sinnott
ΔT = 100 s

0 15 30 45 60
Arc-Minutes

©2020 F. Espenak, www.EclipseWise.com

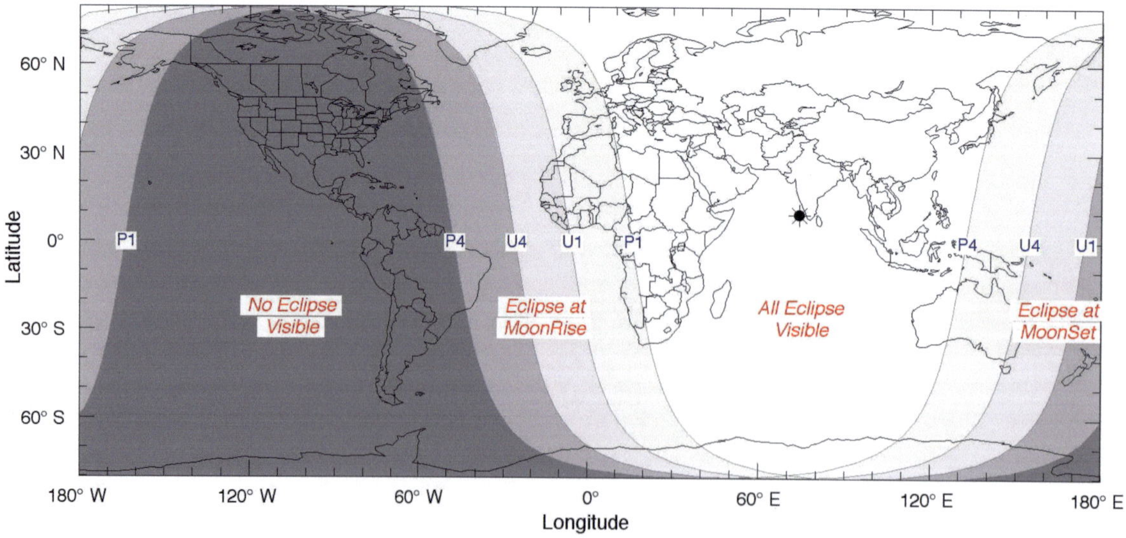

Astropixels Publications

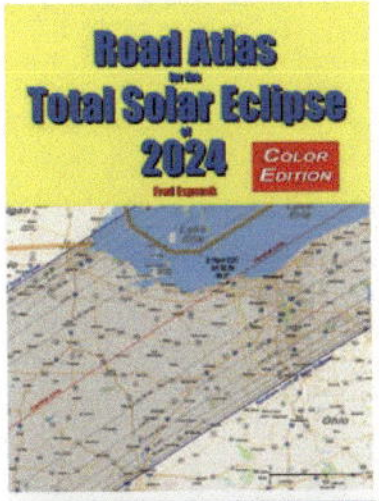

Road Atlas for the Total Solar Eclipse of 2024 (Fred Espenak) contains a comprehensive series of 26 high-resolution maps of the path of totality across the USA, Mexico and Canada. The large scale (1 inch = 22 miles) shows both major and minor roads, towns and cities, rivers, lakes, parks, national forests, wilderness areas and mountain ranges. The duration of totality is plotted in 30-second steps, making it easy to estimate the length of the total eclipse from any location in the eclipse path.

21st Century Canon of Solar Eclipses (Fred Espenak) contains maps and data for all 224 solar eclipses occurring during the 100-year period from 2001 through 2100. Appendix A is comprehensive catalog listing the essential characteristics of each eclipse. Appendix B contains maps depicting the geographic regions of visibility of each eclipse with 12 maps per page. Appendix C has detailed full-page maps of every eclipse from 2017 through 2066. Appendix D plots the track of every central eclipse (total, annular and hybrid) on large-scale maps with countries borders and major cities.

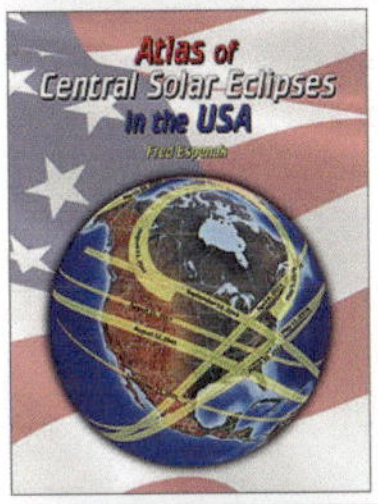

Atlas of Central Solar Eclipses in the USA (Fred Espenak) contains of a series of 499 global maps showing the track of every total and annular solar eclipse across the USA from 1001 through 3000. It is accompanied by a catalog that lists the major characteristics of each eclipse. A set of 20 detailed maps, each covering a 50-year period and centered on the lower 48 states, shows the path of every total and annular eclipse. The maps include state boundaries and major cities. These maps also cover southern Canada and northern Mexico.

Thousand Year Canon of Solar Eclipses 1501 to 2500 (Fred Espenak) contains maps and data for each of the 2,389 solar eclipses occurring over the ten-century period centered on the present era. A comprehensive catalog lists the essential characteristics of each eclipse while a series of global maps show the exact geographic extent of each eclipse.

Thousand Year Canon of Lunar Eclipses 1501 to 2500 (Fred Espenak) contains diagrams, maps and data for each of the 2,424 lunar eclipses occurring over the ten-century period centered on the present era. A comprehensive catalog lists the essential characteristics of each eclipse while a series of diagrams and maps illustrate the Moon-shadow geometry and geographic regions of visibility of each eclipse.

Totality: The Great American Eclipses of 2017 and 2024 (Mark Littmann & Fred Espenak) is the ultimate guide to the most stunning of celestial sights, total eclipses of the Sun The book provides information, photographs, and illustrations to help the public understand and safely enjoy all aspects of these eclipses including how to observe a total eclipse of the Sun, how to photograph and video record an eclipse, why solar eclipses happen, and more. Several chapters focus exclusively on the total eclipses of 2017 and 2024 though the USA.

For complete details on the above ***Astropixels Publications***, visit:

astropixels.com/pubs/

Made in the USA
Las Vegas, NV
03 November 2026

57425081R00105